# LECTURES ON THE PHYSICS OF HIGHLY CORRELATED ELECTRON SYSTEMS V

# Previous Proceedings in the Series of Lectures on the Physics of Highly Correlated Electron Systems

| Year | Title | Publisher | ISBN |
|---|---|---|---|
| 1999 | Fourth Training Course | AIP Conf. Proceedings Vol. 527 | 1-56396-950-5 |
| 1998 | Third Training Course | *not published* | |
| 1997 | Second Training Course | AIP Conf. Proceedings Vol. 438 | 1-56396-789-8 |
| 1996 | First Training Course | *not published* | |

## Other Related Titles from AIP Conference Proceedings

**577** Density Functional Theory and Its Application to Materials
Edited by V. Van Doren, C. Van Alsenoy, and P. Geerlings, July 2001, 0-7354-0016-4

**554** Physics in Local Lattice Distortions: Fundamentals and Novel Concepts; LLD2K
Edited by Hiroyuki Oyanagi and Antonio Bianconi, February 2001, 1-56396-984-X

**551** Atomic Physics 17: XVII International Conference on Atomic Physics; ICAP 2000
Edited by Ennio Arimondo, Paolo De Natale, and Massimo Inguscio, February 2001, 1-56396-982-3

**535** Fundamental Physics of Ferroelectrics 2000: Aspen Center for Physics Winter Workshop
Edited by Ronald E. Cohen, September 2000, 1-56396-959-9

**483** High Temperature Superconductivity
Edited by Stewart E. Barnes, Joseph Ashkenazi, Joshua L. Cohn, and Fulin Zuo, August 1999, 1-56396-880-0

**427** Lectures on Superconductivity in Networks and Mesoscopic Systems
Edited by Carlo Giovannella and Colin J. Lambert, April 1998, 1-56396-750-2

To learn more about these titles, or the AIP Conference Proceedings Series, please visit the webpage http://www.aip.org/catalog/aboutconf.html

# LECTURES ON THE PHYSICS OF HIGHLY CORRELATED ELECTRON SYSTEMS V

Fifth Training Course in the Physics of Correlated Electron Systems and High-Tc Superconductors

Salerno, Italy   30 October –10 November 2000

*EDITOR*
Ferdinando Mancini
*Università degli Studi di Salerno
Salerno, Italy*

Melville, New York, 2001
AIP CONFERENCE PROCEEDINGS ■ VOLUME 580

**Editor:**

Ferdinando Mancini
Dipartimento di Fisica "E. R. Caianiello"
Unità INFM di Salerno
Università degli Studi di Salerno
Via S. Allende
84081 Baronissi (SA)
ITALY

E-mail: mancini@sa.infn.it

Authorization to photocopy items for internal or personal use, beyond the free copying permitted under the 1978 U.S. Copyright Law (see statement below), is granted by the American Institute of Physics for users registered with the Copyright Clearance Center (CCC) Transactional Reporting Service, provided that the base fee of $18.00 per copy is paid directly to CCC, 222 Rosewood Drive, Danvers, MA 01923. For those organizations that have been granted a photocopy license by CCC, a separate system of payment has been arranged. The fee code for users of the Transactional Reporting Service is: 0-7354-0019-9/01/$18.00.

© 2001 American Institute of Physics

Individual readers of this volume and nonprofit libraries, acting for them, are permitted to make fair use of the material in it, such as copying an article for use in teaching or research. Permission is granted to quote from this volume in scientific work with the customary acknowledgment of the source. To reprint a figure, table, or other excerpt requires the consent of one of the original authors and notification to AIP. Republication or systematic or multiple reproduction of any material in this volume is permitted only under license from AIP. Address inquiries to Office of Rights and Permissions, Suite 1NO1, 2 Huntington Quadrangle, Melville, N.Y. 11747-4502; phone: 516-576-2268; fax: 516-576-2450; e-mail: rights@aip.org.

L.C. Catalog Card No. 2001092174
ISBN 0-7354-0019-9
ISSN 0094-243X
Printed in the United States of America

# Contents

**Preface** .................................................. vii

**Strong-Coupling Theory of High-Temperature Superconductivity** ............. 1
*A. S. Alexandrov*
I. Introduction .......................................... 1
II. Migdal-Eliashberg Theory ............................. 4
III. Polaron Dynamics, Bipolarons, and Charged Bose-Gas ... 20
IV. Bipolarons and High-$T_c$ Cuprates: Some Recent Results ... 43
V. Conclusions ........................................... 88
VI. References ............................................ 90

**Transport Properties in Superconducting Layered Systems** ............. 97
*L. Maritato*
I. Introduction .......................................... 97
II. Transport Properties of Type-II Superconductors ....... 98
III. Magnetic Field Dependence of Pinning Mechanisms in BSCCO Thin Films ............................................ 103
IV. H-T Phase Diagram of Conventional Superconducting/Magnetic Artificial Multilayers ................................. 108
V. HTS Artificial Superlattices .......................... 115
VI. Conclusions .......................................... 119
VII. References ........................................... 119

**Spin-Fluctuation Pairing in Strongly Correlated Systems** ................ 121
*N. M. Plakida*
I. Introduction ......................................... 121
II. Models with Strong Correlations ...................... 130
III. Superconducting Pairing in the t-J Model ............. 137
IV. Superconducting Pairing in the Hubbard Model ......... 161
V. Conclusions .......................................... 182
VI. References ........................................... 184

**Non-Abelian Bosonization and WZNW Models** ............................. 189
*A. M. Tsvelik and R. Citro*
I. Introduction ......................................... 189
II. The Chiral Anomaly ................................... 194
III. Anomalous Commutators ................................ 195
IV. Kac-Moody Algebras ................................... 199
V. Sugawara Hamiltonian for Wess-Zumino-Novikov-Witten Model ................................................. 202
VI. Knizhnik-Zamolodchikov (KZ) Equations ................ 205
VII. WZNW Model in the Lagrangian Formulation and Non-Abelian Bosonization ............................. 207
VIII. A Non-Trivial Determinant ............................ 209

| IX. | SU(2)$_1$ WZNW Model: A Gaussian Model | 211 |
| X. | Non-Abelian Bosonization of Interacting Fermions with Spin: The Hubbard Model | 213 |
| XI. | The Kondo Chain | 217 |
| XII. | Conformal Symmetry and General Properties of Correlation Functions for WZNW Models | 221 |
| XIII. | Conclusions | 226 |
| XIV. | References | 227 |

Author Index .......................................................... 229

## PREFACE

The ensuing volume contains the lectures delivered at the "Fifth Training Course in the Physics of Correlated Electron Systems and High-Tc Superconductors", held in Vietri sul Mare (Salerno) Italy, in November 2000.

Following the tradition of previous years, the meeting was devoted to promote the formation of young scientists by means of training through research. Differently from usual workshops, the idea was to bring together for two weeks senior and young researchers in a close and informal atmosphere, paying special attention and effort to an active participation of the young researchers and to introduce them to some specific problems.

The course consisted of four lectures every morning, held by Professors S. Alexandrov, L. Maritato, N. M. Plakida and A. M. Tsvelik, and afternoon activities aimed principally to increase discussions between the students and lecturers, the solving of particular exercises and oral exposition by the students. The outcome of this type of course was an active participation of the students and a large interchange of ideas due to the long afternoon time destined to discussions.

It is hoped that both the meeting, which brought together leaders in the field as well as bright and eager beginners, and the present volume may be useful as an up-to-date book for researchers interested in the field.

I wish to acknowledge and thank those whose support made the course and the proceedings possible. The main sponsor of the event is the European Commission, under the TMR Programme. For additional support I gratefully acknowledge the University of Salerno. Finally, I wish to thank Professor Maria Marinaro, President of the International Institute of Advanced Scientific Studies, who hosted the event in the wonderful and warm venue of Vietri sul Mare.

Salerno, 6 June 2001                                            Ferdinando Mancini

# Strong-Coupling Theory of High-Temperature Superconductivity

A.S. Alexandrov*

*Department of Physics, Loughborough University,
Loughborough LE11 3TU, UK*

**Abstract.** The observation of high-temperature superconductivity in complex layered cuprates by Bednorz and Müller must now rate as one of the greatest experimental discoveries of the last century. Identifying and understanding the microscopic origin of high-temperature superconductivity stands as one of the greatest theoretical challenges of this century. These lectures describe an approach, based on the extension of the BCS theory to the strong-coupling regime with small polarons and bipolarons. We discuss the canonical Migdal-Eliashberg theory of strongly coupled electrons and phonons (or any bosons) and its breakdown in the strong-coupling regime. Then, starting from the dynamic properties of a single polaron, we show how the multi-polaron problem is reduced to a weakly interacting charged Bose gas. Finally we discuss a few recent applications of the bipolaron theory to cuprates, in particular the essential interaction in oxides, the parameter-free expression for the superconducting critical temperature, upper crtitical field, symmetry of the order parameter, the London penetration depth, vortex structure, theory of tunnelling and Andreev reflection, normal and superconducting gaps, angle-resolved photoemission, and stripes.

## I INTRODUCTION

In 1986 Bednorz and Müller [1] discovered the onset of possible superconductivity at exceptionally high temperature in a black ceramic material comprising four elements: Lanthanum, Barium, Copper and Oxygen. Within the next decade many more complex copper oxides (cuprates) were synthesised including the mercury-cuprate compounds which, to date, have the highest confirmed critical temperature for a superconducting transition, some $T_c$=135K at room pressure and approximately 160K under high (applied) pressure. These discoveries could undoubtedly result in large scale commercial applications for cheap and efficient electricity production, providing long lengths of superconducting wires operating above the liquid nitrogen temperature (c.a. 80K) can be routinely manufactured. The new phenomenon initiated by Bednorz and Müller has broken all constraints on the maximum $T_c$ predicted by the conventional theory of low-temperature superconducting metals and their alloys.

Following Kamerlingh-Onnes' discovery of superconductivity in elemental mercury in 1911, subsequent work revealed that many metals and alloys displayed similar superconducting properties, the transition temperature of the alloy $Nb_3Ge$ at 23K being the highest recorded prior to the discoveries in the High-$T_c$ cuprates. Despite intense efforts worldwide, no adequate explanation of the superconductivity phenomenon appeared until the work by Bardeen, Cooper and Schrieffer (BCS) in 1957 [2]. By that time the frictionless flow (i.e. superfluidity) of liquid $^4$He had been discovered below a temperature of some 2.17K. It has been known that the helium atom $^4$He having two protons and two neutrons, is a Bose particle, or boson, while its isotope $^3$He, with two protons and only one neutron, is a fermion. An assembly of bosons obey Bose-Einstein statistics, which allows all of them to occupy a single quantum state. Fermions obey the Pauli exclusion principle and Fermi-Dirac statistics, which dictate that two identical particles must not occupy the same quantum state.

F. London suggested in 1938 that the remarkable superfluid properties of $^4$He were intimately linked to Bose-Einstein 'condensation' of the entire assembly of Bose particles [3]. Nine years later, Bogoliubov [4] and Landau [5] explained how Bose statistics can lead to the frictionless flow of a liquid. The bosons in the lowest energy state within the Bose-Einstein condensate thus form a coherent macromolecule communicating with each other by means of quantum mechanical forces. As soon as one Bose particle in the Bose liquid meets an obstacle to its flow, for example in the form of an impurity, all the others do not allow their condensate partner to be scattered to leave the condensate. A crucial demonstration that superfluidity was linked to Bose particles and Bose-Einstein condensation came after experiments on liquid $^3$He, whose atoms were fermions (two protons+one neutron), which failed to show the characteristic superfluid transition within a reasonable wide temperature interval around the critical temperature for the onset of superfluidity in $^4$He. In sharp contrast, $^3$He becomes a superfluid only below a very low temperature of some 0.0026 K; here we have a superfluid formed from pairs of two $^3$He fermions below this temperature.

The three orders-of-magnitude difference between the critical (superfluidity) temperatures of $^4$He and $^3$He kindles the view that the Bose-Einstein condensation might represent a 'smoking gun' of high temperature superconductivity. 'Unfortunately', electrons are fermions. Therefore, it is not surprising at all that the first proposal for high temperature superconductivity, made by Ogg Jr in 1946 [6], involved the pairing of individual electrons. If two electrons are chemically coupled together the resulting combination is a boson with a total spin $S = 0$. Thus an ensemble of such two-electron entities can, in principle, be condensed into the Bose-Einstein superfluid or superconducting condensate. This idea became very attractive as a natural explanation for superconductivity when Schafroth in 1955 showed that a gas of charged bosons with the charge 2e per particle would expel the magnetic flux (the venerable Meissner-Ochsenfeld effect), which is a characteristic feature of bulk superconductivity [7].

However, with one or two exceptions, the Ogg-Schafroth picture was condemned and practically forgotten because it neither accounted quantitatively for the critical parameters of 'old' (i.e. low- $T_c$) superconductors, nor did it explain the microscopic nature of the attractive force which could overcome the natural Coulomb repulsion between two electrons which constitute a Bose pair. In addition, the same theory which yields a rather precise estimate of the critical temperature of $^4$He leads to an utterly unrealistic result for superconductors, $viz$, $T_c = 10000K$ with the atomic density of electron pairs about $10^{22}$ per $cm^3$, and with the effective mass of each boson twice of the electron mass, $m = 2m_e \simeq 2 \times 10^{-27}g$.

The failure of this 'bosonic' picture of individual electron pairs became fully transparent when Bardeen, Cooper and Schrieffer [2] proposed that two electrons in a superconductor indeed formed a pair, but of a very large (practically macroscopic) dimension of about 10000 times the average inter-electron spacing. The BCS theory derived from an early demonstration by Fröhlich [8] that conduction electrons in states near the Fermi energy could attract each other weakly on account of their interaction with vibrating ions of a crystal lattice. Cooper then showed that electron pairs are stable only due to their quantum interaction with other pairs. The final BCS theory showed that in a small interval round the Fermi energy, the electrons are correlated into pairs in the momentum space. These Cooper pairs would strongly overlap in the real space, in sharp contrast with the model of non-overlapping (local) pairs discussed earlier by Ogg and Schafroth. Highly successful for metals and alloys with a low $T_c$, the BCS theory led the vast majority of theorists to the conclusion that there could be no possibility of superconductivity above 30K, which implied that Nb$_3$Ge already had the highest $T_c$. While the Ogg-Schafroth phenomenology led to unrealistically high values of $T_c$, the BCS theory left perhaps only limited hope for the discovery of new materials which could be superconducting at room temperatures or, at least, at liquid nitrogen temperatures.

Now it is realised [9–11], that the Ogg-Schafroth and BCS descriptions are actually two opposite extremes of the same problem of the electron-lattice interaction. By extending the BCS theory towards the strong interaction between electrons and ions, a Bose liquid of electron pairs surrounded by the lattice deformation (i.e. *bipolarons*) was naturally predicted with the conclusion that high temperature superconductivity could exist in the crossover region of the electron-lattice interaction strength from the BCS-like to bipolaronic superconductivity [11]. Compared with the early Ogg-Schafroth view, two fermions (now polarons) are bound into a bipolaron by a lattice deformation. Such bipolaronic states are much heavier since they are 'dressed' by the same lattice deformation. As a result the superconducting critical temperature, being proportional to the inverse mass of a bipolaron [9], was reduced in comparison with an 'ultra-hot', local-pair, Ogg-Schafroth superconductivity, but occured much higher than the BCS theory prediction. Quite remarkably Bednorz and Müller noted in their original publication and subsequently in their Nobel Price lecture [12], that in their ground-breaking search for High-$T_c$ superconductivity, they were stimulated and guided by the polaron model. Their expectation

[12] was that if 'an electron and a surrounding lattice distortion with a high effective mass can travel through the lattice as a whole, and a strong electron-lattice coupling exists an insulator could be turned into a high temperature superconductor'.

In these lectures I discuss the extension of the BCS theory to the strong-coupling regime with small polarons and bipolarons at a fairly basic level. First the canonical Migdal-Eliashberg theory of coupled electrons and phonons and its breakdown in the strong-coupling regime are discussed in Section 2. I show how the multi-polaron problem is reduced to a weakly interacting charged Bose gas in Section 3. A few recent applications of the bipolaron theory to cuprates are discussed in Section 4, in particular the essential interaction, a parameter-free expression for the superconducting critical temperature, upper crtitical field, symmetry of the order parameter, the London penetration depth, vortex structure, theory of tunnelling, Andreev reflection, angle-resolved photoemission, normal and superconducting gaps, and stripes.

## II  MIGDAL-ELIASHBERG THEORY

The theory of ordinary metals and superconductors is based on 'Migdal's' theorem [13], which showed that the contribution of the diagrams with 'crossing' phonon lines (so called 'vertex' corrections) is small. This is true if the parameter $\lambda \omega_D / \mu$ is small, where $\lambda$ is dimensionless (BCS) electron-phonon coupling constant, $\mu$ the Fermi energy in a rigid lattice (i.e. in the absence of the electron-phonon interaction), and $\omega_D$ is the Debye phonon frequency ($\hbar = c = k_B = 1$). Neglecting the vertex corrections Migdal calculated the renormalised electron mass as $m^* = m(1 + \lambda)$ (near the Fermi level) [13] and by breaking the gauge symmetry Eliashberg extended the Migdal theory to describe the BCS superconducting state at intermediate values of $\lambda$ [14].

### A  Generic Hamiltonian

The Hamiltonian describing interacting electrons on a lattice and phonons is derived by the use of the local-density and harmonic approximations (see, for example, Ref. [9]) as

$$H = H_0 + H_{e-ph} + H_{e-e}, \tag{1}$$

where

$$H_0 = \sum_{\mathbf{k},n,s} \xi_{\mathbf{k},n,s} c^\dagger_{\mathbf{k},n,s} c_{\mathbf{k},n,s} + \sum_{\mathbf{q},\nu} \omega_{\mathbf{q},\nu} (d^\dagger_{\mathbf{q},\nu} d_{\mathbf{q},\nu} + 1/2) \tag{2}$$

describes independent electron Bloch bands and phonons, $\xi_{\mathbf{k},n,s} = E_{\mathbf{k},n,s} - \mu$ and $\omega_{\mathbf{q},\nu}$ are the electron and phonon energy spectrum, respectively. Because the electron-phonon interaction leads to the pairing of electrons it is convenient to consider an open system with a fixed chemical potential $\mu$ rather than to fix the number of electrons $N_e$ to avoid some artificial difference between odd and even $N_e$. The ground state of an open system has the minimum expectation value of $H - \mu N_e$. That is why the electron energy is related to $\mu$.

The part of the electron-phonon interaction, which is linear in phonon operators, $d_{\mathbf{q},\nu}$ can be written as

$$H_{e-ph} = \frac{1}{\sqrt{2N}} \sum_{\mathbf{k},\mathbf{q},n,n',\nu,s} \gamma_{n,n'}(\mathbf{q},\mathbf{k},\nu) \omega_{\mathbf{q},\nu} c^\dagger_{\mathbf{k},n,s} c_{\mathbf{k}-\mathbf{q},n',s} d_{\mathbf{q},\nu} + H.c., \quad (3)$$

where $N$ is the number of unit cells in a crystal. Low energy physics is often described within a single band approximation with the (dimensionless) matrix element $\gamma(\mathbf{q})$ depending only on the momentum transfer $\mathbf{q}$. If

$$\gamma_{n,n'}(\mathbf{q},\mathbf{k},\nu) = \gamma(\mathbf{q}) \quad (4)$$

we call $H_{e-ph}$ as the Fröhlich interaction. Terms of $H_{e-ph}$ quadratic and higher order in phonon operators are small. They play some role only for those phonons which are not coupled with electrons by the linear interaction ($\gamma = 0$). The correlation energy of a homogeneous electron system is written as

$$H_{e-e} = \frac{1}{2} \sum_{\mathbf{q}} V_c(\mathbf{q}) \rho^\dagger_{\mathbf{q}} \rho_{\mathbf{q}} \quad (5)$$

where $V_c(\mathbf{q})$ is the Fourier component of the Coulomb interaction, and

$$\rho^\dagger_{\mathbf{q}} = \sum_{\mathbf{k},s} c^\dagger_{\mathbf{k},s} c_{\mathbf{k}+\mathbf{q},s} \quad (6)$$

is the density fluctuation operator. The operator $c_{\mathbf{k},s}$ annihilates an electron with the momentum $\mathbf{k}$ and spin $s$. For doped semiconductors and amorphous metals a random potential should be also included in $H$.

The harmonic approximation is sufficient for low temperature kinetics and thermodynamics of solids. Further corrections especially those of third and fourth order in ionic displacements, are known as anharmonic terms, and are of importance for the thermal expansion and lattice thermal conductivity.

## B  Effect of the electron-phonon interaction on electrons in a metal

Electrons in ordinary metals screen the bare ion-ion Coulomb repulsion. The residual short range dynamical matrix has the sound wave linear dispersion of eigenfrequencies in the long wave limit. The lowest contribution to the electron self-energy is given by two second order diagrams, Fig. 2.1 a,b. The diagram Fig. 2.1b is proportional to $|\gamma(\mathbf{q})|^2$ with $\mathbf{q} \equiv 0$, which is zero for acoustic phonons.

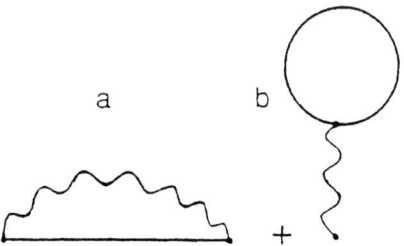

**FIGURE 2.1.** Second order electron self-energy.

The summation of all 'noncrossing' diagrams, Fig. 2.2 yields for the electron self-energy

$$\Sigma(\mathbf{k},\epsilon) = \frac{2i}{(2\pi)^4 N} \int d\mathbf{q} d\omega E_p G(\mathbf{k}-\mathbf{q},\epsilon-\omega) D(\mathbf{q},\omega). \tag{7}$$

Here

$$2E_p = |\gamma(\mathbf{q})|^2 \omega_\mathbf{q}, \tag{8}$$

and the electron and phonon Green's functions (GF) are

$$G(\mathbf{k},\omega) = \frac{1}{\omega - \xi_\mathbf{k} - \Sigma(\mathbf{k},\omega)}, \tag{9}$$

and

$$D(\mathbf{q},\omega) = \frac{\omega_\mathbf{q}^2}{\omega^2 - \omega_\mathbf{q}^2 + i\delta} \tag{10}$$

(spin and phonon mode quantum numbers are omitted, $\delta = +0$).

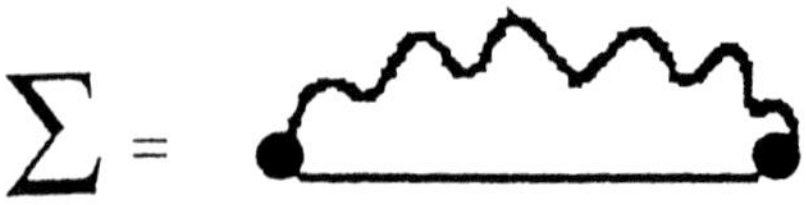

**FIGURE 2.2.** Electron self-energy in the Migdal approximation.

The electron-phonon interaction $E_p$ is generally not small being of the order of the Fermi-energy. Therefore one has to consider fourth and higher order diagrams with the crossing phonon lines as in Fig. 2.3, which are absent from Fig. 2.2. As shown by Migdal their contribution is adiabatically small ($\sim s/v_F$) compared with Eq.(7) ($s$ and $v_F = k_F/m$ are sound and Fermi velocities, respectively, $k_F$ is the Fermi momentum, and $m$ is the band mass). The dimensionless coupling constant

$$\lambda = 2E_p N(0) \tag{11}$$

might be not small:

$$\lambda \simeq 1. \tag{12}$$

Here

$$N(0) = \frac{1}{N} \sum_{\mathbf{k}} \delta(\xi_{\mathbf{k}} - \mu) \tag{13}$$

is the density of states (DOS) per cell at the Fermi level.

**FIGURE 2.3.** Adiabatically small corrections to the electron self-energy.

A new quasiparticle spectrum renormalised by the interaction is determined as a pole of the electron GF:

$$\tilde{E}_{\mathbf{k}} = \tilde{\mu} + v_F(k - k_F) + \delta E_{\mathbf{k}} \tag{14}$$

where

$$\delta E_{\mathbf{k}} = \Sigma(\mathbf{k}, \tilde{E}_{\mathbf{k}} - \tilde{\mu}) - \Sigma(k_F, 0), \tag{15}$$

and $\tilde{\mu} = \mu + \Sigma(k_F, 0)$ is the renormalised Fermi energy. The adiabatic condition allows one to simplify the self energy further by replacing the exact GF in the integral Eq.(7) by the free one. This is appropriate for all quasiparticle energies with the exception of a very narrow region ($<< \omega_D s/v_F$) near the Fermi level, where the difference between $G$ and $G^{(0)}$ appears to be essential for the damping. As a result

$$\delta E_{\mathbf{k}} = \frac{2iE_p}{(2\pi)^4} \int d\mathbf{q}d\omega \left( G^{(0)}(\mathbf{k} - \mathbf{q}, \tilde{\xi} - \omega) - G^{(0)}(k_F - \mathbf{q}, -\omega) \right) D(\mathbf{q}, \omega). \tag{16}$$

To simplify the calculations we take $E_p$ independent of the momentum, apply the Debye approximation for acoustic phonons $\omega_\mathbf{q} = sq$ for $q < q_D$, where $q_D \simeq \pi/a$ is the Debye momentum, and consider a half-filled band, $k_F \simeq \pi/2a$ with the energy independent DOS near the Fermi level $N(0) = m_e a^2 / 4\pi$ ($a$ is the lattice constant).

The main contribution to the integral Eq.(16) comes from the momentum region close to the Fermi-surface:

$$|\mathbf{k} - \mathbf{q}| \simeq k_F. \tag{17}$$

This makes it convenient to introduce a new variable $k' = |\mathbf{k} - \mathbf{q}|$ instead of the angle $\Theta$ between $\mathbf{k}$ and $\mathbf{q}$ and extend the integration to $\pm\infty$ with the variable $\xi = v_F(k' - k_F)$. Thus the angular integration in Eq.(16) becomes:

$$\int d\Theta \sin\Theta \, (...) \sim \int_{-\infty}^{\infty} d\xi \frac{\tilde{\xi}}{[\tilde{\xi} - \omega - \xi + i\delta sgn(\xi)][\omega + \xi - i\delta sgn(\xi)]}. \tag{18}$$

This integral is non-zero only if $\tilde{\xi} > \omega > 0$ or $\tilde{\xi} < \omega < 0$. It is $-2\pi i$ in the first region and $2\pi i$ in the second one. Taking into account that $D$ is an even function of $\omega$ one obtains:

$$\delta E_\mathbf{k} = \frac{2E_p}{(2\pi)^2 v_F N} \int_0^{q_D} dq\, q \int_0^{|\tilde{\xi}|} d\omega\, sgn(\tilde{\xi}) \frac{\omega_q^2}{\omega^2 - \omega_q^2 + i\delta}. \tag{19}$$

The real and imaginary parts of Eq.(19) determine respectively the renormalised spectrum and the life time of quasiparticles:

$$\Re(\delta E_\mathbf{k}) = \frac{E_p}{4\pi^2 v_F N} \int_0^{q_D} dq\, q\, \omega_q \ln\left|\frac{\omega_q - \tilde{\xi}}{\omega_q + \tilde{\xi}}\right|, \tag{20}$$

$$\Im(\delta E_\mathbf{k}) = \frac{E_p}{4\pi v_F N} \int_0^{q_m} dq\, q\, \omega_q\, sgn(\tilde{\xi}) \tag{21}$$

with $q_m = |\tilde{\xi}|/s$ if $|\tilde{\xi}| < \omega_D$ and $q_m = q_D$ if $|\tilde{\xi}| > \omega_D$. For the excitations far away from the Fermi surface with $|\tilde{\xi}| >> \omega_D$

$$\Re(\delta E_\mathbf{k}) = -\lambda \frac{\omega_D^2}{2\tilde{\xi}}. \tag{22}$$

Here $\omega_D = sq_D$ is the Debye frequency. For low-energy excitations with $|\tilde{\xi}| << \omega_D$

$$\Re(\delta E_\mathbf{k}) = -\lambda \tilde{\xi}, \tag{23}$$

which means an increase of the effective mass of the excitation due to the electron-phonon interaction

$$\tilde{\xi} = \frac{k_F}{m^*}(k - k_F). \qquad (24)$$

The renormalised effective mass is

$$m^* = (1 + \lambda)m. \qquad (25)$$

Thus the excitation spectrum of a metal has two different regions with two different values of the effective mass. The thermodynamic properties of a metal at low temperature $T << \omega_D$ involve $m^*$, but the optical properties in a frequency range $\nu >> \omega_D$ are determined by the high-energy excitations, where according to Eq.(22) corrections are small and the mass is equal to the band mass $m$.

The damping has just the opposite behavior. The integral Eq.(21) yields

$$\Im(\delta E_\mathbf{k}) = \frac{sgn(\tilde{\xi})\pi\lambda\omega_D}{3} \qquad (26)$$

if $|\tilde{\xi}| > \omega_D$, and

$$\Im(\delta E_\mathbf{k}) = \frac{sgn(\tilde{\xi})\pi\lambda|\tilde{\xi}|^3}{3\omega_D^2} \qquad (27)$$

for $|\tilde{\xi}| << \omega_D$.

These expressions describe the rate of decay of the quasiparticles due to emission of phonons. In the immediate neighborhood of the Fermi surface $|\tilde{\xi}| << \omega_D$ the decay is small compared with the quasiparticle energy $|\tilde{\xi}|$ even for a relatively strong coupling $\lambda \sim 1$, so that the concept of well-defined quasiparticles has a definite meaning. Hence within the Migdal approximation the electron-phonon interaction does not destroy the Fermi-liquid behavior of electrons. The Pauli exclusion principle is responsible for the stability of the Fermi liquid. In the intermediate-energy region $|\tilde{\xi}| \sim \omega_D$, however, the decay is comparable with the energy. Here the quasiparticle spectrum looses its meaning. In the high-energy region $|\tilde{\xi}| >> \omega_D$ the decay remains the same in absolute value but again becomes small in comparison with $|\tilde{\xi}|$ and the quasiparticle concept recovers its meaning.

## C  Broken gauge symmetry and the BCS ground state

The Migdal approach is justified if the ground state is stable versus a phase transition. Fröhlich [8] realised that the electron-phonon interaction leads to an attraction between two electrons and Cooper [15] discovered that any attraction between *degenerate* electrons leads to their pairing; it does not matter how weak it is. Pairs are bosons which undergo a Bose-Einstein condensation at some critical temperature $T_c$. The condensed state is described by the classical field, which is an average of the product of two annihilation field operators $F \sim <\psi\psi>$ or two creation operators $F^+ \sim <\psi^\dagger\psi^\dagger>$. This average is macroscopically large below $T_c$ in an open system. The appearance of the 'anomalous' averages brakes the gauge symmetry of the bare Hamiltonian, Eq.(1) and cannot be described perturbatively. Following Bardeen, Cooper and Schrieffer [2] one has to go beyond the Migdal approximation, Fig. 2.2 including the anomalous averages in the ground state. This can be done selfconsistenly using the same self-energy diagram Fig. 2.2 as for the normal state but with the matrix electron GF [14]. At finite temperature the diagram technique can be formulated for the 'temperature' GF defined by Matsubara [16] as

$$g(\mathbf{k},\tau) = -\langle\langle T_\tau c_\mathbf{k}(\tau) c_\mathbf{k}^\dagger \rangle\rangle, \tag{28}$$

with $c_\mathbf{k}(\tau) = exp(H\tau) c_\mathbf{k} exp(-H\tau)$ and $0 < \tau < 1/T$ a 'thermodynamic' time. The double angular brackets correspond to the quantum as well as statistical average with the Gibbs distribution:

$$\langle\langle ... \rangle\rangle = \sum_\nu e^{\frac{\Omega - E_\nu}{T}} \langle \nu | ... | \nu \rangle \tag{29}$$

where $\Omega$ is the thermodynamic potential and $|\nu\rangle$ the eigenstates of $H - \mu N$ with the eigenvalues $E_\nu$. Because the thermodynamic time is restricted by $1/T$ the temperature GF is expanded in the Fourier series:

$$g(\mathbf{k},\tau) = T \sum_{\omega_n} e^{-i\omega_n \tau} g(\mathbf{k},\omega_n) \tag{30}$$

with the discrete Matsubara frequencies $\omega_n = \pi T(2n+1)$, $n = 0, \pm 1, \pm 2, .....$ For free electrons:

$$g^{(0)}(\mathbf{k},\omega_n) = \frac{1}{i\omega_n - \xi_\mathbf{k}} \tag{31}$$

and for phonons:

$$d^{(0)}(\mathbf{q},\omega_n - \omega_{n'}) = -\frac{\omega_\mathbf{q}^2}{[\omega_n - \omega_{n'}]^2 + \omega_\mathbf{q}^2} \tag{32}$$

To take into account the Cooper pairing of two electrons with the opposite momentum and spin one can introduce, following Gor'kov [17] and Nambu [18] the matrix GF:

$$\hat{g}(\mathbf{k},\tau) = -\begin{pmatrix} \langle\langle T_\tau c_{\mathbf{k},\uparrow}(\tau) c_{\mathbf{k},\uparrow}^\dagger \rangle\rangle & \langle\langle T_\tau c_{\mathbf{k},\uparrow}(\tau) c_{-\mathbf{k},\downarrow} \rangle\rangle \\ \langle\langle T_\tau c_{-\mathbf{k},\downarrow}^\dagger(\tau) c_{\mathbf{k},\uparrow}^\dagger \rangle\rangle & \langle\langle T_\tau c_{-\mathbf{k},\downarrow}^\dagger(\tau) c_{-\mathbf{k},\downarrow} \rangle\rangle \end{pmatrix}. \tag{33}$$

The matrix self-energy

$$\hat{\Sigma}(\mathbf{k},\omega_n) = \left(\hat{g}^{(0)}(\mathbf{k},\omega_n)\right)^{-1} - \hat{g}^{-1}(\mathbf{k},\omega_n) \tag{34}$$

with $\hat{g}^{(0)}(\mathbf{k},\omega_n) = (i\omega_n \tau_0 - \xi_\mathbf{k} \tau_3)^{-1}$. Here $\tau_{0,1,2,3}$ is the set of the Pauli matrices:

$$\tau_0 = \begin{pmatrix} 1 & 0 \\ 0 & 1 \end{pmatrix},$$

$$\tau_1 = \begin{pmatrix} 0 & 1 \\ 1 & 0 \end{pmatrix},$$

$$\tau_2 = \begin{pmatrix} 0 & -i \\ i & 0 \end{pmatrix},$$

$$\tau_3 = \begin{pmatrix} 1 & 0 \\ 0 & -1 \end{pmatrix}.$$

The generalized equation for the matrix $\hat{\Sigma}$ is given by the same diagram as in the normal state, Fig. 2.2, with substitution $\gamma(\mathbf{q})\tau_3$ instead of $\gamma(\mathbf{q})$ and summation over Matsubara frequencies instead of integration:

$$\hat{\Sigma}(\mathbf{k},\omega_n) = -T \sum_{\omega_{n'}} \int \frac{d\mathbf{q}}{(2\pi)^3} \gamma^2(\mathbf{q}) \omega_{\mathbf{q}} \tau_3 \hat{g}(\mathbf{k}-\mathbf{q},\omega_{n'})\tau_3 d(\mathbf{q},\omega_n - \omega_{n'}) \quad (35)$$

The most important difference of Eq.(35) compared with the normal state Eq.(7) is the possibility of obtaining a finite value of the off-diagonal matrix elements $\sim <cc>$ within the selfconsistent solution. In the second ($\hat{g} \to \hat{g}^{(0)}$) or in any finite order of the perturbation theory there are no anomalous averages from Eq.(35). That means that within the perturbation theory there is no superconducting phase transition. However if one solves Eq.(35) selfconsistently, one obtains the finite anomalous averages.

To illustrate this we adopt a particular momentum dependence of the interaction constant as in the previous section and an approximate form of the phonon Green's function:

$$d(\mathbf{q},\omega_n - \omega_{n'}) \simeq -1 \quad (36)$$

if $|\omega_n - \omega_{n'}| < \omega_D$ and zero otherwise. If there is no current in the system the phase of the order parameter can be chosen to be zero. In this case $\hat{\Sigma}$ is a sum of three Pauli matrices $\tau_{0,1,3}$ with the coefficients $(1-Z)i\omega_n$, $\Delta$ and $\chi$ respectively, which are the functions of frequency and momentum:

$$\hat{\Sigma}(\mathbf{k},\omega_n) = (1-Z)i\omega_n \tau_0 + \Delta \tau_1 + \chi \tau_3. \quad (37)$$

Thus

$$\hat{g}^{-1}(\mathbf{k},\omega_n) = Zi\omega_n \tau_0 - \Delta \tau_1 - \tilde{\xi}\tau_3 \quad (38)$$

and

$$\hat{g}(\mathbf{k},\omega_n) = -\frac{Zi\omega_n + \Delta \tau_1 + \tilde{\xi}\tau_3}{Z^2\omega_n^2 + \tilde{\xi}^2 + \Delta^2} \quad (39)$$

with $\tilde{\xi} = \xi + \chi$. Substitution of Eq.(37-39) in the master equation (35) yields

$$(1-Z)i\omega_n = -\lambda T \int d\tilde{\xi} \sum_{\omega_{n'}} \frac{i\omega_{n'}Z}{Z^2\omega_{n'}^2 + \tilde{\xi}^2 + \Delta^2} = 0 \qquad (40)$$

$$\chi = -\lambda T \int d\tilde{\xi} \sum_{\omega_{n'}} \frac{\tilde{\xi}}{Z^2\omega_{n'}^2 + \tilde{\xi}^2 + \Delta^2} = 0 \qquad (41)$$

$$\Delta = \lambda T \int d\tilde{\xi} \sum_{\omega_{n'}} \frac{\Delta}{Z^2\omega_{n'}^2 + \tilde{\xi}^2 + \Delta^2} \qquad (42)$$

It follows from Eq. (40,41) that $Z = 1$ and $\chi = 0$. Applying the formula for $tanh$ to the sum in Eq.(42) we obtain the familiar BCS equation for the order parameter

$$1 = \frac{\lambda}{2} \int \frac{d\xi}{\sqrt{\xi^2 + \Delta^2}} tanh \frac{\sqrt{\xi^2 + \Delta^2}}{2T}, \qquad (43)$$

where the integration is restricted by the region $|\xi| < \omega_D$ because of the approximation, Eq.(36) for $d(\mathbf{q}, \omega_n - \omega_{n'})$.

There is no direct physical meaning of poles of the temperature GF. To derive the one-particle excitation spectrum one has to calculate the real-time GF determined at finite temperature as

$$G(\mathbf{k}, t) = -i\langle\langle T_t c_\mathbf{k}(t) c_\mathbf{k}^\dagger\rangle\rangle, \qquad (44)$$

with the real time $t$. One can use the retarded $G^R$ and advanced $G^A$ Green functions:

$$G^R(\mathbf{k}, t) = -i\Theta(t)\langle\langle[c_\mathbf{k}(t)c_\mathbf{k}^\dagger]\rangle\rangle \qquad (45)$$

$$G^A(\mathbf{k}, t) = i\Theta(-t)\langle\langle[c_\mathbf{k}(t)c_\mathbf{k}^\dagger]\rangle\rangle \qquad (46)$$

where [...] is an anticommutator. They are analytical in the upper or lower halfplane of $\omega$ correspondingly. There is a simple connection between the Fourier components of $G$ and $G^{R,A}$

$$G^{R,A}(\mathbf{k}, \omega) = \Re G(\mathbf{k}, \omega) \pm i coth(\frac{\omega}{2T}) Im G(\mathbf{k}, \omega) \qquad (47)$$

from one hand and between those of $G^{R,A}$ and $g$ from another hand,

$$G^R(\mathbf{k}, i\omega_n) = g(\mathbf{k}, \omega_n) \qquad (48)$$

for $\omega_n > 0$ and

$$g(\mathbf{k}, -\omega_n) = g^*(\mathbf{k}, \omega_n). \qquad (49)$$

In our case the temperature GF

$$g(\mathbf{k}, \omega) = \frac{u_{\mathbf{k}}^2}{i\omega_n - \epsilon_{\mathbf{k}}} + \frac{v_{\mathbf{k}}^2}{i\omega_n + \epsilon_{\mathbf{k}}}. \tag{50}$$

where $u_{\mathbf{k}}^2, v_{\mathbf{k}}^2 = (\epsilon_{\mathbf{k}} \pm \xi_{\mathbf{k}})/2\epsilon_{\mathbf{k}}$ and $\epsilon_{\mathbf{k}} = \sqrt{\xi_{\mathbf{k}}^2 + \Delta^2}$. The analytical continuation of this expression to the upper half-plane yields

$$G^R(\mathbf{k}, \omega) = \frac{u_{\mathbf{k}}^2}{\omega - \epsilon_{\mathbf{k}} + i0^+} + \frac{v_{\mathbf{k}}^2}{\omega + \epsilon_{\mathbf{k}} + i0^+}. \tag{51}$$

and with Eq.(47) one obtains

$$G(\mathbf{k}, \omega) = \Re \left( \frac{u_{\mathbf{k}}^2}{\omega - \epsilon_{\mathbf{k}}} + \frac{v_{\mathbf{k}}^2}{\omega + \epsilon_{\mathbf{k}}} \right) - i\pi \tanh(\frac{\omega}{2T}) \left( u_{\mathbf{k}}^2 \delta(\omega - \epsilon_{\mathbf{k}}) + v_{\mathbf{k}}^2 \delta(\omega + \epsilon_{\mathbf{k}}) \right). \tag{52}$$

If $T = 0$, $\tanh(\frac{\omega}{2T}) = sgn(\omega)$ and

$$G(\mathbf{k}, \omega) = \frac{u_{\mathbf{k}}^2}{\omega - \epsilon_{\mathbf{k}} + i0^+} + \frac{v_{\mathbf{k}}^2}{\omega + \epsilon_{\mathbf{k}} - i0^+}. \tag{53}$$

The poles of GF are the same as the BCS quasiparticles:

$$\epsilon_{\mathbf{k}} = \sqrt{\xi_{\mathbf{k}}^2 + \Delta^2(0)}. \tag{54}$$

Thus the Migdal-Eliashberg theory reproduces the BCS results if a similar approximation for the attraction between electrons is made. The critical temperature and the BCS gap are adiabatically small ($\sim \omega_D$) compared with the Fermi energy and one can worry about the 'crossing' diagrams as in Fig. 2.3, which are neglected in the master equation, Eq.(35). However, the BCS state is essentially the same as the normal one outside the narrow momentum region around the Fermi surface. The outside region contribute mainly to the integrals in the crossing diagrams, which makes them small in the BCS state as well.

Within a more general consideration the equation (35) takes properly into account the phonon spectrum, retardation and realistic matrix element of the electron-phonon interaction. In particular it is useful in a study of the effect of the Coulomb repulsion on the pairing. There is no adiabatic parameter for this interaction. Nevertheless in a qualitative analysis one can adopt the same contribution to the electron self-energy from the Coulomb interaction as from phonons replacing $\gamma^2(\mathbf{q})\omega_\mathbf{q} d(\mathbf{q}, \omega_n - \omega_{n'})$ in Eq.(35) for the Fourier component of the Coulomb potential $4\pi e^2/q^2$. The Coulomb interaction is nonretarded for the frequencies less or compared with the Fermi energy, so the equation for the order parameter becomes

$$\Delta(\omega_n) = T \int d\xi \sum_{\omega_{n'}} K(\omega_n - \omega_{n'}) \frac{\Delta(\omega_{n'})}{\omega_{n'}^2 + \xi^2 + \Delta^2(\omega_{n'})}, \tag{55}$$

where the kernel $K$ is given by

$$K(\omega_n - \omega_{n'}) = \lambda\Theta(\omega_D - |\omega_n - \omega_{n'}|) - \mu_c\Theta(\mu - |\omega_n - \omega_{n'}|) \tag{56}$$

with $\mu_c$ the product of the Fourier component of the Coulomb potential and the normal density of states at the Fermi level. At $T = T_c$ one can neglect the second power of the order parameter in Eq.(55) and integrating over $\xi$ obtain

$$\Delta(\omega_n) = \pi T_c \sum_{\omega_{n'}} K(\omega_n - \omega_{n'})\frac{\Delta(\omega_{n'})}{|\omega_{n'}|}. \tag{57}$$

To solve this equation we adopt the BCS-like parametrization of the kernel

$$K(\omega_n - \omega_{n'}) \simeq \lambda\Theta(2\omega_D - |\omega_n|)\Theta(2\omega_D - |\omega_{n'}|) \\ - \mu_c\Theta(2\mu - |\omega_n|)\Theta(2\mu - |\omega_{n'}|) \tag{58}$$

and replace the summation by the integration

$$\pi T_c \sum \to \int_{\pi T_c}^{\infty} d\omega \tag{59}$$

because $T_c \ll \omega_D, \mu$. The solution can be found in the form

$$\Delta(\omega) = \Delta_1\Theta(2\omega_D - |\omega|) + \Delta_2\Theta(2\mu - |\omega|)\Theta(|\omega| - 2\omega_D) \tag{60}$$

with constant but different values of the order parameter below ($\Delta_1$) and above ($\Delta_2$) the cut-off energy $2\omega_D$. Substitution of Eq.(58,60) into Eq.(57) yields for $\Delta_{1,2}$

$$\Delta_1\left[1 - (\lambda - \mu_c)\ln\frac{2\omega_D}{\pi T_c}\right] + \Delta_2\mu_c\ln\frac{\mu}{\omega_D} = 0 \tag{61}$$

$$\Delta_1\mu_c\ln\frac{2\omega_D}{\pi T_c} + \Delta_2\left[1 + \mu_c\ln\frac{\mu}{\omega_D}\right] = 0 \tag{62}$$

The condition of the existence of a nontrivial solution of these coupled equations gives $T_c$

$$T_c = \frac{2\omega_D}{\pi}exp\left(-\frac{1}{\lambda - \mu_c^*}\right) \tag{63}$$

where

$$\mu_c^* = \frac{\mu_c}{1 + \mu_c\ln(\mu/\omega_D)} \tag{64}$$

is the Coulomb pseudopotential [19]. This is a remarkable result. It shows that even a large Coulomb repulsion $\mu_c > \lambda$ may not destroy Cooper pairs because its contribution is suppressed down to the value $\sim 1/\ln\frac{\mu}{\omega_D} \ll 1$. The retarded attraction mediated by phonons acts well after two electrons meet each other. This time delay is sufficient for two electrons to be separated by the relative distance at which the Coulomb repulsion is small.

In the normal state the Coulomb correlations lead to a damping of excitations of the order $\xi^2/\mu$, which is relevant only in a narrow region around the Fermi surface of the order $\omega_D\sqrt{m/M}$. Outside this region the damping due to the Fröhlich interaction dominates.

For metals and their alloys the empirical McMillan [20] formula for $T_c$ is adopted:

$$T_c = \frac{\omega_D}{1.45} exp\left(-\frac{1.04(1+\lambda)}{\lambda - \mu_c^*(1+0.62\lambda)}\right) \quad (65)$$

which works well for low-$T_c$ materials even if the estimated $\lambda$ is large ($> 1$) as in $Pb$. However already in materials with a moderate $T_c \sim 20K$ as in $A-15$ compounds ($Nb_3Sn, V_3Si$) the discrepancy between the values of $\lambda$ estimated with Eq.(65) and the direct band-structure calculations exceeds by several times the limit allowed by the experimental and computation accuracy [21].

In original papers Migdal [13] and Eliashberg [14] restricted the region of the applicability of their approach by the value of the coupling $\lambda < 1$. We have shown that the proper extension of the BCS theory to the strong coupling region $\lambda > 1$ inevitably involves small polaron formation [11].

## D  Breakdown of the Migdal-Eliashberg theory in the strong-coupling regime

The same theory, applied to phonons, yields the renormalised phonon frequency $\tilde{\omega} = \omega(1-2\lambda)^{1/2}$ [13] with an instability at $\lambda = 0.5$. Because of this instability both Migdal [13] and Eliashberg [14] restricted the applicability of their approach to $\lambda \leq 1$.

It was then shown that if the adiabatic Born-Oppenheimer approach is properly applied to a metal, there is only negligible renormalisation of the phonon frequencies of the order of the adiabatic ratio, $\omega/\mu \ll 1$ for *any* value of $\lambda$ [22]. The conclusion was that the standard electron-phonon interaction could be applied to electrons for any value of $\lambda$ but it should not be applied to renormalise phonons. As a result, many authors used the Migdal-Eliashberg theory with $\lambda$ much larger than 1 [23].

However, starting from the infinite coupling limit, $\lambda = \infty$ and applying the inverse $(1/\lambda)$ expansion technique [24,25] we showed [11,26,25] that the many-electron system collapses into small polaron regime at $\lambda \sim 1$ independent of the adiabatic ratio (see section 3). This regime is beyond the Migdal-Eliashberg theory. It cannot be reached by summation of the standard Feynman-Dyson perturbation diagrams even including *all* vertex corrections, because of the broken translational symmetry, as first discussed by Landau [27] for a single electron and by us [26] for the many-electron system. During last years quite a few numerical and analytical studies have confirmed this conclusion [28–32,9,33–44] (and references theirein). On the other hand a few others (see, for example, [45,46]) still argue that the breakdown of the Migdal-Eliashberg theory might happen only at $\lambda \geq \mu/\omega \gg 1$. Indeed numerical study of the finite bandwidth effects [47] and some analytical calculations of the vertex corrections to the vertex function [48,49,46] with the standard Feynman-Dyson perturbation technique confirm the second conclusion.

Here I compare the Migdal solution of the Holstein Hamiltonian with the exact one in the extreme adiabatic regime, $\omega/\mu \to 0$, to show that the ground state of the system is a self-trapped insulating state with a broken translational symmetry at $\lambda \geq 1$ rather than a Fermi liquid expected with the Migdal approach.

While the vertex corrections and the finite bandwidth are technical issues, playing no role in the extreme adiabatic limit [13,48,47], the starting basic assumption of the canonical Migdal-Eliashberg theory is that the electron and phonon Green functions (GF) are translationally invariant. As a result one takes $G(\mathbf{r}, \mathbf{r}', \tau) = G(\mathbf{r} - \mathbf{r}', \tau)$ with Fourier component $G(\mathbf{k}, \omega_n)$ prior to solving the Dyson equations. This assumption excludes the possibility of local violation of the translational symmetry due to the lattice deformation in any order of the Feynman-Dyson perturbation theory. This is similar to the absence of the anomalous (Bogoliubov) averages in any order of the perturbation theory (subsection 2.3). To enable electrons to relax into the lowest polaronic states, one has to introduce an infinitesimal translationally non-invariant potential, which should be set equal to zero only in the final solution for the GF [26]. As in the case of the off-diagonal superconducting order parameter, a small translational-symmetry-breaking potential drives the system into a new ground state at sufficiently large coupling, $\lambda \sim 1$, independent of the adiabatic ratio. Setting it equal to zero in the solution of the equations of motion restores the translational symmetry but in a new polaronic band rather than in the bare electron band, which turns out to be an excited state.

In particular, let us consider the extreme adiabatic limit of the Holstein chain [50], which has the simple analytical solution:

$$H = -t \sum_{<ij>} c_i^\dagger c_j + H.c. + 2(\lambda k t)^{1/2} \sum_i x_i c_i^\dagger c_i$$
$$+ \sum_i \left( -\frac{1}{2M} \frac{\partial^2}{\partial x_i^2} + \frac{k x_i^2}{2} \right), \tag{66}$$

where $t$ is the nearest neighbour hopping integral, $c_i^\dagger, c_i$ are the electron operators, $x_i$ is the normal coordinate of the molecule (site) $i$, and $k = M\omega^2$ with $M$ the ion mass [51]. We first consider the two-site case (zero dimensional limit), $i, j = 1, 2$ with one electron, and than generalise the result for the infinite lattice with many electrons. The transformation $X = (x_1 + x_2)$, $\xi = x_1 - x_2$ allows us to eliminate the coordinate $X$, which is coupled only with the total density $(n_1 + n_2 = 1)$, leaving the following Hamiltonian to solve in the extreme adiabatic limit $M \to \infty$:

$$H = -t(c_1^\dagger c_2 + c_2^\dagger c_1) + (\lambda k t)^{1/2}\xi(c_1^\dagger c_1 - c_2^\dagger c_2) + \frac{k\xi^2}{4}. \tag{67}$$

The solution is

$$\psi = (\alpha c_1^\dagger + \beta c_2^\dagger)|0>, \tag{68}$$

where

$$\alpha = \frac{t}{[t^2 + ((\lambda k t)^{1/2}\xi + (t^2 + \lambda k t\xi^2)^{1/2})^2]^{1/2}}, \tag{69}$$

$$\beta = -\frac{(\lambda k t)^{1/2}\xi + (t^2 + \lambda k t\xi^2)^{1/2}}{[t^2 + ((\lambda k t)^{1/2}\xi + (t^2 + \lambda k t\xi^2)^{1/2})^2]^{1/2}}, \tag{70}$$

and the energy

$$E = \frac{k\xi^2}{4} - (t^2 + \lambda k t\xi^2)^{1/2}. \tag{71}$$

In the extreme adiabatic limit the displacement $\xi$ is classical, so the ground state energy, $E_0$ and the ground state displacement $\xi_0$ are obtained by minimising Eq.(71) with respect to $\xi$. If $\lambda \geq 0.5$ one obtains

$$E_0 = -t(\lambda + \frac{1}{4\lambda}), \tag{72}$$

and

$$\xi_0 = \left[\frac{t(4\lambda^2 - 1)}{\lambda k}\right]^{1/2}. \tag{73}$$

The symmetry-breaking ('order') parameter is

$$\Delta \equiv \beta^2 - \alpha^2 = \frac{[2\lambda + (4\lambda^2 - 1)^{1/2}]^2 - 1}{[2\lambda + (4\lambda^2 - 1)^{1/2}]^2 + 1}. \tag{74}$$

If, however, $\lambda < 0.5$ the ground state is translationally invariant with $E_0 = -t, \xi = 0, \beta = -\alpha$, and $\Delta = 0$. Precisely this state is the 'Migdal' solution of the Holstein model. Indeed, in the Migdal approximation GF is diagonal in the $\mathbf{k}$ representation, $G(\mathbf{k}, \mathbf{k}', \tau) = G(\mathbf{k}, \tau)\delta_{\mathbf{k},\mathbf{k}'}$. The site operators can be transformed into momentum space as

$$c_k = N^{-1/2} \sum_j c_j \exp(ikaj), \tag{75}$$

with $k = 2\pi n/Na$, $-N/2 < n \leq N/2$. Than the off-diagonal GF with $k = 0$ and $k' = \pi/a$ of the two-site chain ($N = 2$) at $\tau = -0$ is given by

$$G(k, k', -0) = \frac{i}{2}\langle(c_1^\dagger - c_2^\dagger)(c_1 + c_2)\rangle. \tag{76}$$

Calculating this average one obtains

$$G(k, k', -0) = \frac{i}{2}(\alpha^2 - \beta^2), \tag{77}$$

which should vanish in the Migdal theory. Hence, this theory only provides symmetric (translationally invariant) solution with $|\alpha| = |\beta|$. When $\lambda > 0.5$ this solution is *not* the ground state of the system, Fig. 2.4. The system collapses into a localised adiabatic polaron, trapped on the 'right' (or on the 'left') site due to a finite local lattice deformation $\xi_0$. On the other hand, when $\lambda < 0.5$, the Migdal solution is the *only* solution with $\xi_0 = 0$. Thus the Migdal-Eliashberg constraint [13,14] on the applicability of their approach is perfectly correct irrespective of the phonon frequency renormalisation.

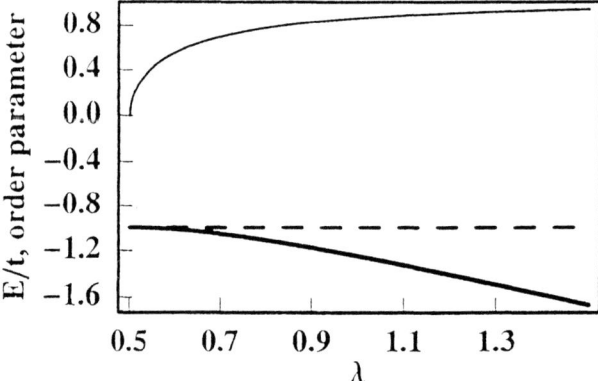

**FIGURE 2.4.** The ground state energy (in units of $t$, solid line) and the order parameter (thin solid line) of the adiabatic Holsten model. The Migdal solution is shown as the dashed line.

The generalisation to the multi-polaron system on the infinite lattice of any dimension is straightforward in the extreme adiabatic regime. The adiabatic solution of the infinite one-dimensional (1D) chain with one electron was obtained by Rashba [52] in the continuous approximation, and by Holstein [50] and Kabanov and Mashtakov [28] for a discrete lattice. The last authors also studied the Holstein two-dimensional (2D) and three-dimensional (3D) lattices in the adiabatic limit. According to Ref. [28] the self-trapping of a single electron occures for any value of $\lambda$ in a 1D Holstein chain, and at $\lambda \geq 0.875$ and $\lambda \geq 0.92$ in 2D and 3D, respectively. The radius of the self-trapped adiabatic polaron, $r_p$, is readily derived from its continous wave function [52]

$$\psi(x) \sim 1/\cosh(\lambda x/a). \tag{78}$$

It becomes smaller than the lattice constant, $r_p = a/\lambda$ for $\lambda \geq 1$. That is why in the strong-coupling ($\lambda \geq 1$) adiabatic regime the multi-polaron system remains in the self-trapped insulating state no matter how many polarons it has. The only instability which might occur in this regime is the formation of on-site self-trapped bipolarons, if the Holstein on-site attractive interaction, $2\lambda zt$, is larger than the repulsive Hubbard $U$ [53]. Actually, this instability can be seen in the second order Feynamn-Dyson diagramm containing the polarisation loop, as explained in Ref. [54]. On-site bipolarons forme charge ordered insulating state due to weak repulsion between them [55]. The exact analytical and numerical proof of this statement as well as different polaronic and bipolaronic configurations in the adiabatic Holstein model was reviewed by Aubry [56]. For example, the asymptotically exact many-particle ground state of the half-filled Holstein model in the strong-coupling limit ($\lambda \gg 1$) is [55]

$$\psi = \prod_{j \in B} c^\dagger_{j,\uparrow} c^\dagger_{j,\downarrow} |0\rangle \tag{79}$$

for any value of the adiabatic ratio, $\omega/zt$ [55,56]. Here $j$ are $B$-sites of the bipartite lattice $A+B$. It is a charged ordered insulating state, rather than the Fermi liquid, expected in the Migdal approximation at any value of $\lambda$ in the extreme adiabatic limit, $\omega \to 0$. If the Coulomb repulsion (Hubbard $U$) is sufficiently strong to prevent the bipolaron formation, than at half filling every site is occupied by one polaron. The deformation barrier for their tunneling to a neighboring site disappears, so one could erroneously believe that the system should be metallic. However, to keep polarons unbound into on-site bipolarons, the Hubbard $U$ should be larger than $2\lambda zt$, i.e. larger than the bandwidth, $U > 2zt$, if $\lambda > 1$. It is well known that in this regime the system is a Mott-Hubbard insulator, rather than an ordinary metal. Hence, the Migdal-Eliashberg theory cannot be applied at $\lambda \geq 1$, no matter what a strength of the Coulomb interaction, the number of electrons and the dimension of the lattice are.

The non-adiabtic corrections (phonons) allow polarons and bipolarons propagate as Bloch states in a new (narrow) band (see section 3). Thus, under certain conditions [44] the multi-polaron system might be metallic with polaronic (or bipolaronic) carriers rather than bare electrons. However, there is a qualitative difference between the ordinary Fermi liquid and the polaronic one. In particular, the renormalized (effective) mass of electrons is independent of the ion mass $M$ in ordinary metals (where the Migdal adiabatic approximation is believed to be valid), because $\lambda$ does not depend on the isotope mass. However, the polaron effective mass $m^*$ will depend on $M$. This is because the polaron mass $m^* = m \exp(A/\omega)$ [57], where $m$ is the band mass in the absence of the electron-phonon interaction, and $A$ is a constant. Hence, there is a large isotope effect on the carrier mass in polaronic metals, in contrast to the zero isotope effect in ordinary metals. Recently, this effect has been experimentally found in cuprates [58] and manganites [59].

The essential physics of strongly coupled electrons and phonons has been understood with the $1/\lambda$ expansion technique [25,9], which starts with the exact solution in the extreme limit $\lambda = \infty$ and allows for the summation of all multiphonon diagrams in any order of the small parameter $1/\lambda$ for any value of the adiabatic ratio $\omega/t$ (section 3). It predicts a breakdown of the Migdal-Eliashberg theory at $\lambda \simeq 1$ independent of the adiabatic ratio, in agreement with the exact solution of the extreme adiabatic Holstein model discussed here. There are also other studies of the same problem, which do not relay on the standard Feynman-Dyson perturbation theory in powers of $\lambda$. In particular, Takada *et al* [33,36,60] applied the gauge-invariant self-consistent method neglecting the vector term in the Ward identity. Benedetti and Zeyher [39] applied the dynamical mean-field theory in infinite dimensions. As in the $1/\lambda$ expansion technique, both approaches avoided the problem of the broken translational symmetry by using the nondispersive vertex and Green functions as the starting point. As a result they arrived at the same correct conclusion about the applicability of the Migdal approach (in Ref. [39] the critical value of $\lambda$ was found to be 1.3 in the adiabatic limit). Meanwhile, some authors [46,45], who relay on the perturbation expansion in powers of $\lambda$, fail to notice the self-trapping transition, erroneously claiming that the Migdal-Eliashberg theory could be applied at practically any $\lambda$ in the extreme adiabatic regime.

It becomes clear that the Migdal-Eliashberg theory cannot be applied if the (BCS) e-ph coupling constant, $\lambda$, is about 1 or larger even in the adiabatic regime of the strongly coupled many-electron system. There is a transition into the self-trapped state due to the broken translational symmetry, Fig. 2.4. The transition appears at $0.5 < \lambda < 1.3$ depending on the lattice dimensionality. This conclusion is correct for any electron-phonon interaction conserving the on-site electron occupation numbers. In particular, Hiramoto and Toyozawa [61] calculated the strength of the deformation potential, which transforms electrons into small polarons and bipolarons. Their continuous approach is sufficient for a qualitative estimate if the Debye wavenumber $q_D \sim \pi/a$ is introduced as an upper limit cut-off in all sums in the momentum space. Hiramoto and Toyozawa found that the transition between two electrons and a small bipolaron occurs at $\lambda \simeq 0.5$, that is half of the critical value of $\lambda$ at which the transition from electron to small polaron takes place in the extreme adiabatic limit ($sq_D << zt$, s the sound velocity). The effect of the adiabatic ratio $sq_D/zt$ on the critical value of $\lambda$ was found to be negligible. The radius of the acoustic polaron and bipolaron is the lattice constant, so the critical $\lambda$ does not depend on the number of electrons in this case either.

# III POLARON DYNAMICS, BIPOLARONS, AND CHARGED BOSE-GAS

The basic features of small polarons were well recognised a long time ago by Tjablikov [62], Yamashita and Kurosava [63], Sewell [64], Holstein [50], Lang and Firsov [24], Kudinov and Firsov [65] and others, and described in several review papers and textbooks [66,67,34,68,69,9]. The main feature is the exponential reduction of the bandwidth at intermediate and large values of the electron-phonon coupling, $\lambda$, resulting in a coherent small polaron tunneling at low temperatures and a thermally activated hopping at high temperatures. The polaronic bandwidth decreases with increasing temperature. A crossover from the polaronic Bloch states to the incoherent hopping takes place at temperatures $T \simeq \omega/2$ or even higher, where $\omega$ is the characteristic phonon frequency. The numerical solution for several vibrating molecules coupled with one or two electrons [29] revealed an agreement of the numerical bandwidth with the analytical Holstein results at large $\lambda$ both in the nonadiabatic, $\omega \geq t$ and adiabatic, $\omega \leq t$ regimes ($t$ is the hopping integral). For a multi-polaron system a $1/\lambda$ perturbation theory has been developed [11,25], which allowed us to extend the BCS theory to the strong-coupling regime $\lambda > 1$ and predict the transition to a Bose liquid of $2e$ charged bipolarons in the crossover region of intermediate values of the BCS coupling constant $\lambda$ [11]. The renormalised phonon frequencies were obtained [28,25] in agreement with the numerical results [29]. The theory has been applied to cuprates [70,9,71] and more recently to manganites [72] providing a description of many unusual properties of these materials ranging from high-$T_c$ superconductivity in cuprates (section 4) to colossal magnetoresistance (CMR) and ferromagnetism in doped manganites.

In this section the polaron dynamics, damping and bipolarons are discussed in more detail to show that small polarons (and bipolarons) exist in the itinerant (Bloch) states at zero temperature no matter which values the parameters of the translationally invariant electron-phonon system take. The Fröhlich interaction leads to relatively light polarons with the atomic size of the wave function and a large size of the phonon cloud.

We discuss the $t/\omega$ - $\lambda$ 'phase' diagram with polaronic and bipolaronic domains, bipolaron configuration in cuprates, and map the multi-polaron problem onto the charged Bose gas (CBG).

## A  Polaron band

The canonical approach to a small polaron problem is based on the canonical displacement (Lang-Firsov) transformation [24] of the electron-phonon Hamiltonian, Eq.(1) allowing for the summation of all diagrams including the vertex corrections. In the site (Wannier) representation for electrons, the Hamiltonian is

$$H = \sum_{i,j} T_{ij} c_i^\dagger c_j + \sum_{\mathbf{q},j} \omega_\mathbf{q} \hat{n}_j \left( u_j(\mathbf{q}) d_\mathbf{q} + h.c. \right)$$
$$+ \sum_\mathbf{q} \omega_\mathbf{q} (d_\mathbf{q}^\dagger d_\mathbf{q} + \frac{1}{2}) + \sum_{i,j} V_{ij} \hat{n}_i \hat{n}_j, \qquad (80)$$

with the bare hopping integral $T_{ij} = \delta_{s,s'} T(\mathbf{m})$ and the matrix element of the electron-phonon interaction

$$u_i(\mathbf{q}) = \frac{1}{\sqrt{2N}} \gamma(\mathbf{q}) e^{i\mathbf{q} \cdot \mathbf{m}}. \qquad (81)$$

Here $i = (\mathbf{m}, s)$, $j = (\mathbf{n}, s')$, $\hat{n}_i = c_i^\dagger c_i$, and $c_i, d_\mathbf{q}$ are the electron (hole) and phonon operators, respectively, $V_{ij}$ is the (direct) Coulomb repulsion. The term with $i = j$ should be excluded from the last sum.

As long as $\lambda > 1$ the kinetic energy remains smaller than the interaction energy and a self-consistent treatment of the many-electron system strongly coupled with phonons is possible with the '$1/\lambda$' expansion technique [25]. This possibility stems from the fact, known for a long time, that there is an exact solution for a single electron in the strong-coupling limit $\lambda \to \infty$. Following Lang and Firsov one can apply the canonical transformation $e^S$ to diagonalise the Hamiltonian. The diagonalisation is exact if $T(\mathbf{m}) = 0$ (or $\lambda = \infty$):

$$\tilde{H} = e^S H e^{-S}, \tag{82}$$

where

$$S = \sum_{\mathbf{q},i} \hat{n}_i \left[ u_i(\mathbf{q}) d_{\mathbf{q}} - H.c. \right] \tag{83}$$

The electron operator transforms as

$$\tilde{c}_i = c_i exp\left( -\sum_{\mathbf{q}} u_i(\mathbf{q}) d_{\mathbf{q}} - H.c. \right) \tag{84}$$

and the phonon one as:

$$\tilde{d}_{\mathbf{q}} = d_{\mathbf{q}} + \sum_i \hat{n}_i u_i^*(\mathbf{q}). \tag{85}$$

It follows from Eq.(85) that the Lang-Firsov canonical transformation shifts ions to new equilibrium positions. In a more general sense it changes the boson vacuum. As a result,

$$\tilde{H} = \sum_{i,j} \hat{\sigma}_{ij} c_i^\dagger c_j - E_p \sum_i \hat{n}_i + \sum_{\mathbf{q}} \omega_{\mathbf{q}}(d_{\mathbf{q}}^\dagger d_{\mathbf{q}} + 1/2) + \frac{1}{2} \sum_{i \neq j} v_{ij} \hat{n}_i \hat{n}_j, \tag{86}$$

where

$$\hat{\sigma}_{ij} = T(\mathbf{m} - \mathbf{n}) \delta_{s,s'} \exp\left( \sum_{\mathbf{q}} [u_i(\mathbf{q}) - u_j(\mathbf{q})] d_{\mathbf{q}} - H.c. \right) \tag{87}$$

is a renormalised hopping integral depending on the phonon variables, and

$$v_{ij} = V_{ij} - \frac{1}{N} \sum_{\mathbf{q}} |\gamma(\mathbf{q})|^2 \omega_{\mathbf{q}} \cos[\mathbf{q} \cdot (\mathbf{m} - \mathbf{n})] \tag{88}$$

is the the interaction of polarons owing to their Coulomb repulsion and local lattice deformation.

In the strong coupling limit $\lambda \to \infty$ one can neglect the hopping term of the transformed Hamiltonian. The rest has analytically determined eigenstates and eigenvalues. The eigenstates $|\tilde{N}\rangle = |n_i, n_{\mathbf{q}}\rangle$ are classified with the polaron $n_{\mathbf{m},s}$ and phonon $n_{\mathbf{q}}$ occupation numbers and the energy levels are:

$$E = -E_p \sum_i n_i + \frac{1}{2} \sum_{i \neq j} v_{ij} n_i n_j + \sum_{\mathbf{q}} \omega_{\mathbf{q}} (n_{\mathbf{q}} + 1/2) \qquad (89)$$

with $n_i = 0, 1$ and $n_{\mathbf{q}} = 0, 1, 2, 3, ....\infty$.

Hence, the Hamiltonian Eq.(1) in zero order with respect to the hopping describes localised polarons and independent phonons which are vibrations of ions relative to new equilibrium positions depending on the polaron occupation numbers. The phonon frequencies remain unchanged in this limit. The middle of the electronic band falls by the polaronic level shift $E_p$ as a result of a potential well created by the lattice deformation,

$$E_p = \frac{1}{2N} \sum_{\mathbf{q}} |\gamma(\mathbf{q})|^2 \omega_{\mathbf{q}}. \qquad (90)$$

First we limit our discussion to a single-polaron problem with no polaron-polaron interaction. The effect of the interaction (including also the direct Coulomb repulsion) such as the bipolaron formation and screening are discussed later.

With the finite hopping term polarons tunnel in a narrow band owing to the degeneracy of the zero order Hamiltonian with respect to the site position of a single polaron in a regular lattice. To see it one can apply the perturbation theory using $1/\lambda$ as a small parameter with $\lambda \equiv E_p/zt$ ($z$ is the coordination lattice number and $t$ the nearest-neighbour hopping integral). The proper (Bloch) set of $N$ degenerate zero order eigenstates of the lowest energy level $(-E_p)$ of the unperturbed Hamiltonian is

$$|\mathbf{k}, 0\rangle = \frac{1}{\sqrt{N}} \sum_{\mathbf{m}} c_{\mathbf{m}}^{\dagger} \exp(i\mathbf{k} \cdot \mathbf{m})|0\rangle, \qquad (91)$$

$|0\rangle$ is the vacuum. By applying the textbook perturbation theory one readily calculates the lowest energy levels of the polaron in a crystal. Up to the second order in the hopping integral the result is

$$E(\mathbf{k}) = -E_p + \epsilon_{\mathbf{k}}$$
$$- \sum_{\mathbf{k}', n_{\mathbf{q}}} \frac{|\langle \mathbf{k}, 0| \sum_{i,j} \hat{\sigma}_{i,j} c_i^{\dagger} c_j |\mathbf{k}', n_{\mathbf{q}}\rangle|^2}{\sum_{\mathbf{q}} \omega_{\mathbf{q}} n_{\mathbf{q}}} \qquad (92)$$

with $|\mathbf{k}', n_{\mathbf{q}}\rangle$ the exited states of the unperturbed Hamiltonian with one electron and at least one real phonon. The second term in Eq.(92), which is linear with respect to the bare hopping $t$, determines the small polaron band dispersion as

$$\epsilon_{\mathbf{k}} = \sum_{\mathbf{m}} t(\mathbf{m}) e^{-g^2(\mathbf{m})} \exp(-i\mathbf{k} \cdot \mathbf{m}), \qquad (93)$$

with the band-narrowing factor (at zero temperature)

$$g^2(\mathbf{m}) = \frac{1}{2N} \sum_{\mathbf{q}} |\gamma(\mathbf{q})|^2 [1 - \cos(\mathbf{q} \cdot \mathbf{m})] \qquad (94)$$

The third term in Eq. (92), quadratic in $t$, yields a negative k-*independent* correction to the polaron level shift of the order of $1/\lambda^2$. The origin of this correction, which is much larger than the fist order in $t$ contribution (containing a small exponent) is understood from Fig. 3.1. The polaron localised in the potential well of the depth $E_p$ on the site $\mathbf{m}$ hops onto a neighbouring site $\mathbf{n}$ with no deformation around and comes back. As any second order correction this transition shifts the energy down by an amount $\sim -t^2/E_p$. It has little to do with the polaron effective mass and the polaron tunneling mobility because the lattice deformation around $\mathbf{m}$ does not follow the electron. The electron hops 'forth and back' many times (about $e^{g^2}$) 'waiting' for a sufficient lattice deformation to appear around the site $\mathbf{n}$. Only after it 'creates' the deformation around $\mathbf{n}$ the electron tunnels onto the next site together with the deformation.

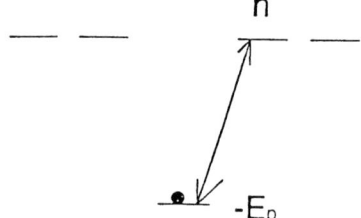

FIGURE 3.1. 'Forth and back' virtual transitions of polaron without any transfer of the lattice deformation from one site to another. These transitions shift the midle of the band further down with no real charge delocalization.

## B Damping of the polaron band

The polaron band is exponentially narrow, Eq.(93). Hence, one can raise a question concerning its existence in real solids. At zero temperature the perturbation term of the transformed Hamiltonian conserves the momentum because all off-diagonal matrix elements vanish,

$$\langle \mathbf{k}, 0 | \sum_{i,j} \hat{\sigma}_{i,j} c_i^\dagger c_j | \mathbf{k}', 0 \rangle = 0 \qquad (95)$$

if $\mathbf{k} \neq \mathbf{k}'$. The absorption or emission of a single high-frequency phonon is forbidden by the energy conservation because the polaron half-bandwidth $w \leq \omega$. Hence, there is no damping of the polaron band at T= 0 no matter how strong the interaction $\lambda$ and how small the adiabatic ratio $\omega/t$ are. However, the polaron bandwidth depends on temperature. For high temperatures $T \gg \omega/2$ the band shrinks exponentially with increasing temperature from $w = zte^{-g^2}$ (at $T = 0$) down to [66,24,9]

$$w \simeq zt \exp\left(-\frac{2E_p T}{\omega^2}\right). \tag{96}$$

here $t$ is the value of $T(\mathbf{a})$ for the nearest neighbours. On the other hand, the scattering of polarons within their narrow band becomes more important with the increasing temperature owing to the simultaneous emission and absorption of phonons, Fig. 3.2.

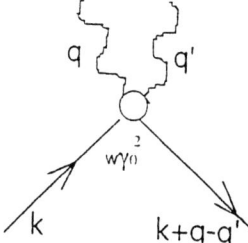

**FIGURE 3.2.** Two-phonon scattering responsible for the damping of the polaron band.

These incoherent events tend to destroy the coherent polaron tunneling within the band. The corresponding scattering rate is given by the Fermi Golden rule as

$$\frac{1}{\tau} = 2\pi \langle \sum_{\mathbf{q},\mathbf{q}'} |\langle \mathbf{k}+\mathbf{q}-\mathbf{q}', n_\mathbf{q}-1, n_{\mathbf{q}'}+1| \sum_{i,j} \hat{\sigma}_{i,j} c_i^\dagger c_j |\mathbf{k}, n_\mathbf{q}, n_{\mathbf{q}'}\rangle|^2 \delta(\epsilon_\mathbf{k} - \epsilon_{\mathbf{k}+\mathbf{q}-\mathbf{q}'})\rangle, \tag{97}$$

Expanding $\hat{\sigma}_{ij}$ operators in the powers of the phonon creation and annihilation operators one estimates the matrix element of the two-phonon scattering as

$$|\langle \mathbf{k}+\mathbf{q}-\mathbf{q}', n_\mathbf{q}-1, n_{\mathbf{q}'}+1| \sum_{i,j} \hat{\sigma}_{i,j} c_i^\dagger c_j |\mathbf{k}, n_\mathbf{q}, n_{\mathbf{q}'}\rangle| \sim \frac{1}{N} w\gamma_0^2 \sqrt{n_\mathbf{q}} \sqrt{n_{\mathbf{q}'}+1}. \tag{98}$$

Substituting this estimate into Eq.(18) and using the definition of the density of states in the polaron band

$$N_p(\xi) \equiv \frac{1}{N} \sum_{\mathbf{k}} \delta(\xi - \epsilon_\mathbf{k}) \simeq \frac{1}{2w} \tag{99}$$

one obtains

$$\frac{1}{\tau} \simeq w\gamma_0^4 n_\omega(1+n_\omega), \tag{100}$$

with the momentum independent $\gamma(\mathbf{q}) = \gamma_0$ and the phonon distribution function $n_\omega = [exp(\omega/T) - 1]^{-1}$. The polaron band is well defined if

$$\frac{1}{\tau} < w, \tag{101}$$

which is satisfied for a wide temperature range

$$T \leq \frac{\omega}{ln\gamma_0^4} \tag{102}$$

below about half of the characteristic phonon frequency for the relevant values of $\gamma_0^2$. The incoherent thermally activated hopping dominates in the polaron motion at higher temperatures where the polaronic states cannot be classified by their momenta.

## C  Small Holstein polaron and 'small Fröhlich polaron'

The analytical $1/\lambda$ expansion allows us to analyse both small Holstein polaron (SHP) with a short-range interaction and a lighter small polaron with a long-range Fröhlich interaction, i.e mobile small Fröhlich polaron (SFP) [73]. To illustarte the point we express the electron-phonon interaction in terms of real displacements $\xi_\mathbf{n}$ as [42]

$$H_{e-ph} = -\sum_{\mathbf{n},i} f(\mathbf{m} - \mathbf{n})\xi_\mathbf{n}\hat{n}_i. \tag{103}$$

Here $\xi_\mathbf{n} = \sum_\mathbf{q}(2NM\omega_\mathbf{q})^{-1/2} \exp(i\mathbf{q} \cdot \mathbf{n})d_\mathbf{q}^\dagger + H.c.$ is a normal coordinate at site $\mathbf{n}$, and $f(\mathbf{m} - \mathbf{n}) = N^{-1} \sum_\mathbf{q} \gamma(\mathbf{q})(M\omega_\mathbf{q}^3)^{1/2} \exp[i\mathbf{q} \cdot (\mathbf{n} - \mathbf{m}))]$ is the force between the electron at site $\mathbf{m}$ and the normal coordinate $\xi_\mathbf{n}$.

In general, there is no simple relation between the polaronic shift $E_p$ and the exponent $g^2$ of the mass enhancement. This relation depends on the form of the electron-phonon interaction. Indeed, for dispersionless phonons, $\omega_\mathbf{q} = \omega$ one obtains

$$E_p = \frac{1}{2M\omega^2} \sum_\mathbf{m} f^2(\mathbf{m}), \tag{104}$$

while

$$g^2 = \frac{1}{2M\omega^3} \sum_\mathbf{m} \left[f^2(\mathbf{m}) - f(\mathbf{m})f(\mathbf{m} + \mathbf{a})\right]. \tag{105}$$

where $\mathbf{a}$ is the lattice vector.

The effective mass renormalisation is $m^*/m = e^{g^2}$, where $m$ is the bare band mass and $1/m^* = \partial^2 E(\mathbf{k})/\partial k^2$ with $k \to 0$.

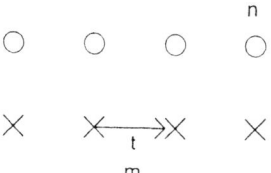

**FIGURE 3.3.** One-dimensional model of the small Fröhlich polaron on the chain interacting with all ions of another chain.

If the interaction is local, $f(\mathbf{m}) = \kappa \delta_{\mathbf{m},0}$ (Holstein model), then $g^2 = E_p/\omega$. In general, one has $g^2 = \gamma E_p/\omega$ with a numerical coefficient $\gamma = 1 - \sum_{\mathbf{m}} f(\mathbf{m}) f(\mathbf{m} + \mathbf{a})/\sum_{\mathbf{n}} f^2(\mathbf{n})$ less than unity [73].

To calculate $\gamma$ one can introduce a one-dimensional lattice model with a long-range Coulomb interaction between the electron and ions, Fig. 3.3 [42].

The electron in a Wannier state on a site $\mathbf{m}$ of the infinite chain ($\times$) interacts with the vibrations of *all* ions of another chain ($\circ$) polarised in the direction perpendicular to the chains. The corresponding force is given by

$$f(\mathbf{m} - \mathbf{n}) = \frac{\kappa}{(|\mathbf{m} - \mathbf{n}|^2 + 1)^{3/2}}, \quad (106)$$

The distance along the chains $|\mathbf{m} - \mathbf{n}|$ is measured in units of the lattice constant $a$, the inter-chain distance is also $a$. Here and further on we take $a = 1$. For this long-range interaction one obtains $E_p = 1.27\kappa^2/(2M\omega^2)$, $g^2 = 0.49\kappa^2/(2M\omega^3)$ and $g^2 = 0.39 E_p/\omega$. The effective mass renormalisation is much smaller than in the dispersionless Holstein model, roughly as $m^*_{SFP} \propto (m^*_{SHP})^{1/2}$.

Not only the small polaron mass strongly depends on the radius of the electron-phonon interaction, but also does the range of the applicability of the analytical Lang-Firsov theory. The theory appears almost exact in a wide region of parameters for the Fröhlich interaction. The polaron mass in a wide region of the adiabatic parameter and coupling has been recently calculated [42] with the continuous-time path-integral Quantum Monte Carlo (QMC) algorithm. This method is free from any systematic finite-size, finite-time-step and finite-temperature errors and allows for the *exact* (in the QMC sense) calculation of the ground-state energy and the effective mass of the lattice polaron for any electron-phonon interaction.

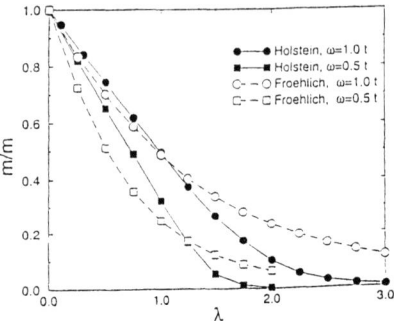

**FIGURE 3.4.** Inverse effective polaron mass in units of $1/m = 2ta^2$.

At large $\lambda$ ($> 1.5$) we found SFP to be much lighter than SHP, while the large Fröhlich polaron (i.e. at $\lambda < 1$) is *heavier* than the large Holstein polaron with the same binding energy, Fig. 3.4. The mass ratio $m_{FP}^*/m_{HP}^*$ is a non-monotonic function of $\lambda$. The effective mass of small *and* large Fröhlich polarons, $m_{FP}^*(\lambda)$ is well fitted by a single exponent, which is $e^{0.73\lambda}$ for $\omega = t$ and $e^{1.4\lambda}$ for $\omega = 0.5\,t$, which is not the case for the Holstein polaron. The exponents are remarkably close to those obtained with the Lang-Firsov transformation, $e^{0.78\lambda}$ and $e^{1.56\lambda}$, respectively. Hence, in the case of the Fröhlich interaction the transformation is perfectly accurate even in the adiabatic regime, $\omega/t \leq 1$ for *any* coupling strength.

Another interesting point is that the size of SFP and the length over which the distortion spreads are *different*. In the extreme strong-coupling limit, the Lang-Firsov transformation is exact, and the polaron is entirely localised on one site **m**. Hence, the size of its wave function is the atomic size. On the other hand, the ion displacements, proportional to the displacement force $f(\mathbf{m} - \mathbf{n})$, spread over a large distance. Their amplitude at a site **n** falls with the distance as $1/|\mathbf{m} - \mathbf{n}|^3$ in our one-dimensional model. The polaron cloud (i.e lattice distortion) can be more extended than the polaron itself (see, also Ref. [63,74,73]). Such polaron tunnels with a larger probability than the nondispersive Holstein polaron due to a smaller *relative* lattice distortion around two neighboring sites. It can be equally called a 'large descreat strong-coupling Fröhlich' polaron, if the lattice distortion is included in the definition of its size. On the other hand, historically one referes to a 'small' polaron, as a quasiparticle well described by the $1/\lambda$ expansion technique. With this definition polarons on a lattice are small for any value of the long-range electron-phonon interaction.

The model, Eq.(106), contains only one phonon mode polarised along c-axis, so that, let say, c-component of the field from a c-polarised dipole falls of with the distance as $1/r^3$. An *isotropic* Fröhlich interaction might be longer-ranged than ours giving rise to $1/r^2$ law. Hence, the polaron mass might be even lighter than shown in Fig. 3.4. The fact that the Lang-Firsov transformation is perfectly accurate for the long-range interaction in a wide region of the parameters, allows us to generalise this result. Including all phonon polarisations in a three-dimensional lattice we obtain $m^*_{SFP} \sim (m^*_{SHP})^\gamma$ with the constant $\gamma = \sum_\mathbf{q} \gamma^2(\mathbf{q})[1 - \cos(\mathbf{q} \cdot \mathbf{m})]/\sum_\mathbf{q} \gamma^2(\mathbf{q})$. Calculating the constant with the Fröhlich matrix element ($\gamma(\mathbf{q}) \sim 1/q$) we find $\gamma = 0.57$ in the cubic lattice, and $\gamma = 0.44$, $\gamma = 0.255$ in the cuprate lattice for the apex and in-plane oxygen hole, respectively, in a fair agreement with the numerical result.

A lighter mass of SFP compared with the nondispersive SHP is a generic feature of any dispersive electron-phonon interaction. As an example a short-range interaction with dispersive acoustic phonons ($\gamma(\mathbf{q}) \sim 1/q^{1/2}, \omega_\mathbf{q} \sim q$) also leads to a lighter polaron in the strong-coupling regime compared with the nondispersive SHP. Actually, Holstein [50] pointed in his original paper that the dispersion is a vital ingredient of the theory.

## D  Electron Green function in the polaronic regime

We have shown that the problem of a fermion on a lattice coupled with the bosonic field of lattice vibrations has an exact solution in terms of the coherent (Glauber) states in the extremely strong-coupling limit, $\lambda = \infty$ for any type of electron-phonon interaction conserving the on-site occupation numbers of fermions. For the finite coupling the $1/\lambda$ perturbation diagrammatic technique can be applied. The expansion parameter actually is $1/2z\lambda^2$ [67,74,75,25], so the analytical perturbation theory might have a wider region of applicability than one can expect from a naive variational estimate ($z$ is the lattice coordination number). However, it is not clear how fast the expansion converges. The exact numerical diagonalisation of vibrating clusters, variational calculations [29,31,32,35,37,41,40], dynamical mean-field approach in infinite dimensions [39], and Quantum-Monte-Carlo simulations [42] revealed that the ground state energy (the polaron binding energy $E_p$) is not very sensitive to the parameters. On the contrary, the effective mass, the bandwidth and the polaron density of states strongly depend on the adiabatic ratio $\omega_0/t$ in case of a short-range (Holstein) interaction. While the perturbation theory is almost exact in the nonadiabatic regime *for all values* of the coupling constant, there is no agreement in the adiabatic region, $\omega_0/t << 1$, where the first order perturbation expression *overestimates* polaron mass by a few orders of magnitude [29]. Here I would like to remind that $\omega_0$ is the characteristic phonon frequency and $t$ is the nearest-neighbour hopping integral, so that the dimensionless (BCS) coupling constant is $\lambda = E_p/(zt)$. A much lower effective mass of the adiabatic small polaron in the intermediate coupling regime compared with that estimated by the first order perturbation theory is a result of a poor convergence of perturbation expansion owing to the appearance of the familiar double-well potential [50] in adiabatic limit. The tunnelling probability is extremely sensitive to the shape of this potential.

As discussed above the range of applicability of the analytical theory depends also on the radius of interaction. While the analytical approach is applicable only if $\omega_0 \geq$

$t$ for short-range interaction, the theory appears almost exact in a substantially wider region of the parameters for a long-range (Fröhlich) interaction.

Now we can use this result to calculate the single-particle Green function of a fermion on a lattice coupled with the bosonic field via long-range Fröhlich interaction.

QMC result [42] allows us to average the hopping term in the transformed Hamiltonian, Eq.(86) with respect to phonons both in the nonadiabatic and intermediate regimes ($\omega_0 \geq t/2$) no matter what the value of Fröhlich interaction is. Then we can apply canonical trasformation together with the averaged Hamiltonian to derive single-particle Green's function (GF). The interaction is completely integrated out from the averaged Hamiltonian,

$$\tilde{H} = H_p + H_{ph} \tag{107}$$

where the 'free' polaron part is given by

$$H_p = \sum_{\mathbf{k}} \xi(\mathbf{k}) c_\mathbf{k}^\dagger c_\mathbf{k}, \tag{108}$$

(we omit the spin index), and the free phonon part is

$$H_{ph} = \sum_{\mathbf{q}} \omega_\mathbf{q} (d_\mathbf{q}^\dagger d_\mathbf{q} + 1/2). \tag{109}$$

Here $\xi(\mathbf{k}) = Z'E(\mathbf{k}) - \mu$ is the renormalised polaron-band dispersion with the chemical potential $\mu$, which includes polaron binding energy $(-E_p)$, $E(\mathbf{k}) = \sum_\mathbf{m} t(\mathbf{m}) exp(-i\mathbf{k} \cdot \mathbf{m})$ is the bare disperison in a rigid lattice, and the mass-renormalisation exponent is

$$Z' = \frac{\sum_\mathbf{m} t(\mathbf{m}) e^{-g^2(\mathbf{m})} \exp(-i\mathbf{k} \cdot \mathbf{m})}{\sum_\mathbf{m} t(\mathbf{m}) \exp(-i\mathbf{k} \cdot \mathbf{m})}. \tag{110}$$

Quite generally one finds $Z' = exp(-\gamma E_p/\omega)$, where the numerical coefficient $\gamma$ might be as small as 0.4 and even smaller in cuprates with the nearest neighbour oxygen-oxygen distance less than the lattice constant (see subsection 3.3)

Applying Lang-Firsov canonical transformation the Fourier component of the retarded GF,

$$G_R(\mathbf{k}, \omega) = -i \sum_\mathbf{m} \int_0^\infty e^{-i(\mathbf{k}\cdot\mathbf{m}-\omega t)} \langle 0|c_0(t) c_\mathbf{m}^\dagger(0)|0\rangle dt \tag{111}$$

is expressed as a convolution of the Fourier components of the coherent retarded polaron GF, $G_p(\mathbf{n}, t)$, and the multiphonon correlation function $\sigma(\mathbf{n}, t)$:

$$G_R(\mathbf{k}, \omega) = \frac{1}{2\pi} \sum_\mathbf{m} e^{-i\mathbf{k}\cdot\mathbf{m}} \int_{-\infty}^\infty d\omega' G_p(\mathbf{m}, \omega') \sigma(\mathbf{m}, \omega - \omega'). \tag{112}$$

Here

$$G_p(\mathbf{m},\omega) = -i \int_0^\infty dt e^{i\omega t} \langle \tilde{0}|e^{iH_p t} c_{\mathbf{0}} e^{-iH_p t} c_{\mathbf{m}}^\dagger|\tilde{0}\rangle, \tag{113}$$

and

$$\sigma(\mathbf{m},\omega) = \int_{-\infty}^\infty dt e^{i\omega t} \langle \tilde{0}|e^{iH_{ph} t} X_{\mathbf{0}} e^{-iH_{ph} t} X_{\mathbf{m}}^\dagger|\tilde{0}\rangle \tag{114}$$

with $X_{\mathbf{m}} = \exp\left(\sum_{\mathbf{q}} u_i(\mathbf{q}) d_{\mathbf{q}} - H.c.\right)$, and $|\tilde{0}\rangle$ the ground state of $\tilde{H}$.

In the following we consider the dispersionless phonons, $\omega_{\mathbf{q}} = \omega_0$, and the Fröhlich interaction with $\gamma(\mathbf{q}) \sim 1/q$. The straightforward calculations [69,9] yield

$$G_p(\mathbf{m},\omega) = \frac{1}{N} \sum_{\mathbf{k'}} \frac{e^{i\mathbf{k'}\cdot\mathbf{m}}}{\omega - \xi(\mathbf{k'}) + i\delta}, \tag{115}$$

and

$$\sigma(\mathbf{m},\omega) = -2\pi i Z \sum_{l=0}^\infty \frac{1}{l!} \left(\frac{1}{2N} \sum_{\mathbf{q}} |\gamma(\mathbf{q})|^2 e^{i\mathbf{q}\cdot\mathbf{m}}\right)^l \delta(\omega - l\omega_0). \tag{116}$$

Here

$$Z = \exp(-\frac{1}{2N} \sum_{\mathbf{q}} |\gamma(\mathbf{q})|^2) \tag{117}$$

and $\delta = +0$. The convolution of Eq.(18) and Eq.(19) yields

$$G_R(\mathbf{k},\omega) = \sum_{l=0}^\infty G_R^{(l)}(\mathbf{k},\omega), \tag{118}$$

where

$$G_R^{(l)}(\mathbf{k},\omega) = Z \sum_{\mathbf{q}_1,\ldots\mathbf{q}_l} \frac{|\gamma(\mathbf{q}_1)|^2 \times |\gamma(\mathbf{q}_2)|^2 \times \ldots \times |\gamma(\mathbf{q}_l)|^2}{(2N)^l l!(\omega - \omega_0 l - \xi(\mathbf{k} + \mathbf{q}_1 + \ldots \mathbf{q}_l) + i\delta)} \tag{119}$$

with $\delta = +0$.

Obviously, Eq.(118) is in the form of a perturbative multiphonon expansion. A term with an index $l$ corresponds to a transition from the initial state $\mathbf{k}$ of the polaron band to the final state $\mathbf{k} + \mathbf{q_1} + \ldots \mathbf{q_l}$ with the emission of $l$ optical phonons.

The Green function of a polaronic carrier, Eq.(118) comprises two different contributions. The first ($l = 0$) coherent $\mathbf{k}$-dependent term arises from the polaron band tunneling,

$$G_R^{(0)} = \frac{Z}{\omega - \xi(\mathbf{k}) + i\delta}. \tag{120}$$

The spectral weight of the coherent part is strongly (exponentially) suppressed as $Z = \exp(-E_p/\omega_0)$ while the effective mass might only be slightly enhanced, $\xi_{\mathbf{k}} = Z'E_{\mathbf{k}} - \mu$, because $Z \ll Z' < 1$ in the case of the Fröhlich interaction.

The second incoherent phonon-assisted contribution with $l \geq 1$ describes the excitations accompanied by emission of phonons. We notice that its spectral density spreads over a wide energy range of about twice the polaron level shift $E_p$. On the contrary the coherent term shows an angular dependence in the energy window of the order of polaron bandwidth, $2Z'zt$. Interestingly, there is some $\mathbf{k}$ dependence of the *incoherent* background as well, if the matrix element of electron-phonon interaction depends on $q$ (see also Ref. [76]).

## E  Polaron-polaron interaction and screening

Polarons interact not only with phonons but also with each other. The range of the deformation surrounding (Fröhlich) polarons is quite large, so the polaron deformation fields are overlapped at finite density. Hence, one can worry about the effect of the overlap on their stability. Taking into account both the long-range attraction of polarons owing to their lattice deformations *and* the direct Coulomb repulsion, the residual long-range interaction has been found to be rather weak and repulsive [9]. The Fourier component of the residual polaron-polaron interaction, $v(\mathbf{q})$, comprising the direct Coulomb repulsion and the attraction mediated by phonons, is given by

$$v(\mathbf{q}) = \frac{4\pi e^2}{\epsilon_\infty q^2} - |\gamma(\mathbf{q})|^2 \omega_{\mathbf{q}}. \qquad (121)$$

In the long-wave limit ($q \ll \pi$) the Fröhlich interaction dominates in the attractive part, so we have

$$|\gamma(\mathbf{q})|^2 \omega = \frac{4\pi e^2(\epsilon_\infty^{-1} - \epsilon_0^{-1})}{q^2}, \qquad (122)$$

where $\epsilon_\infty$ and $\epsilon_0$ are the high frequency and the static dielectric constants of the host ionic insulator. At large distances the polaron-polaron interaction is repulsive,

$$v_{ij} = \frac{e^2}{\epsilon_0 |\mathbf{m} - \mathbf{n}|}. \qquad (123)$$

Optical phonons nearly nullify the bare Coulomb repulsion in ionic solids if $\epsilon_0 \gg 1$, which is normally the case in oxides. Hence, there is no effect of the overlapping deformations on the small polaron stability.

In the absence of bipolarons (see below) one can apply the canonical random phase approximation to calculate the dielectric response function of polarons,

$$\epsilon(\mathbf{q}, \Omega) = 1 - 2v(\mathbf{q}) \sum_{\mathbf{k}} \frac{n_{\mathbf{k}+\mathbf{q}} - n_{\mathbf{k}}}{\Omega - \epsilon_{\mathbf{k}} + \epsilon_{\mathbf{k}+\mathbf{q}}}. \qquad (124)$$

This expression describes the response of small polarons to a perturbation of a frequency $\Omega \leq w$, when phonons in the polaronic cloud are not excited. In the static limit at large distances (or $q \to 0$) we obtain the usual Debye screening with a rather small Debye radius owing to a heavy mass. Actually, for a temperature larger than the polaronic halfbandwidth one can expand the polaron distribution function as

$$n_{\mathbf{k}} \simeq \frac{n}{2}\left(1 - \frac{(2-n)\epsilon_{\mathbf{k}}}{2T}\right) \tag{125}$$

with n the density of polarons, to get

$$\epsilon(q, 0) = 1 + \frac{q_s^2}{q^2}, \tag{126}$$

where $q_s = (2\pi e^2 n(2-n)/T\epsilon_0)^{1/2}$. However, already for a finite but rather low frequency $\Omega \geq w$ the polaron response is dynamic

$$\epsilon(\mathbf{q}, \Omega) = 1 - \frac{\omega_p^2(\mathbf{q})}{\Omega^2} \tag{127}$$

with the temperature dependent polaron plasma frequency

$$\omega_p^2(\mathbf{q}) = 2v(\mathbf{q}) \sum_{\mathbf{k}} n_{\mathbf{k}}(\epsilon_{\mathbf{k}+\mathbf{q}} - \epsilon_{\mathbf{k}}), \tag{128}$$

proportional to the inverse temperature at $T >> w$.

Considering the electron-phonon interaction in a multi-polaron system one has to take into account the dynamic properties of the response function. One can (erroneously) believe that the long-range Fröhlich interaction becomes a short range (Holstein) due to screening. This is not correct. Replacing the bare electron-phonon interaction $\gamma(\mathbf{q})$ by a screened one, $\gamma_{sc}(\mathbf{q}, \omega)$ as shown in Fig. 3.5 we obtain

$$\gamma_{sc}(\mathbf{q}, \omega) = \frac{\gamma(\mathbf{q})}{\epsilon(\mathbf{q}, \omega)}. \tag{129}$$

In the long-wave limit the response of polarons at the optical phonon frequency is dynamic, because $\omega >> qv$ ($v$ is the characteristic group velocity of polarons). Also their (renormalised) plasma frequency, $\omega_p(\mathbf{q})$, is lower than the optical phonon frequency due to the large static dielectric constant, enhanced effective mass, and relatively low density of polarons. Therefore, a singular behaviour of $\gamma(\mathbf{q}) \sim 1/q$ is unaffected by the screening. The optical phonon frequency remains almost unchanged as well [25]. Polarons are slow enough and *cannot* screen the high-frequency crystal field oscillations. As a result, the interaction with the high-frequency optical phonons in ionic polaron solids remains long-range.

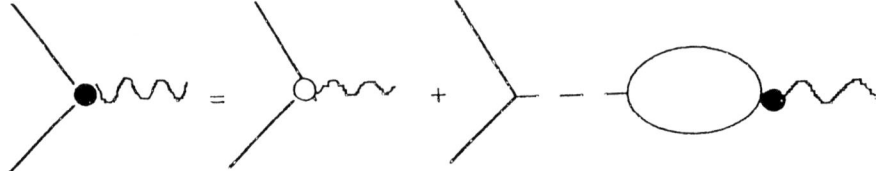

**FIGURE 3.5.** Electron-phonon vertex (dark circle) screened by the Coulomb interaction (dashed line).

Another important point is a possibility of the Wigner crystallization of the (bi)polaronic liquid [9]. Because the residual long-range repulsion is relatively weak, a relevant dimensionless parameter $r_s = m^*e^2/\epsilon_0(4\pi n/3)^{1/3}$ is not very large in doped cuprates. The Wigner crystallization appears around $r_s \simeq 100$ or larger, which corresponds to the atomic density of polarons $n \leq 10^{-6}$ with $\epsilon_0 = 30$ and $m^* = 5m_e$. This estimate tells us that (bi)polaronic carriers in cuprates are in the liquid state (see also section 4).

## F  Bipolarons and charged Bose gas

The Fröhlich interaction together with a short-range deformation potential and/or Jahn-Teller distorion can easily overcome the Coulomb repulsion at a distance about the lattice constant. Then, (owing to a narrow band) polarons form real space small bipolarons rather than the Cooper pairs. We can estimate the characteristic parameters $\lambda$ and $t/\omega$ of the bipolaronic instability. The characteristic attractive potential is $V = zt(\lambda - \mu)$, where $\mu$ is the dimensionless Coulomb pseudopotential. A bound state of two polarons appears if [77]

$$V \geq \frac{\pi^2}{8m^*}. \tag{130}$$

Substituting the polaron mass, $m^* = \exp(\gamma\lambda zt/\omega)/2t$, we find

$$\frac{t}{\omega} \leq (\gamma z\lambda)^{-1} \ln\left[\frac{\pi^2}{4z(\lambda - \mu)}\right]. \tag{131}$$

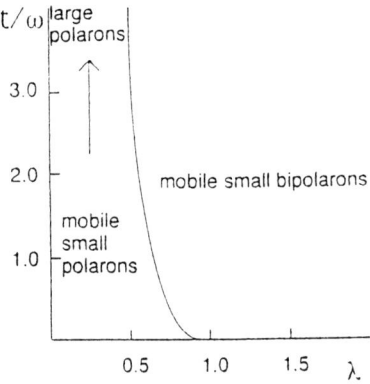

**FIGURE 3.6.** $'t/\omega - \lambda'$ diagram with a small bipolaron (BEC) domain, a large polaron (BCS) domain, and a region of unbound small polarons for $z = 6, \gamma = 0.4$, and the Coulomb pseudopotential $\mu = 0.5$.

The corresponding 'phase' diagram is shown in Fig. 3.6. Bipolarons are formed at $\lambda \geq \mu + \pi^2/4z$ in the nonadiabatic and intermediate regime $t/\omega \simeq 1$. In the case of the Fröhlich interaction there is no sharp transition between small and large polarons as one can see in Fig. 3.4. However, due to the fact that the Lang-Firsov transformation is practically exact in the whole region of coupling for nonadiabatic and intermediate regime (up to $t/\omega = 2$), the carriers are small polarons *independent* of the value of $\lambda$ in this regime. It means that the radius of their wave function is about the atomic size and they tunnel together with the entire phonon cloud no matter how 'thin' the cloud is. Our estimates are fully confirmed by the numerical simulations of ionic perovskite lattices [78] which established the existence of stable *intersite* bipolarons in doped cuprates.

## G  Bipolaron anisotropic flat bands in high-$T_c$ copper oxides

Consideration of particular lattice structures shows that small inter-site bipolarons can be perfectly mobile even when the electron-phonon coupling is strong and the bipolaron binding energy is large. Here we analyse the important case of copper based high-$T_c$ oxides. The existance of the 'parent' Mott insulators suggest that high-$T_c$ superconductors are in fact doped semiconductors with narrow electron bands. Therefore, different types of bipolarons can be found with computer simulation techniques based on the minimization of the ground state energy without the kinetic energy term in Eq.(86).

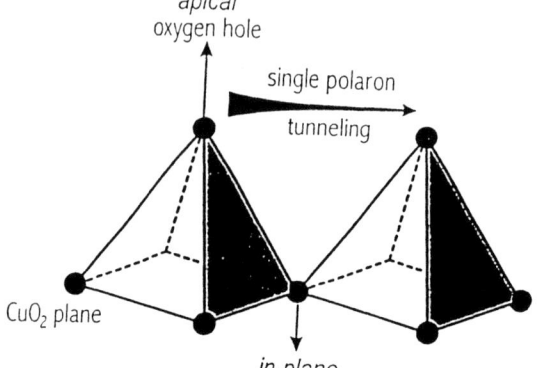

**FIGURE 3.7.** Apex bipolaron tunneling in perovskites. Reprinted from [109], Copyright 2000, with permission from Elsevier Science.

The intersite pairing of the in-plane oxygen hole with the *apex* one is energetically favorable in the perovskite structures with the binding energy $\Delta = 0.119 eV$ for $La_2CuO_4$ [78]. Obviously, this apex bipolaron can tunnel from one cell to another via a direct *single polaron* tunneling from one apex oxygen to its apex neighbour as shown in Fig. 3.7.

Oxides are strongly polarizable materials, so the coupling with optical phonons dominates in the electron-phonon interaction [8],

$$\gamma(\mathbf{q}) = -\frac{i\sqrt{8\pi\alpha}}{\sqrt{\Omega}(2m\omega)^{1/4}q}. \tag{132}$$

The lattice polarization is coupled with the electron density, therefore the interaction is diagonal in the site representation and the coupling constant does not depend on the particular orbital. The canonical displacement transformation eliminates an essential part of the electron-phonon interaction. The transformed Hamiltonian is given by

$$\tilde{H} = SHS^{-1} = (T_p - E_p)\sum_{i(p)} n_{i(p)} + (T_d - E_d)\sum_{i(d)} n_{i(d)} + \sum_{i \neq j} \hat{\sigma}_{ij} c_i^\dagger c_j$$
$$+ \sum_{\mathbf{q}} \omega_{\mathbf{q}}(d_{\mathbf{q}}^\dagger d_{\mathbf{q}} + 1/2) - \frac{1}{2}\sum_{\mathbf{q},i,j} \left(2\omega_{\mathbf{q}} u_i(\mathbf{q}) u_j^*(\mathbf{q}) - V_{ij}\right) \hat{n}_i \hat{n}_j. \tag{133}$$

The first oxygen ($p$) and the second copper ($d$) diagonal terms include the polaronic level shift, which is the same for oxygen and copper ions

$$E_p = E_d = \sum_{\mathbf{q}} |u_j(\mathbf{q})|^2 \omega_{\mathbf{q}}. \tag{134}$$

The transformed hopping term involves phonon operators

$$\hat{\sigma}_{ij} = T_{ij} exp \left( \sum_{\mathbf{q}} u_i^*(\mathbf{q}) d_{\mathbf{q}}^\dagger - h.c. \right) exp \left( \sum_{\mathbf{q}} u_j(\mathbf{q}) d_{\mathbf{q}} - h.c. \right). \quad (135)$$

There are two major effects of the electron-phonon interaction. One is the band narrowing due to a phonon cloud surrounding the hole. In case of a large charge transfer gap $E_g \gg \omega$ the bandwidth narrowing factor is the same for the direct $t_{pp'}$ and the second order via copper $t_{pp'}^{(2)}$ oxygen-oxygen transfer, Fig. 3.8a,b

$$t_{pp'} \equiv \langle 0|\hat{\sigma}_{pp'}|0\rangle = T_{pp'} e^{-g_{pp'}^2} \quad (136)$$

$$t_{pp'}^{(2)} \equiv \sum_\nu \frac{\langle 0|\hat{\sigma}_{pd}|\nu\rangle\langle \nu|\hat{\sigma}_{dp'}|0\rangle}{E_0 - E_\nu} \simeq \frac{T_{pd}^2}{E_g} e^{-g_{pp'}^2}, \quad (137)$$

where $|\nu\rangle, E_\nu$ are eigenstates and eigenvalues of the transformed Hamiltonian, Eq.(133) without the third hopping term, $|0\rangle$ the phonon vacuum, and the reduction factor is

$$g_{pp'}^2 = \frac{1}{2N} \sum_{\mathbf{q}} |\gamma(\mathbf{q})|^2 \left(1 - cos[\mathbf{q} \cdot (\mathbf{m}_p - \mathbf{m}_{p'})]\right). \quad (138)$$

These expressions are the result of the straightforward calculations described below. The direct hopping is given by

$$t_{pp'} = T_{pp'} \langle 0|exp \left( \sum_{\mathbf{q}} u_p^*(\mathbf{q}) d_{\mathbf{q}}^\dagger - h.c. \right) exp \left( \sum_{\mathbf{q}} u_{p'}(\mathbf{q}) d_{\mathbf{q}} - h.c. \right) |0\rangle. \quad (139)$$

With the help of $e^{A+B} = e^A e^B e^{-[AB]/2}$ one obtains

$$t_{pp'} = T_{pp'} e^{-g_{pp'}^2} \langle 0|exp \left( \sum_{\mathbf{q}} u_p^*(\mathbf{q}) d_{\mathbf{q}}^\dagger \right) exp \left( -\sum_{\mathbf{q}} u_{p'}(\mathbf{q}) d_{\mathbf{q}}^\dagger \right) |0\rangle, \quad (140)$$

where

$$g_{ij}^2 = \frac{1}{2} \sum_{\mathbf{q}} \left( |u_i(\mathbf{q})|^2 + |u_j(\mathbf{q})|^2 - 2u_i^*(\mathbf{q}) u_j(\mathbf{q}) \right). \quad (141)$$

The bracket in Eq.(140) is equal to unity. Then Eq.(136) follows from Eq.(141) using the definition of $u_j(\mathbf{q})$.

Taking into account that $E_\nu - E_0 = E_g + \sum_{\mathbf{q}} \omega_{\mathbf{q}} n_{\mathbf{q}}$, the second order indirect hopping, Eq.(137) is written as

$$t_{pp'}^{(2)} = i \int_0^\infty dt e^{-iE_g t} \langle 0|\hat{\sigma}_{pd}(t) \hat{\sigma}_{dp'}|0\rangle, \quad (142)$$

where

$$\hat{\sigma}_{pd}(t) = T_{pd} exp\left(\sum_{\mathbf{q}} u_p^*(\mathbf{q},t) d_\mathbf{q}^\dagger - h.c.\right) exp\left(\sum_{\mathbf{q}} u_d(\mathbf{q},t) d_\mathbf{q} - h.c.\right). \quad (143)$$

Here $u_j(\mathbf{q},t) \equiv u_j(\mathbf{q})exp(i\omega_\mathbf{q}t)$ and $n_\mathbf{q} = 0,1,2,...$ the phonon occupation numbers. Calculating the bracket in Eq.(142) one obtains

$$\langle ... \rangle = e^{-g_{pd}^2} e^{-g_{dp'}^2} exp\left(-\sum_\mathbf{q}[u_p(\mathbf{q}) - u_d(\mathbf{q})][u_d^*(\mathbf{q}) - u_{p'}^*(\mathbf{q})]e^{-i\omega_\mathbf{q}t}\right). \quad (144)$$

If $\omega_\mathbf{q}$ is q-independent the integral in Eq.(142) is calculated by the expansion of the exponent in Eq.(144):

$$t_{pp'}^{(2)} = \frac{T_{pd}^2}{E_g} e^{-g_{pd}^2} e^{-g_{dp'}^2} \sum_{k=0}^{\infty} \frac{(-1)^k \left(\sum_\mathbf{q}[u_p(\mathbf{q}) - u_d(\mathbf{q})][u_d^*(\mathbf{q}) - u_{p'}^*(\mathbf{q})]\right)^k}{k!(1 + k\omega/E_g)}. \quad (145)$$

Then Eq.(137) is obtained in the limit $E_g >> \omega$.

Substitution of Eq.(132) into Eq.(138) yields

$$g_{pp'}^2 = \frac{E_p}{\omega}\left(1 - \frac{Si(q_d m)}{q_d m}\right), \quad (146)$$

if the Debye approximation for the Brillouin zone is applied. Here $Si(x) = \int_0^x sin(t)dt/t$, $m = a/\sqrt{2}$ and $m = a$ for the in-plane and for the apex reduction factor, respectively. For cuprates with $q_d \simeq 0.7\text{Å}^{-1}$ and $a \simeq 3.8\text{Å}$ one obtains $g_{pp'}^2 \simeq 0.2 E_p/\omega$ and $g^2 \simeq 0.3 E_p/\omega$. Because the nearest neighbor oxygen-oxygen distance in copper oxides is less than the lattice constant the calculations yield a remarkably lower value of $g_{pp'}^2 \simeq 0.2 E_p/\omega$ than one can expect with the Holstein estimate ($= E_p/\omega$).

The other effect of the electron-phonon coupling is the attraction between two polarons given by the last term in Eq.(133). For the Fröhlich interaction the polaron level shift in $La_{2-x}Sr_xCuO_4$ is estimated to be as large as $E_p \simeq 0.6eV$, section 4. Then with $\omega = 0.06eV$ one obtains $g_{pp'}^2 \simeq 2$. As a result a large attraction between two polarons of the order of $2E_p \geq 1eV$ is possible accompanied by only one order of magnitude mass enhancement. Such a posssibility is a result of the particular lattice structure and the matrix element dispersion.

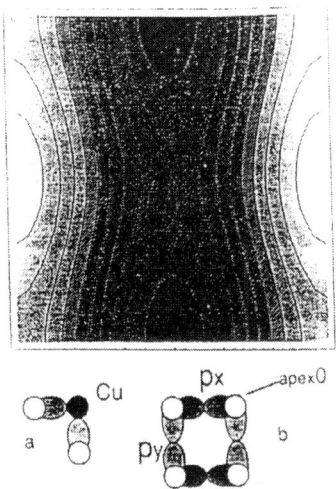

**FIGURE 3.8.** Counterplot of the 'x'-biplaron dispersion $E_{\mathbf{k}}^x$. Dark regions correspond to the bottom of the center-of-mass band. $E_{\mathbf{k}}^y$ energy surfaces are obtained by $\pi/2$ rotation. Three-band (in-plane) model (a) and two-band apex orbitals (b).

The *bipolaron* hopping integral $t$ is obtained by projecting the Hamiltonian, Eq.(133) onto the reduced Hilbert space containing only empty or doubly occupied elementary cells, and averaging the result with the phonon density matrix. The wave function of the apex bipolaron localised, let say, in the cell **m** is written as

$$|\mathbf{m}\rangle = \sum_{i=1}^{4} A_i c_i^\dagger c_{apex}^\dagger |0\rangle, \qquad (147)$$

where $i$ denotes the $p_{x,y}$ orbitals and spins of the four plane oxygen ions in the cell **m**, Fig. 3.7 and $c_{apex}^\dagger$ is the creation operator for the hole on one of the three apex oxygen orbitals with the spin, which is the same or opposite to the spin of the plane hole, depending on the total spin of the bipolaron. The probability amplitudes $A_i$ are normalised by the condition $|A_i| = 1/2$ if four plane orbitals $p_{x1}, p_{y2}, p_{x3}$ and $p_{y4}$ are involved, or by $|A_i| = 1/\sqrt{2}$ if only two of them are relevant.

The matrix element of the Hamiltonian Eq.(6.19) of the first order with respect to the transfer integral responsible for the bipolaron tunneling to the nearest neighbour cell **m** + **a** is

$$t = \langle \mathbf{m}|\tilde{H}|\mathbf{m}+\mathbf{a}\rangle = |A_i|^2 T_{pp'}^{apex} e^{-g^2}, \qquad (148)$$

where $T^{apex}_{pp'}e^{-g^2}$ is the single polaron hopping between two apex ions. As a result the hole bipolaron energy spectrum in a tight binding approximation consists of two bands $E^{x,y}$ formed by the overlap of $p_x$ and $p_y$ *apex polaron* orbitals, respectively, Fig. 3.8:

$$E^x_{\mathbf{k}} = -tcos(k_x) + t'cos(k_y), \quad (149)$$

$$E^y_{\mathbf{k}} = t'cos(k_x) - tcos(k_y) \quad (150)$$

where the in-plane lattice constant is taken to be $a = 1$, $t$ is the renormalized hopping integral, Eq.(148) between $p$ orbitals of the same symmetry elongated in the direction of the hopping ($pp\sigma$) and $t'$ is the renormalized hopping integral in the perpendicular direction ($pp\pi$). Their ratio $t/t' = T^{apex}_{pp'}/T'^{apex}_{pp'} = 4$ as follows from the tables of hopping integrals in solids.

Two different bands are not mixed because $T^{apex}_{p_x,p_y} = 0$ for the nearest neighbors. The random potential does not mix them either if it varies smoothly on the lattice scale. Hence, we can distinguish '$x$' and '$y$' bipolarons with a lighter effective mass in $x$ or $y$ direction, respectively. The apex $z$ bipolaron, if formed is ca. four times less mobile than $x$ and $y$ bipolarons. The bipolaron bandwidth is of the same order as the polaron one, which is a specific feature of the inter-site bipolarons. For a large part of the Brillouin zone near $(0, \pi)$ for '$x$' and $(\pi, 0)$ for '$y$' bipolaron, one can adopt the effective mass approximation

$$E^{x,y}_{\mathbf{k}} = \frac{k_x^2}{2m_{x,y}} + \frac{k_y^2}{2m_{y,x}} \quad (151)$$

with $k_{x,y}$ taken relative to the band bottom positions and $m_x = 1/t$, $m_y = 4m_x$.

## H  Bipolaronic Hamiltonian and Charged Bose Gas

In the subspace, in which there no single polarons one can rewrite the Hamiltonian in terms of the creation $b_i^\dagger = c^\dagger_{\mathbf{m}\uparrow}c^\dagger_{\mathbf{m},\downarrow}$ and annihilation (singlet) bipolaron operators as

$$H_b = \sum_{\mathbf{m}\neq\mathbf{m'}} \left( -t_{\mathbf{m},\mathbf{m'}} b_{\mathbf{m}}^\dagger b_{\mathbf{m'}} + \frac{1}{2}\bar{v}_{\mathbf{m},\mathbf{m'}} n_{\mathbf{m}} n_{\mathbf{m'}} \right), \quad (152)$$

where $n_{\mathbf{m}} = b_{\mathbf{m}}^\dagger b_{\mathbf{m}}$ is the bipolaron occupation number operator, and $\bar{v}_{\mathbf{m},\mathbf{m'}}$ the bipolaron-bipolaron repulsion including the direct Coulomb repulsion (four times of the polaron-polaron one, Eq.(121)) and the hard-core short-range repulsion. The latter appears due to the fact that the bipolaron operators have the pseudospin commutation rules [55].

In the case of *inter − site* bipolarons, which are the bound states of two small polarons on neighbouring sites, there are two additional points. First of all an inter-site bipolaron can be formed with a nonzero spin $S = 1$ (triplet state) and the energy $J > 0$ above the singlet state $S = 0$.

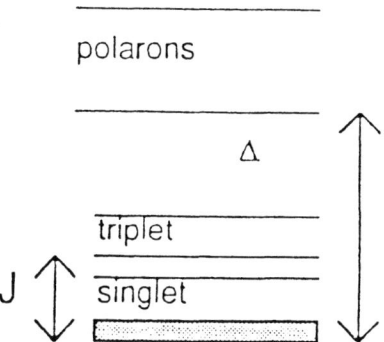

**FIGURE 3.9.** Singlet and triplet bipolaron (bosonic) bands separated by the exchange energy $J$, and the single polaron (fermionic) band. $2\Delta$ is the singlet bipolaron binding energy.

This should be taken into account by introducing additional spin quantum numbers $S = 1; l = 0, \pm 1$ in the definition of $b_m$. The second point is that in a simple square or cubic lattice the inter-site bipolaron tunnels via the next neighbor hopping of a single polaron rather than via simultaneous two-particle tunnelling in the case of on-site bipolaron. This 'crab-like' tunnelling results in a bipolaron bandwidth of the same order as the polaron one. In the perovskite structures the *apex* bipolaron is the ground state. In general, $H_b$ in the form Eq.(152) is applied not only to on-site or inter-site bipolarons but also to more extended non-overlapping pairs if m includes the spin, and $t_{m,m'}$ is considered as a phenomenological parameter. The site index m should be generally considered as a position of the centre of mass of the bipolaron.

As a result the low-energy band structure of strongly coupled electrons and phonons consists of two bosonic bands (singlet and triplet), separated by the singlet-triplet exchange energy $J$, Fig. 3.9. If the bipolaron binding energy is large, $\Delta > T$, single polarons exist only as thermally excited non-degenerate fermions. We argue that many features of spin and charge excitations of metal oxides can be described with this simple model, section 4.

The population of singlet and triplet bands is controlled by the chemical potential $\mu = Tlny$, where $y$ is determined from the thermal equilibrium of singlet and triplet bipolarons (if $T > T_c$) as

$$\ln\left(\frac{1 - ye^{-2t/T}}{1-y}\right) + 3\ln\left(\frac{1 - ye^{(-2t-J)/T}}{1 - ye^{-J/T}}\right) = \frac{2n_B t}{T}, \tag{153}$$

and $y = 1$ if $T < T_c$. Here $n_B$ is the total number of pairs per cell and the density of states is assumed to be energy independent within the bands.

When the carrier density is small, $n_B \ll 1$ (as in cuprates), bipolaronic operators are almost bosonic. The hard core interaction does not play any role in this dilute limit, so only the Coulomb repulsion is relevant. This repulsion is significantly reduced due to the large static dielectric constant in oxides, $\epsilon_0 \gg 1$. Hence, the Hamiltonian $H_b$ is that of the charged Bose gas on a lattice.

The charged Coulomb Bose-gas (CBG) is a fundamental reference system in many-particle physics with a superfluid phase transition. It has been studied by several authors and recently became of particular interest motivated by the bipolaron theory of high temperature superconductivity. Schafroth [7] demonstrated that an ideal gas of charged bosons exihibits the Meissner-Ochsenfeld effect (expulsion of a magnetic field) below the ideal Bose-gas condensation temperature. Later on the one-particle excitation spectrum at $T = 0$ was calculated by Foldy [79], who worked at zero temperature using the Bogoliubov approach. The Bogoliubov method leads to the result that the ground state of the system has a negative correlation energy whose magnitude increases with the density of bosons. Perhaps more interesting is the fact that the elementary excitations of the system have, for small momenta, energies characteristic of plasma oscillations which pass over smoothly for large momenta to the energies characteristic of single particle excitations. Further investigations have been carried out at or near $T_c$, the transition temperature for the gas. These works has been concerned with the critical exponents [80,81], and the change in the transition temperature from that of the ideal gas [81,80]. The RPA dielectric responce function and screening in CBG have been studied in the high-density limit [82], including a low-dimensional (2d) CBG. We have found the crtitical magnetic field of the Bose-Einstein condensation of CBG [83,86]. The superfluid properties of charged bosons as well as their excitation spectrum and the response function have been studied by the use of the Bogoliubov-de Genes (BdG) type equations, fully taking into account the interaction of quasiparticles with the condensate [85].

When we are interested in the long-wave, low-frequency responce, the effective mass approximation of the bipolaron dispersion, Eq.(151) is applied. In this case one can use the Hamiltonian for a system of charged bosons on an oppositely charged background (to ensure charge neutrality) in an external field as

$$H = \int d\mathbf{r} \psi^\dagger(\mathbf{r}) \left[ -\frac{(\hbar \nabla - ie^* \mathbf{A}/c)^2}{2m^{**}} - \mu \right] \psi(\mathbf{r})$$
$$+ \frac{1}{2} \int d\mathbf{r} \int d\mathbf{r}' V(\mathbf{r} - \mathbf{r}') [\psi^\dagger(\mathbf{r}) \psi^\dagger(\mathbf{r}') \psi(\mathbf{r}') \psi(\mathbf{r}) - 2n_B \psi^\dagger(\mathbf{r}) \psi(\mathbf{r})], \qquad (154)$$

where $m^{**}, e^* = 2e, n_B, \mu$ are the mass, charge, average density and the chemical potential of bosons, and $V(\mathbf{r}) = e^{*2}/\epsilon_0 |\mathbf{r}|$ is the Coulomb repulsion.

# IV  BIPOLARONS AND HIGH-$T_C$ CUPRATES: SOME RECENT RESULTS

There is a number of anomalous properties of high-$T_c$ cuprates, which seem to be intrinsic and universal. The isotope effect on the carrier mass [87], unusually high values of the static dielectric constants [88], and the non-Fermi liquid and non-BCS spectral densities seen in the angle-resolved photoemission (ARPES) [89–91], superconductor- insulator-normal metal (SIN) and SIS tunnelling spectra [92–98] are among those. The isotope effect and the high values of the static dielectric constants as well as the optical spectroscopy [99] suggest that the electron-phonon interaction is more than sufficient to bind carriers and phonons into small polarons and bipolarons in highly polarisable cuprates [9,100](see below). The bipolaron theory [9] provides a natural explanation of the normal state pseudogap (as a half of the bipolaron binding energy, $\Delta_p$) [101,102], NMR linewidth [101], normal state kinetics [9], tunnelling [103] and ARPES [105,106]. The theory accounts for two distinct energy gaps (coherent and incoherent) [107] as observed in the recent tunnelling [96–98] and Andreev reflection experiments [108,96], for the parameter-free fit of the critical temperature , upper critical field , specific heat, London penetration depth, and the symmetry of the order parameter (see Ref. [109] for recent review).

In this section we discuss a few of these applications.

## A  Essential interaction in cuprates

Assessing the role of different interactions in doped cuprates one has to take into account that these materials are highly polarizable ionic lattices, so that the Fröhlich electron-phonon interaction with optical phonons should be strong. Indeed, our estimate shows that the polaron binding energy, $E_p$ is about 1 eV. Hence, the Fröhlich interaction should play the dominant role together with the direct Coulomb repulsion. In the case of the Fröhlich interaction with optical phonons the exact expression for $E_p$ depends only on the *experimentally* measured high-frequency, $\epsilon_\infty$ and static $\epsilon_0$ dielectric costants of the host insulator as [110]

$$E_p = \frac{1}{2\kappa} \int_{BZ} \frac{d^3 q}{(2\pi)^3} \frac{4\pi e^2}{q^2}, \tag{155}$$

where $\kappa^{-1} = \epsilon_\infty^{-1} - \epsilon_0^{-1}$, and the size of the integration region (the Brillouin zone, BZ) is determined by the lattice constants $a, b, c$. As shown in the Table, the canonical Fröhlich interaction alone gives the binding energy of two holes ($2E_p$) almost by one order of magnitude larger than the magnetic interaction $J \sim 0.15$ eV in the $t-J$ model. The data in the Table represent the *lower* boundaries of the polaron binding energy, since the short-wave phonons due to the deformation potential and/or Jahn-Teller interaction are not included in Eq.(155).

The polaron shift about 1eV provides small polarons no matter which of criteria of their formation is applied. The bare half-bandwidth is about 1 eV or less in cuprates, so even a naive variational criterion of the small polaron formation ($E_p > D$) is satisfied (here $D$ is the half-bandwidth in a rigid lattice). Moreover, in the nonadabatic and intermediate regime the variational criterion of the small polaron formation fails. The optical phonon frequencies are high in oxides, about $\omega \simeq 0.1$ eV. Hence, the adiabatic ratio is of the order of unity, $z\omega/D \simeq 1$ ($z$ is the coordination number). Hence, there is no well-defined double-well potential, and the variational criterion cannot be applied. In this regime the correct criterion of the small polaron formation is determined by the convergency of the $1/\lambda$ perturbation expansio (section 3).

TABLE. Polaron shift $E_{sp}$ due to Fröhlich interaction. Data from *Handbook of Optical Constants of Solids*, edited by E.D. Palik (Academic, New York, 1997) and *Landolt-Börnstein*. The value $\epsilon_\infty = 5$ for $WO_3$ is an estimate. [a] Distance between $CuO_2$ planes, [b] T. Ishikawa (private communication), [c] distance between $MnO_2$ planes.

| System | $\epsilon_\infty$ | $\epsilon_0$ | $a \times b \times c$ (Å$^3$) | $E_{sp}$(eV) |
|---|---|---|---|---|
| $SrTiO_3$ | 5.2 | 310 | $3.905^3$ | 0.852 |
| $BaTiO_3$ | 5.1-5.3 | 1499. | $3.992^2 \times 4.032$ | 0.842 |
| $BaBiO_3$ | 5.7 | 30.4 | $4.34^2 \times 4.32$ | 0.579 |
| $La_2CuO_4$ | 5.0 | 30 | $3.8^2 \times 6^a$ | 0.647 |
| $LaMnO_3$ | $3.9^b$ | $16^b$ | $3.86^3$ | 0.884 |
| $La_{2-2x}Sr_{1+2x}$ $-Mn_2O_7$ | $4.9^b$ | $38^b$ | $3.86^2 \times 3.9^c$ | 0.807 |
| NiO | 5.4 | 12 | $4.18^3$ | 0.429 |
| $TiO_2$ | 6-7.2 | 89-173 | $4.59^2 \times 2.96$ | 0.643 |
| MgO | 2.964 | 9.816 | $4.2147^3$ | 0.982 |
| CdO | 5.4 | 21.9 | $4.7^3$ | 0.522 |
| $WO_3$ | 5 | 100-300 | $7.31 \times 7.54 \times 7.7$ | 0.445 |
| NaCl | 2.44 | 5.90 | $5.643^3$ | 0.749 |
| EuS | 5.0 | 11.1 | $5.968^3$ | 0.324 |
| EuSe | 5.0 | 9.4 | $6.1936^3$ | 0.266 |

The exact Monte-Carlo calculations [42] show that the polaron theory based on this expansion describes quantitatively *both* strong-coupling (small polaron) and weak coupling (large polaron) regimes without any restriction on the value of $E_p$ for the long-range electron-phonon interaction with high-frequency phonons.

The effective polaron-polaron attraction, due to the overlap of the deformation fields, about $2E_p$, is more than sufficient to overcome the intersite Coulomb repulsion at short distances, as also confirmed by the first-principles lattice minimization technique [78].

One could believe that complete localisation in a random potential of the heavy polaronic carriers is inevitable in doped cuprates. All states of the polaronic band are localised if $F/W_p \geq 5$, where $F$ is the characteristic fluctuation of the electrostatic energy of the order of $e^2 n_{im}^{1/3}/\epsilon_0$, and $W_p = z/2m^{**}a^2$ the (bi)polaron half-bandwidth. Even with the impurity density as high as $n_{im} = 0.3$ per cell (of the size $a^3$) and $m^{**} = 10 m_e$ one obtains $F/W_p \simeq 0.5$, which is about one order of magnitude below the critical ratio, which is necessary to localise all polaronic states. Hence, small (bi)polarons are the most probable dispersive quasiparticles in cuprates and manganites.

## B  Parameter-free expression of the critical temperature

In contrast with the BCS theory the bipolaron theory allows us [111,112] to 'integrate out' the interaction and express $T_c$ via the static response functions. In the framework of the BCS theory (largely independent of the nature of coupling) the critical temperature is fairly well approximated by McMillan's formula, Eq.(65). which works well for simple metals and their alloys. There are no general restrictions on the BCS value of $T_c$ if the dielectric function formalism is properly applied [113]. Allen and Dynes [114] found that in the strong-coupling limit $\lambda \gg 1$ the critical temperature might be as high as $T_c \simeq \omega \lambda^{1/2}/2\pi$. Nevertheless, applying this kind of theory to high-$T_c$ cuprates is problematic. Since the bare electron bands are narrow, strong correlations are important giving rise to a doped Mott insulator. As a result, the Coulomb pseudopotential and $\lambda$ are ill-defined and polaronic effects are important as in many doped semiconductors [9]. Taking the 'magic' numbers $\lambda = 0.5$, $\mu^* = 0.14$ and the experimental Debye temperature $\omega = 400K$ one obtains $T_c \simeq 2K$ with Eq.(65) - clearly too low to explain high-$T_c$. One could hardly expect that the Coulomb pseudopotential is lower than 0.1 because the Tolmachev logarithm cannot be large in narrow bands. In fact, $\mu^*$ is of the order of the bare Coulomb repulsion, $\mu^* \simeq \mu \simeq 1$. Hence, an estimate of $T_c$ in cuprates within the BCS theory appears to be an exercise in calculating $\mu^*$ rather than $T_c$ itself. Nor can one increase $\lambda$ without accounting for a polaron collapse of the band and bipolaron formation (section 3). As discussed above this appears at $\lambda \simeq 0.5$ for uncorrelated polarons, and even for a smaller value of the bare electron phonon coupling in strongly correlated models [30,31].

In the framework of the bipolaron theory the critical temperature is determined by the bipolaron energy spectrum. Quite generally the bipolaron energy spectrum is a degenerate doublet due to two ($x$ and $y$) oxygen orbitals elongated along the $CuO_2$ planes, Eq.(149,150). The energy band minima are found at the Brillouin zone boundary, $(\pm \pi, 0)$ and $(0, \pm \pi)$ rather than at $\Gamma$ point owing to the opposite sign of the $pp\sigma$ and $pp\pi$ oxygen hopping integrals. Near these points the effective mass approximation is applied as discussed in section 3. If we take into account also a possibility of pair tunnelling between $CuO_2$ planes, than

$$E_{\mathbf{k}}^{x,y} = \frac{\hbar^2 k_{x,y}^2}{2m_x} + \frac{\hbar^2 k_{y,x}^2}{2m_y} + 2t_\perp [1 - \cos(k_z d)] \quad (156)$$

where $k_{x,y}$ are taken relative $(\pm\pi, 0)$ (or $(0, \pm\pi)$) points, $d$ is the interplane distance and $t_\perp$ is the interplane bipolaron hopping integral.

The condensation temperature is calculated by substituting this spectrum into the density sum rule,

$$\sum_{\mathbf{k},i=(x,y)} \left[\exp(E_{\mathbf{k}}^i/T_c) - 1\right]^{-1} = n_B, \quad (157)$$

so one readily obtains $T_c$ as

$$T_c = f \times \frac{3.31(n_B/2)^{2/3}}{(m_x m_y m_c)^{1/3}}. \quad (158)$$

The coefficient $f$ is almost 1 in a wide range of the anisotropy $t_\perp/T_c$ [112] ($m_c = 1/2|t_\perp|d^2$). This expression is rather ambiguous so far because the effective mass tensor as well as the bipolaron density $n_B$ are unknown and doping dependent. Fortunately, one can express the band-structure parameters through the in-plane, $\lambda_{ab} = [m_x m_y/8\pi n_B e^2(m_x + m_y)]^{1/2}$ and out-of-plane penetration depth, $\lambda_c = [m_c/16\pi n_B e^2]^{1/2}$. The bipolaron density is expressed through the in-plane Hall constant ( just above the transition) as

$$R_H = \frac{1}{2en_B} \times \frac{4m_x m_y}{(m_x + m_y)^2}. \quad (159)$$

As a result one obtains

$$T_c = 1.64 \times \left(\frac{eR_H}{\lambda_{ab}^4 \lambda_c^2}\right)^{1/3}. \quad (160)$$

with $T_c$ measured in Kelvin, $eR_H$ in cm$^3$ and $\lambda$ in cm. Hence, our theory yields a parameter-free expression, which unambiguously tells us how close cuprates are to the Bose-Einstein condensation regime. This expression has been compared with the experimental $T_c$ of more than 30 different cuprates, for which both $\lambda_{ab}$ and $\lambda_c$ are measured along with $R_H$ [112]. The theoretical $T_c$ coincides with the experimental one within the experimental error bar for the penetration depth (about $\pm 10\%$) no matter what the doping level is. A few examples are $La_{1.8}Sr_{0.2}CuO_4$ ($\lambda_{ab} = 2000\text{Å}$, $\lambda_c = 25400\text{Å}$, $R_H = 0.8 \times 10^{-3} 1/Ccm^3$), $T_c^{exp}$=36.2K, our theoretical value, Eq.(49) is $T_c$=38K; $YBa_2Cu_3O_7$ ($\lambda_{ab} = 1400\text{Å}$, $\lambda_c = 12600\text{Å}$, $R_H = 1.2 \times 10^{-3} 1/Ccm^3$), $T_c^{exp}$=92.5K, the theoretical value is $T_c$=111 K; $YBa_2Cu_3O_{6.84}$ ($\lambda_{ab} = 1771\text{Å}$, $\lambda_c = 15570\text{Å}$, $R_H = 1.9 \times 10^{-3} 1/Ccm^3$), $T_c^{exp}$=83.7K, the theoretical value is $T_c$=83 K.

One can argue that cuprates belong to a 2D 'XY' universality class with the Kosterlitz-Thouless critical temperature $T_{KT}$ due to a large anisotropy [115–117].

If this is the case then one could not discriminate the Cooper pairs with respect to bipolarons. The Kosterlitz-Thouless temperature, expressed trough the in-plane penetration depth is [117]

$$T_{KT} \simeq \frac{0.9d}{16\pi e^2 \lambda_{ab}^2}. \tag{161}$$

It appears significantly (about twice) higher than the experimental value in most cases. There are quite a few cuprates with about the same in-plane penetration depth and interplane distance, but with essentially different $T_c$. Obviously, this observation is not compatible with Eq.(161). Also many cuprates do not share the critical behaviour of the BCS superfluids nor the universal (3D) $x-y$ properties of neutral superfluids like $^4$He [118,119] but exhibit a critical behaviour of charged bosons. These observations favor the 3D Bose-Einstein condensation of charged bosons as the mechanism of high $T_c$ rather than any low-dimensional phase-fluctuation scenario.

## C  Upper critical field

As noted by Schafroth [7] an ideal charged Bose-gas in a magnetic field cannot be condensed because of the one-dimensional character of particle motion within the lowest Landau level. However, the interacting charged Bose-gas is condensed in a field lower than a certain critical value $H^*$ because the interaction with impurities or between bosons broadens the Landau levels and thereby eliminates the one-dimensional singularity of the density of states [83]. As we discuss below the critical field of BEC has an unusual positive curvature near $T_{c0}$, $H^*(T) \sim (T_c-T)^{3/2}$. At low impurity concentration it diverges at $T \to 0$. The localization can drastically change the low-temperature behavior of $H^*(T)$, so at high concentration of impurities the re-entry effect to the normal state occures.

$H^*$ is determined as the field in which the first nonzero solution of the linearized stationary equation for the macroscopic condensate wave function $\psi_0(\mathbf{r}) = \langle N|\hat{\psi}(\mathbf{r},\tau)|N+1\rangle$, $(N \to \infty, N/V = n_B = const)$ appears:

$$\left[-\frac{1}{2m^{**}}(\nabla - 2ie\mathbf{A}(\mathbf{r}))^2 + U_{imp}(\mathbf{r})\right]\psi_0(\mathbf{r}) = \mu\psi_0(\mathbf{r}), \tag{162}$$

where $U_{imp}(\mathbf{r})$ is the random potentials.

Here we suppose that the particle-particle interaction is taken into account within the Hartree approximation and included in the chemical potential, so the main origin of the broadening of the Landau levels is the impurity scattering. The boson-boson scattering, when it is important, also yields the divergent upward temperature dependence of $H^*$ [70]. The definition of $H^*$ is identical to that of the upper critical field $H_{c2}$ of BCS superconductors. Therefore $H^*$ determines the upper critical field of any "bosonic" superconductor.

In general the energy spectrum of the Hamiltonian Eq.() contains discrete levels (localized states) and a continuous part (delocalized states). The density of delocalized states $\tilde{N}(\epsilon, H) \sim \Im\Sigma(\epsilon)$ and the lowest delocalized energy $E_c$ (the mobility edge, $\tilde{N}(E_c, H) = 0$) can be found with the random phase ("ladder") approximation for the one-particle self-energy:

$$\Sigma(\epsilon) = M^2 \int \frac{N(\epsilon', H) d\epsilon'}{\epsilon - \epsilon' - \Sigma(\epsilon)} \tag{163}$$

where $M^2$ is the squared matrix element for the boson-impurity scattering multiplied by the impurity density, and

$$N(\epsilon, H) = \frac{\sqrt{2}(m^{**})^{3/2}\omega}{4\pi^2} \Re \sum_{N=0}^{\infty} \frac{1}{\sqrt{\epsilon - \omega(N + 1/2)}} \tag{164}$$

is the density of states for a noninteracting system with $\omega = 2eH/m^{**}$.

The solution of Eq.(163) yields the density of states in the lowest Landau level ($N = 0$) as

$$\tilde{N}_0(\epsilon, H) = \frac{\sqrt{6}(m^{**})^{3/2}\omega}{8\pi^2 \sqrt{\Gamma_0}} \left[ \left( \frac{\tilde{\epsilon}^3}{27} + \frac{1}{2} + \sqrt{\frac{\tilde{\epsilon}^3}{27} + \frac{1}{4}} \right)^{1/3} - \left( \frac{\tilde{\epsilon}^3}{27} + \frac{1}{2} - \sqrt{\frac{\tilde{\epsilon}^3}{27} + \frac{1}{4}} \right)^{1/3} \right], \tag{165}$$

and

$$E_c = \frac{\omega}{2} - \frac{3\Gamma_0}{2^{2/3}}. \tag{166}$$

Here $\Gamma_0 = 0.5(2M^2 eH\sqrt{m^{**}}/\pi)^{2/3}$ the characteristic broadening of the lowest Landau level and $\tilde{\epsilon} = (\epsilon - \omega/2)/\Gamma_0$.

Because the singularity of the density of states of upper levels is integrated out one can neglect their quantization using the zero field density of states for $\epsilon > \omega$:

$$N(\epsilon) \simeq \frac{(m^{**})^{3/2}\sqrt{\epsilon}}{\sqrt{2}\pi^2}. \tag{167}$$

The first nontrivial *delocalised* solution of Eq.(162) appears at $\mu = E_c$. Thus the critical curve $H^*(T)$ is determined from the conservation of the number of particles $n_B$ under the condition that the chemical potential coincides with the mobility edge:

$$\int_{E_c}^{\infty} \frac{\tilde{N}_0(\epsilon, H^*) d\epsilon}{exp(\frac{\epsilon - E_c}{T}) - 1} = n_B \left( 1 - (T/T_{c0})^{3/2} - \frac{n_L(T)}{n_B} \right). \tag{168}$$

The left-hand side of Eq.(168) is the number of bosons on the lowest Landau level, while the second term in the right-hand side is the number of bosons on all upper Landau levels, calculated with the classical density of states. The last term ($n_L(T)$) is the number of bosons localised in the random potential below the mobility edge. It can be found by the use of a 'single well-single particle' approximation. Within this approximation *localized charged* bosons obey the Fermi-Dirac statistics because of their Coulomb repulsion in the localised states, so that

$$n_L(T) = \int_{-\infty}^{E_c} dE \frac{\rho_L(E)}{exp[(E-\mu)/T]+1}, \qquad (169)$$

where $\rho_L(E)$ is the density of localized states. It can be approximated as

$$\rho_L(E) = \frac{N_L}{\gamma} \exp(\frac{E-E_c}{\gamma}). \qquad (170)$$

with $\gamma$ of order of a binding energy in a single random potential well and $N_L$ the total number of localized states per unit cell. Substitution of Eq.(165) and Eq.(169) into Eq.(168) yields the expression for the critical field of BEC as

$$H^*(T) = H_d(T_{c0}/T)^{3/2}\left(1 - (T/T_{c0})^{3/2} - \frac{TN_L}{\gamma n}\beta(T/\gamma)\right)^{3/2}, \qquad (171)$$

with

$$\beta(x) = \sum_{k=0}^{\infty} \frac{(-1)^k}{x+k}, \qquad (172)$$

and temperature independent $H_d = \phi_0/2\pi\xi_0^2$. Here $T_{c0} \simeq 3.3 n_B^{2/3}/m^{**}$ is the BEC temperature of an ideal Bose gas with the density $n_B$. The "coherence" length $\xi_0$ is determined by both the mean free path $l = \pi/M^2(m^{**})^2$ and the inter-particle distance:

$$\xi_0 \simeq 0.8(l/n_B)^{1/4}. \qquad (173)$$

$\phi_0 = \pi/e$ is the flux quantum.

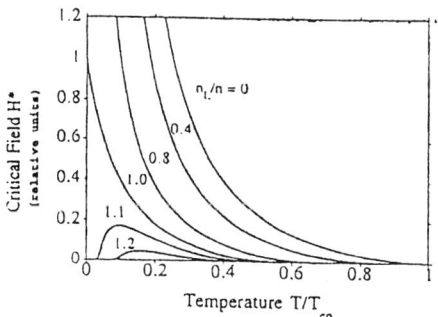

**FIGURE 4.1.** Upper critical field of CBG for different disorder $N_L$.

Using the asymptotic $\beta(x) \simeq (2x)^{-1}$ at temperatures $T > \gamma$ one obtains:

$$H^* = H_d \sqrt{1 - \frac{n_L}{2n_B}} \left(\frac{1}{\tau} - \sqrt{\tau}\right)^{3/2}, \qquad (174)$$

where $\tau \equiv T/T_c$ is the reduced temperature and

$$T_c = T_{c0}\left(1 - \frac{N_L}{2n_B}\right)^{2/3} \qquad (175)$$

is the critical temperature of BEC in a random potential for zero magnetic field.

Thus $H^*(T)$ has the positive "3/2" curvature near $T_c$. This curvature is a universal feature of CBG, which does not depend on a particular scattering mechanism and on approximations made. The number of bosons at the lowest Landau level is proportional to the density of states near the mobility edge $\tilde{N}_0 \sim H/\sqrt{\Gamma(H)}$, where the "width" of the Landau level is also proportional to the same density of states $\Gamma(H) \sim H/\sqrt{\Gamma(H)}$. Hence $\Gamma(H) \sim H^{2/3}$ and the number of condensed bosons is proportional to $H^{2/3}$. On the other hand this number in the vicinity of $T_c$ should be proportional to $T_c - T$ (the total number minus the number of thermally excited bosons). That gives the "3/2" law for $H^*(T)$.

At low temperatures, $T \ll \gamma$, the temperature dependence of $H^*$ turns to be different for different impurity concentrations. If $0 < N_L < n$ the critical field diverges at $T \to 0$:

$$H^* \simeq H_d(T_{co}/T)^{3/2}\left(1 - \frac{N_L}{n_B}\right)^{3/2} \qquad (176)$$

because the number of localized states is smaller then the number of bosons. In this case only the paramagnetic limit restricts the value of $H^*(0)$ if bosons are composed from two fermions with the opposite spins. This limit is not related to the value of $T_c$ and might exceed the BCS paramagnetic limit by many times. If $N_L = n_B$ the critical field reaches its maximum at $T = 0$:

$$H^* \simeq H_d(T_{c0} ln2/\gamma)^{3/2}\left(1 - \frac{\pi^2 T}{8\gamma ln2}\right). \qquad (177)$$

And finally, if $n_B < N_L < 2n_B$ there is a reentry effect to the normal state at temperature below some $T^*$, so $H^* = 0$ for $T < T^*$, Fig. 4.1.

If the number of localized states is large, $N_L > 2n_B$ Bose condensation is impossible, $T_c = 0$.

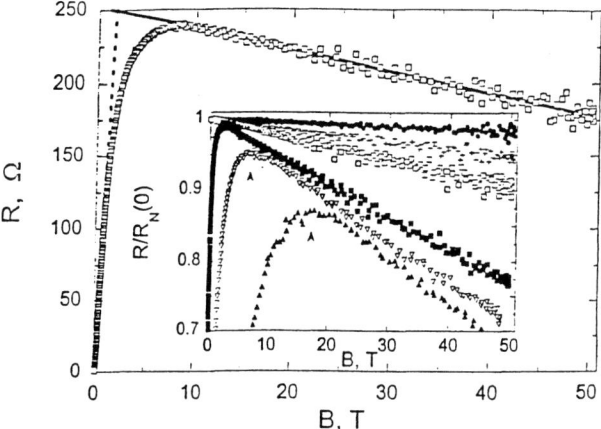

**FIGURE 4.2.** The magnetic field dependence of the out-of-plane resistance of BSCCO-2212 measured at 78K ($T_{c0} = 92$ K). Linear fits to the flux-flow portion of the curve and that attributed to the normal state magnetoresistance are shown by dashed and solid lines respectively. The inset shows the variation with field of the normal state magnetoresistance measured at different temperatures, 115, 103, 98, 90.1, 78, and 57.5 K (from the top) normalised by the value of $R_N(0)$. Reprinted from [120], with permission from EDP Sciences.

In the cuprates superconductors, high magnetic field studies have revealed a non-BCS divergent shape of the upper critical field $H_{c2}(T)$ [120-127]. These studies were performed both in relatively low-$T_c$ cuprates [122,123,125-127] and in some high-$T_c$ compounds in a moderate [124] (below 15 T) and high (up to 60 T) field [120]. The upper critical field was determined from the temperature dependence of c-axis resistivity with some uncertainty due to fluctuations [124]. The uncertainty was removed in the comprehensive study by Gantmakher et al [127] of the in-plane resistivity of high-quality YBa$_2$Cu$_3$O$_{7-\delta}$ crystals, which confirmed the non-BCS upper critical field observed in Ref. [122-126] strongly supporting the bipolaron theory of cuprates [83].

Here I discuss recent study in pulsed magnetic fields of Bi$_2$Sr$_2$CaCu$_2$O$_8$, with $T_{c0} \simeq 91 - 93\,K$, which reveals new features of the c-axis transport [120]. A positive linear magnetoresistance in the flux flow (superconducting) regime and a negative linear magnetoresistance in the normal state have been observed. This allows for determination of the upper critical field, as the point of intersection of these two regimes.

Fig. 4.2 shows a typical measurement of the effect of magnetic field on the out-of-plane resistance of a BSCCO-2212 single crystal below a zero-field critical temperature, $T_{c0}$. There is a low-field regime, $R_{FF}(B,T)$, where a linear field dependence fits the experimental observations rather well, Fig. 4.2.

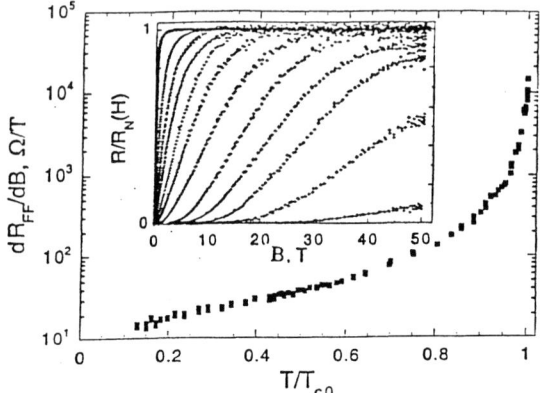

**FIGURE 4.3.** The temperature dependence of the flux-flow resistance slope. Inset: Field dependence of out-of-plane resistance of BSCCO-2212 normalised by its normal state value, $R_N(B)$. The selected traces are obtained (from right to the left) at 16, 20, 25, 30, 35, 45, 52.6, 57.5, 65, 70, 78, and 88.7 K respectively. Reprinted from [120], with permission from EDP Sciences.

A linear field dependence suggests a usual flux-flow regime. Of course, there is no such thing as flux flow resistivity for current flowing along the field direction. Nevertheless, a highly anisotropic structure of our *Bi* samples with alternating quasi-metallic and disordered non-metallic layers favors the current path with the in-plane meanders. Then there is a finite Lorentz force applied to the vortex even in the longitudinal geometry.

It is natural to attribute the high field portion of the curve in Fig. 4.2 (assumed to be above $H_{c2}$) to a normal state magnetoresistance, $R_N(B,T)$, which appears to be *negative* and *linear* in B. The latter is unusual for the longitudinal transport but is also evident in other studies [121,128,129]. It can be explained as the result of the field dependent normal state gap in the bipolaronic system [120]. With these assumptions one can determine the upper critical field, $H_{c2}(T)$, from the intersection of two linear approximations in Fig. 4.2.

This procedure allows us to separate contributions originating from the normal and superconducting states and, in particular, to avoid to large extent an ambiguity due to fluctuations in the crossover region. Referring to Fig. 4.3, the inset shows the field dependence of BSCCO-2212 out-of-plane resistance normalised by its normal state field dependence, $R_N(B)$, thus accounting for its variation with field and temperature. The slope of the flux-flow resistance is inversely proportional to $H_{c2}$

as $R_{FF} = R_N \times B/H_{c2}$.

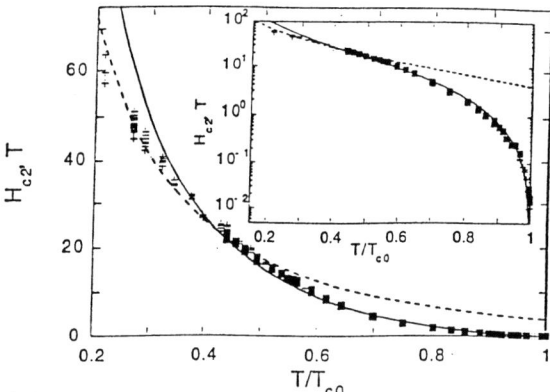

**FIGURE 4.4.** The resistive upper critical field of BSCCO-2212 as a function of temperature obtained from the intersections of the linear extrapolations from the normal and flux-flow regimes (solid circles), and as the ratio of the extrapolated $R_N(T)$ and flux-flow resistance slope (crosses). A fit to the Bose-Einstein condensation field, Eq.(174), is shown by the solid curve, while the dashed line shows a fit to the 'pseudo-upper-critical field' of Ref. [130]. Reprinted from [120], with permission from EDP Sciences.

Indeed, $H_{c2}$ determined from (i) the intersection of the linear fits mentioned above and (ii) that obtained from $R_{FF}(B)$ as $R_N(O,T)(\partial R_{FF}/\partial B)^{-1}$ are almost identical as is seen from Fig. 4.4 where the temperature dependence of $H_{c2}$ is presented together with the theoretical fit using the Bose-Einstein condensation critical field [83] given by Eq. (174).

$H_{c2}(T)$ shows an upward temperature dependence in agreement with the previous results. One can try (unsuccessfully) to fit the data with the pseudo-upper-critical field, $H^* \sim T^{-1}exp(-T/T_0)$ [130] as suggested in Ref. [131]( the dashed line in Fig. 4.4). Therefore, the model which lies behind this equation, which is based on Josephson-coupling as the origin of the anomalous $H_{c2}(T)$, is not supported by the experiment. Some diamagnetism observed above the resistive $T_c(B)$ [132,131] is explained as the *normal* state Landau diamagnetism of singlet bosons [133].

## D  Symmetry of the order parameter

There is strong evidence for a *d*-like order parameter (changing sign when the $CuO_2$ plane is rotated by $\pi/2$) in cuprates reviewed in Ref. [134]. A number of phase-sensitive experiments [135] provide unambiguous evidence in this direction; furthermore, the low temperature magnetic penetration depth [136,137] has been found to be linear in many cuprates as expected for a d-wave BCS superconductor.

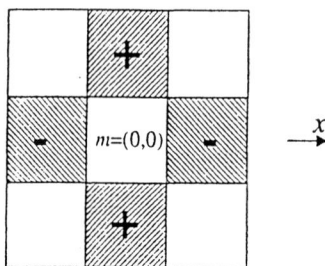

FIGURE 4.5. D-wave condensate wave function, in the real (Wannier) space. The order parameter has different signs in the shaded cells and is zero in the blank cells.
Reprinted from [139], Copyright 1999, with permission from Elsevier Science.

However, SIN (and SIS) tunnelling studies and some high-precision magnetic measurements show the more usual $s$-like shape of the gap function or even reveal an upturn in the temperature dependence of the penetration depth below some characteristic temperature [138].

In the framework of the bipolaron theory the symmetry of the Bose-Einstein condensate should be distinguished from that of the 'internal' wave function of a single bipolaron, i.e. from the symmetry of the single-particle excitation gap. Also the low temperature dependence of the penetration depth, $\lambda(T)$ might *not* be related to the symmetry of the order parameter, but determined by the localisation of bipolarons in the random potential. Both linear positive and negative slopes of $\lambda(T)$ occur depending on the random field profile [139].

If the bipolaron density is low, the bipolaron Hamiltonian can be mapped onto the charged Bose-gas as discussed above. Charged bosons are condensed below $T_c$ into the states of the Brillouin zone with the lowest energy, which are $\mathbf{k} = (\pm \pi, 0)$ and $\mathbf{k} = (0, \pm \pi)$ for the $x$ and $y$ bipolarons, respectively. These four states are degenerate, so the condensate field-operator $\Psi(\mathbf{m})$ in the real (site) space $\mathbf{m} = (m_x, m_y)$ is given by

$$\hat{\Psi}_{\pm}(\mathbf{m}) = N^{-1/2} \sum_{\mathbf{k}=(\pm\pi,0),(0,\pm\pi)} b_{\mathbf{k}} exp(-i\mathbf{k} \cdot \mathbf{m}). \qquad (178)$$

where $b_{\mathbf{k}}$ is the bipolaron (boson) annihilation operator in $\mathbf{k}$ space.

This is a $c$-number below $T_c$, so that the condensate wave function, which is the superconducting order parameter, is given by

$$\Psi_{\pm}(\mathbf{m}) = n_c^{1/2} \left[ \cos(\pi m_x) \pm \cos(\pi m_y) \right], \qquad (179)$$

where $n_c$ is the number of bosons per cell in the condensate. Other combinations of the four degenerate states do not respect time-reversal and (or) parity symmetry. The two solutions, Eq.(179), are physically identical being related by: $\Psi_+(m_x, m_y) = \Psi_-(m_x, m_y + 1)$. They have $d$-wave symmetry changing sign when the $CuO_2$ plane is rotated by $\pi/2$ around $(0,0)$ for $\Psi_-$ or around $(0,1)$ for $\Psi_+$ (Fig. 4.5). The $d$-wave symmetry is entirely due to the bipolaron energy dispersion with four minima at $\mathbf{k} \neq \mathbf{0}$. With the energy minimum located at the $\Gamma$ point of the Brillouin zone the condensate is $s$-like.

## E  London penetration depth

If the total number of bipolarons in one unit cell is $n_B$ of which $n_L$ are in localised states and $n_D$ are in delocalised states then the number in the condensate $n_c$ is

$$n_c = n_B - n_L - n_D \tag{180}$$

and the London penetration depth $\lambda \propto 1/\sqrt{n_c}$. Taking the delocalised bipolarons to be a free three-dimensional gas we have $n_D \propto T^{\frac{3}{2}}$. Thus in the limit of low temperature we can neglect $n_D$ and make the approximation

$$n_c \approx n_B - n_L \tag{181}$$

In this limit $\lambda(T) - \lambda(0)$ is small and so

$$\lambda(T) - \lambda(0) \propto n_L(T) - n_L(0) \tag{182}$$

i.e. $\lambda$ has the same temperature dependence as $n_L$. For small amounts of disorder the delocalised bipolarons may contribute to the low temperature dependence of $\lambda(T)$ as well; for non-interacting bipolarons moving in $d$ dimensions this would give $\lambda \propto T^{d/2}$.

In our picture interacting bosons fill up all the localised single-particle states in a random potential and Bose-condense into the first extended state at the mobility edge, $E_c$. For convenience we choose $E_c = 0$. When two or more charged bosons are in a single localised state of energy $E$ there may be significant Coulomb energy and we take this into account as follows. The localisation length $\xi$ is thought to depend on $E$ via

$$\xi \propto \frac{1}{(-E)^\nu} \tag{183}$$

where $\nu > 0$. The Coulomb potential energy of $p$ charged bosons confined within a radius $\xi$ can be expected to be

$$\text{potential energy} \sim \frac{p(p-1)e^2}{\epsilon_0 \xi}. \tag{184}$$

Thus the total energy of $p$ bosons in a localised state of energy $E$ is taken to be

$$w(E) = pE + p(p-1)\kappa(-E)^\nu \qquad (185)$$

where $\kappa > 0$. We can thus define an energy scale $E_1$:

$$E_1 = \kappa^{\frac{1}{1-\nu}}. \qquad (186)$$

From here on *we choose our units of energy such that $E_1 = 1$*. We take the total energy of charged bosons in localised states to be the sum of the energies of the bosons in the individual potential wells. The partition function $Z$ for such a system is then the product of the partition functions $z(E)$ for each of the wells,

$$z(E) = e^{\alpha p_0^2} \sum_{p=0}^{\infty} e^{-\alpha(p-p_0)^2} \qquad (187)$$

where

$$p_0 = \frac{1}{2}\left\{1 + (-E)^{1-\nu}\right\}, \qquad (188)$$

$$\alpha = \frac{(-E)^\nu}{\theta}, \qquad (189)$$

and

$$\theta = \frac{k_B T}{E_1}. \qquad (190)$$

The average number $n_L$ of bosons in localised states is

$$n_L = \int_{-\infty}^{0} dE \langle p \rangle \rho_L(E), \qquad (191)$$

where the mean occupancy $\langle p \rangle$ of a single localised state is taken to be

$$\langle p \rangle = \frac{\sum_{p=0}^{\infty} p\, e^{-\alpha(p-p_0)^2}}{\sum_{p=0}^{\infty} e^{-\alpha(p-p_0)^2}}, \qquad (192)$$

and $\rho_L(E)$ is the one-particle density of localised states per unit cell below the mobility edge, Eq.(170).

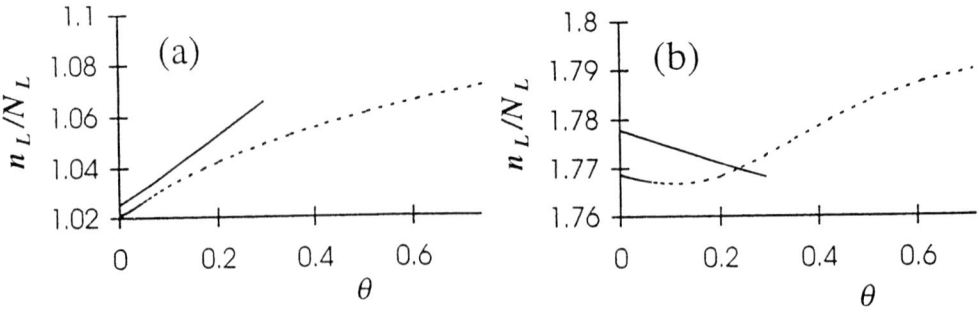

FIGURE 4.6. Dependence of the density of localised bosons $n_L$ on temperature $\theta$ for $\gamma = 20$. (a) $\nu = 1.5$, (b) $\nu = 0.65$. The solid lines correspond to the low temperature predictions from equations 193 and 194, while the dashed lines are derived from an accurate numerical calculation.
Reprinted from [139], Copyright 1999, with permission from Elsevier Science.

We now focus on the temperature dependence of $n_L$ at low temperature ($\theta \ll 1$) for the case where the (dimensionless) width of the impurity tail $\gamma$ is large ($\gamma > 1$). We note that the parameter $\gamma$ is the ratio of the width of the tail to the characteristic Coulomb repulsion, $E_1$.

In the following we consider first the case $\nu > 1$ and then $\nu < 1$. If $\nu > 1$ we can approximate $n_L$ as

$$\frac{n_L}{N_L} \approx 1 + \frac{\nu - 1}{2(2-\nu)\gamma} + \frac{2\theta}{(2-\nu)\gamma}\ln 2. \tag{193}$$

So we expect $n_L$ to be close to the total number of wells $N_L$ and to increase linearly with temperature. Fig. 4.6a compares this analytical formula with accurate numerical calculation for the case $\nu = 1.5$, $\gamma = 20$. We also note that even when $\gamma < 1$, $n_L(\theta)$ will still be linear with the same slope provided that $\theta \ll \gamma$.

If $\nu < 1$ we obtain, keeping only the lowest power of $\theta$ (valid provided $\theta^{\frac{1}{\nu}} \ll \theta$)

$$\frac{n_L}{N_L} = \frac{1}{2} + \frac{\Gamma(2-\nu)\gamma^{1-\nu}}{2} + \frac{1-\nu}{2(2-\nu)\gamma} - \frac{\theta}{\gamma}\ln 2. \tag{194}$$

Hence in this case $n_L$ decreases linearly with increasing temperature (in the low temperature limit).

Fig. 4.6b compares this analytical formula with the numerical calculation for the case $\nu = 0.65$, $\gamma = 20$. We note that such a value of $\nu$ is typical for amorphous semiconductors [140]. A large value of $\gamma$ is also expected in disordered cuprates with their large static dielectric constant.

Fig. 4.7 shows that the low temperature experimental data [138] on the London penetration depth $\lambda$ of $YBCO$ films can be fitted very well by this theory with $\nu < 1$. It is more usual to see $\lambda$ increase linearly with temperature [188,137] and this would correspond to $\nu > 1$ (or to the predominance of the effect of delocalised bipolarons moving in two dimensions).

We believe that $\nu < 1$ is more probable for a rapidly varying random potential while $\nu \geq 1$ is more likely for a slowly varying one. Both $\nu < 1$ and $\nu \geq 1$ are observed in doped semiconductors [140]. Hence, it is not surprising that drastically different low-temperature dependence of the London penetration depth is observed in different samples of doped cuprates. In the framework of our approach $\lambda(T)$ is related to the localisation of carriers at low temperatures rather than to any energy scale characteristic of the condensate. The excitation spectrum of the charged Bose-liquid determines, however, the temperature dependence of $\lambda(T)$ at higher temperatures including an unusual critical behaviour near $T_c$ [9].

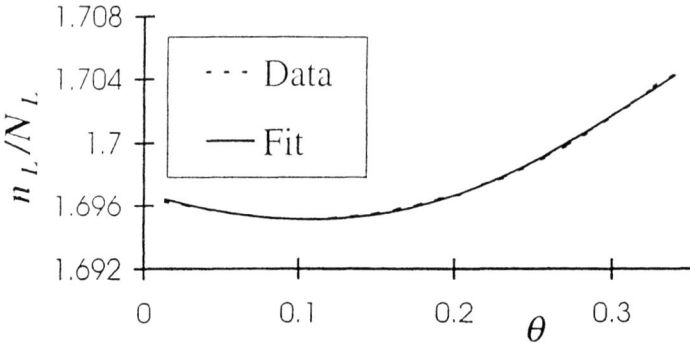

**FIGURE 4.7.** Fit to the London penetration depth obtained by Walter *et al* [138] for a YBCO film. The parameter values from the fit were $E_1 = 74K$, $\gamma = 20$ and $\nu = 0.67$. Reprinted from [139], Copyright 1999, with permission from Elsevier Science.

As a result we argue that the experimentally observed d-wave symmetry of the superfluid in some cuprates is not due to the internal symmetry of Cooper pairs (or bipolarons) but due to the bipolaron dispersion relation. Owing to the symmetry of the oxygen atomic orbitals two (degenerate) bipolaron bands have their minima at the Brillouin zone boundaries rather than at the $\Gamma$ point. Then the condensate wave function is modulated in space changing sign under $\pi/2$ rotations. The mobility edge is found near the bottom of these bands. The energy structure and the localisation of carriers in a random potential are very similar to those in usual semiconductors like $Ge$ or $Si$, where the band minima are not at the $\Gamma$ point. However, the statistics (Bose-Einstein) are drastically different and this leads to the cuprates being high-$T_c$ superconductors. The key parameter of the temperature dependence of the London penetration depth is the exponent $\nu$ of the localisation length. Experimentally, the more usual linear increase of $\lambda$ with temperature is observed in samples which either have no disorder or a shallow and smooth random potential profile. Here we expect $\nu \geq 1$ and (or) a substantial contribution from delocalised bipolarons moving principally in two dimensions, and thus that $\lambda$ increases linearly with temperature, as observed [137]. On the other hand heavy-ion bombardment [138] introduces rather deep and narrow potential wells for which we might expect $\nu < 1$. This would explain the upturn in the temperature dependence of $\lambda$ in the disordered films [138].

## F  Single vortex structure

Let us now formulate and solve the equations describing a single vortex in CBL at T= 0K [141].

The equation of motion for the field operator, $\psi$, is derived using the Hamiltonian, Eq. (154). If the density is relatively high, so the dimensionless Coulomb repulsion

$r_s = m^{**}e^{*2}/\epsilon_0(4\pi n_B/3)^{1/3}$ is not very large, one can expect that the occupation numbers of one-particle states are not very much different from those in the ideal Bose-gas. In particular, one state remains to be macroscopically occupied at T= 0K. Then, following Bogoliubov [4] one can separate the large matrix element $\psi_0$ from $\psi$ by treating the rest $\tilde{\psi}$ as small fluctuations,

$$\psi(\mathbf{r},t) = \psi_0(\mathbf{r},t) + \tilde{\psi}(\mathbf{r},t). \qquad (195)$$

The anomalous average $\psi_0(\mathbf{r},t) = \langle \psi(\mathbf{r},t) \rangle$ is equal to $(n_c)^{1/2}$ in a homogeneous system, where $n_c$ is the condensate density.

Substituting the Bogoliubov displacement transformation, Eq.(2) into the equation of motion and collecting $c - number$ terms of $\psi_0$, we obtain a set of the BdG-type equations [85]. The macroscopic condensate wave function (the order parameter) obeys the following equation

$$i\frac{\partial}{\partial t}\psi_0(\mathbf{r},t) = \left[ -\frac{(\nabla - ie^*\mathbf{A}/c)^2}{2m^{**}} - \mu \right] \psi_0(\mathbf{r},t) + \int d\mathbf{r}' V(\mathbf{r}-\mathbf{r}')[\psi_0^*(\mathbf{r}',t)\psi_0(\mathbf{r}',t) - n_B]\psi_0(\mathbf{r},t). \qquad (196)$$

There are also higher order terms in $r_s$ proportional to the density of bosons, $\tilde{n} \simeq 0.2 n_B r_s^{3/4}$, pushed up from the condensate by the Coulomb interaction at T= 0K [79]. They are negligible even for the intermediate value of $r_s \simeq 1$. With the same accuracy one can take $n_c = n_B$.

The integro-differential equation (196) is essentially different from the Ginsburg-Landau [142] and the Gross-Pitaevskii [143,144] equations, describing the order parameter in the BCS and neutral superfluids, respectively. As recognised by London [145] and discussed in detail by Vinen [146] superconducting metals and neutral superfluids ($^4$He) have many features in common with their characteristic properties resulting from the macroscopic quantum coherence and a short-range (local) interaction. While CBL shares the quantum coherence owing to the Bose-Einstein condensate (BEC), the long-range (nonlocal) interaction leads to a few peculiarities. It is worthwhile to emphasise that one *should* apply a bare (unscreened) Coulomb repulsion in Eq.(196) to avoid the double-counting of the interaction in the perturbation theory diagrams [85]. Actually, the coherence length appears to be just the same as the screening radius (see below).

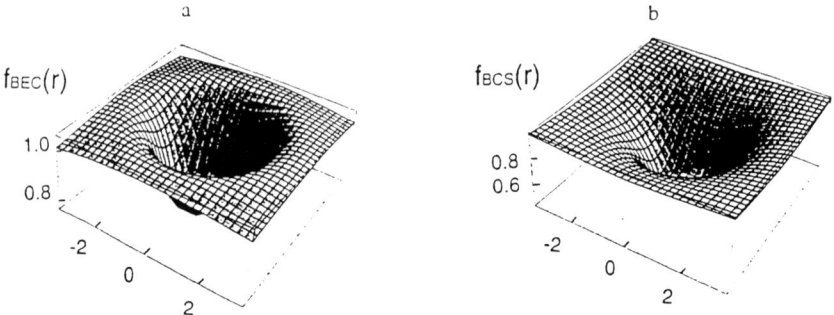

**FIGURE 4.8.** Vortex core profile in the charged Bose liquid, $f_{BEC}$, (a) compared with the vortex in the BCS superconductors, $f_{BCS}$, (b).

Under the stationary conditions, Eq. (196) supplemented by the current density, $\mathbf{j}(\mathbf{r}) = (e^*/m^{**})\Re[\psi_0^*(\mathbf{r})(-i\hbar\nabla - e^*\mathbf{A}/c)\psi_0(\mathbf{r})]$, and by the Maxwell equations, is reduced to a set of three nonlinear differential equations:

$$\frac{1}{\kappa^2 \rho}\frac{d}{d\rho}\rho\frac{df}{d\rho} - \frac{1}{f^3}\left(\frac{dh}{d\rho}\right)^2 - \phi f = 0, \tag{197}$$

$$\frac{1}{\kappa^2 \rho}\frac{d}{d\rho}\rho\frac{d\phi}{d\rho} = 1 - f^2, \tag{198}$$

and

$$-\frac{1}{\rho}\frac{d}{d\rho}\frac{\rho}{f^2}\frac{dh}{d\rho} = h. \tag{199}$$

As usual [147] these equations are written in a form which introduces the dimensionless quantities: $|\psi_0| = n_c^{1/2} f$, $\mathbf{r} = \lambda(0)\rho$, $e^*\xi(0)\lambda(0)\text{curl}\mathbf{A} = \mathbf{h}$ for the order parameter, length and magnetic field, respectively. A new feature is the electric field potential given by

$$\phi = \frac{1}{e^*\phi_c}\int d\mathbf{r}' V(\mathbf{r} - \mathbf{r}')[|\psi_0(\mathbf{r}')|^2 - n] \tag{200}$$

with a new fundamental unit $\phi_c = 1/2e^*m^{**}\xi(0)^2$. The coherence length is about the same as the RPA screening radius at T= 0K, $\xi(0) = (2^{1/2}m^{**}\omega_{p0})^{-1/2}$, where $\omega_{p0} = (4\pi n_B e^{*2}/\epsilon_0 m^*)^{1/2}$ is the zero-temperature plasma frequency [79]. The London penetration depth is $\lambda(0) = (m^{**}/4\pi n_B e^{*2})^{1/2}$. Cylindrical symmetry is assumed.

There are six boundary conditions in a single-vortex problem. Four of them are the same as in the BCS superconductor: $h = dh/\rho = 0$, $f = 1$ for $\rho = \infty$ and the flux quantization condition, $dh/d\rho = -pf^2/\kappa\rho$ for $\rho = 0$, where $p$ is an integer. The remaining two conditions are derived from the global charge neutrality: $\phi = 0$ for $\rho = \infty$ and

$$\phi(0) = \int_0^\infty \rho \ln(\rho)(1 - f^2)d\rho \tag{201}$$

for the electric field at the origin, and $\mu = 0$.

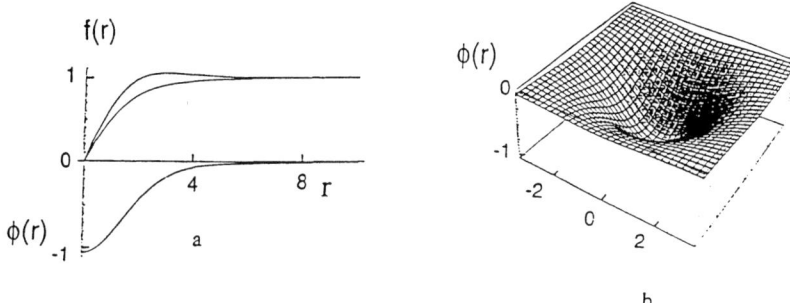

**FIGURE 4.9.** Electric field potential $\phi$ as a function of the distance (measured in units of $\xi(0)$) (lower curve) together with the CBL (upper curve) and BCS order parameters (a); its profile is shown in Fig. 4.9b.

The most realistic regime of CBL is an extreme type II with a very large Ginsburg-Landau parameter, $\kappa = \lambda(0)/\xi(0) \gg 1$ (see below). In this regime Eq.(6) is reduced to the London equation with the familiar solution $h = pK_0(\rho)/\kappa$, where $K_0(\rho)$ is the Hankel function of imaginary argument of zero order. For the region $\rho \leq p$, where the order parameter and the electric field differ from unity and zero, respectively, one can use the flux quantization condition to integrate out the magnetic field. That leaves us with the two parameter-free equations written for $r = \kappa\rho$ as

$$\frac{1}{r}\frac{d}{dr}r\frac{df}{dr} - \frac{p^2 f}{r^2} - \phi f = 0, \tag{202}$$

and

$$\frac{1}{r}\frac{d}{dr}r\frac{d\phi}{dr} = 1 - f^2. \tag{203}$$

They are satisfied by regular solutions of the form $f = c_p r^p$ and $\phi = \phi(0) + (r^2/4)$ when $r \to 0$. The constants $c_p$ and $\phi(0)$ are determined by complete numerical integration of Eq.(202) and Eq.(203). The numerical results for $p = 1$ are shown in Fig. 4.8 and Fig. 4.9 with $c_1 \simeq 1.5188$ and $\phi(0) \simeq -1.0515$.

In the region $p \ll r < p\kappa$ the solutions are $f = 1 + (4p^2/r^4)$ and $\phi = -p^2/r^2$. In this region the solution for $f$ differs qualitatively from the BCS order parameter [147], $f_{BCS} = 1 - (p^2/r^2)$ (see also Fig. 4.8), which is a result of the local charge redistribution caused by the magnetic field. Different from the BCS superconductor, where the total density of electrons remains constant across the sample, CBL allows for the flux penetration by redistributing the density of bosons within the coherence volume.

This leads to an increase of the order parameter compared with the homogeneous case ($f > 1$) in the region close to the vortex core. Inside the core the order parameter is suppressed, Fig. 4.8, as in the BCS superconductors. The resulting electric field, Fig. 4.9b, (together with the magnetic field) acts as an additional centrifugal force increasing the steepness ($c_p$) of the order parameter compared with the BCS superfluid where $c_1 \simeq 1.1664$, Fig. 4.9a.

The breakdown of the local charge neutrality is due to the absence of any equilibrium *normal* state solution in CBL below $H_{c2}(T)$ line. Both superconducting ($\Delta \neq 0$) and normal ($\Delta = 0$) solutions are allowed at any temperature in the BCS superconductors. Then the system decides which of two phases (or their mixture) is energetically favorable, so that the local charge neutrality is respected providing the minimum of free energy. In contrast, an equilibrium normal state solution of the complete set of the BdG equations (with $\psi_0 = 0$) does not exist in CBL because it does not respect the density sum rule below $H_{c2}(T)$ [83]. Hence, there are no different phases to mix, and the only way to aquire a flux in the thermal equilibrium is to redistribute the local density of bosons at the expence of their Coulomb energy. This energy determines the vortex free energy $\mathcal{F} = E_v - E_0$, which is the difference of the energy of CBL with, $E_v$, and without, $E_0$, the magnetic flux:

$$\mathcal{F} = \int d\mathbf{r} \left[ \frac{1}{2m^*} |(\hbar \nabla - ie^* \mathbf{A}/c)\psi_0|^2 + \frac{e^*}{2} \phi_c \phi (|\psi_0|^2 - n) + \frac{(curl \mathbf{A})^2}{8\pi} \right]. \quad (204)$$

By the use of Eq.(4) and Eq.(6) it can be written in the dimensionless form as

$$F = 2\pi \int_0^\infty [h^2 - \frac{1}{2}\phi(1 + f^2)]\rho d\rho. \quad (205)$$

In the large $\kappa$ limit the main contribution comes from the region $p/\kappa < \rho < p$, where $f \simeq 1$ and $\phi \simeq -p^2/\kappa^2 \rho^2$; the energy is thus the same as that in the BCS superconductor, $F \simeq 2\pi p^2 \ln(\kappa)/\kappa^2$. It is seen that the most stable solution is the formation of the vortex with one flux quantum, and the lower critical field is the same as in the BCS superconductor, $h_{c1} \simeq \ln(\kappa)/2\kappa$. However, differently from the BCS superconductors, where the Ginsburg-Landau phenomenology is microscopically derived by Gor'kov [148] in the temperature region close to $T_c$, the CBL vortex structure is derived here in the low temperature region. The zero temperature solution, Fig. 4.9, is actually applied in a wide temperature range well below the Bose-Einstein condensation temperature, where the depletion of the condensate remains small.

Let us also estimate the size and the electric field of the vortex core by the use of the material parameters typical for oxides such as $m^{**} = 10m_e$, $e^* = 2e$, $n_B = 10^{21} cm^{-3}$ and $\epsilon_0 = 10^3$. With these parameters I obtain $\xi(0) \simeq 0.48$nm, $\lambda(0) \simeq 265$nm, the Ginsburg-Landau ratio $\kappa \simeq 552$ and $\phi(0) \simeq -8.4$mV. Owing to a large dielectric constant the Coulomb repulsion remains weak even for heavy bosons, $r_s \simeq 0.46$. The actual size of the charged domain is about $4\xi$, Fig. 4.9a, i.e. about five to ten lattice spacings.

## G  SIN tunneling and Andreev reflection

There is convincing experimental evidence that the pairing of carriers takes place well above T$_c$ in underdoped cuprates (for a review see Ref. [9]), the clearest being the uniform susceptibility [149,150] and tunnelling [94]. The gap in the tunnelling and photoemission is almost temperature independent below T$_c$ [93,94] and exists above T$_c$ [94,96,89,90]. Kinetic [151] and thermodynamic [152] data suggest that the gap opens in both charge and spin channels at any relevant temperature in a wide range of doping. A plausible explanation is that the normal gap is half of the bipolaron binding energy [101,9], although alternative explanations have also been proposed. The temperature and doping dependence of the gap still remains a subject of controversy. Reflection experiments, in which an incoming electron from the normal side of a normal/superconducting contact is reflected as a hole along the same trajectory [153], revealed a much smaller gap edge than the bias at the tunnelling conductance maxima in a few underdoped cuprates [108]. Recent intrinsic tunnelling measurements on a series of Bi '2212' single crystals [97] showed distinctly different behaviour of the superconducting and normal state gaps with the magnetic field. Such coexistance of two distinct gaps in cuprates is not well understood [154].

In the framework of the bipolaron theory we can consider a simplified model, which describes the temperature dependence of the gap and tunnelling spectra in cuprates, and account for two different energy scales in the electron-hole reflection [107]. The assumption is that the attraction potential in cuprates is large compared with the (renormalised) Fermi energy of polarons. The model is a generic one-dimensional Hamiltonian including the kinetic energy of carriers in the effective mass $(m^*)$ approximation and a local attraction potential, $V(x - x') = -U\delta(x - x')$ as

$$H = \sum_s \int dx \psi_s^\dagger(x) \left( -\frac{1}{2m^*} \frac{d^2}{dx^2} - \mu \right) \psi_s(x) - U \int dx \psi_\uparrow^\dagger(x) \psi_\downarrow^\dagger(x) \psi_\downarrow(x) \psi_\uparrow(x), \quad (206)$$

where $s = \uparrow, \downarrow$ is the spin. The first band to be doped in cuprates is the oxygen band inside the Hubbard gap as established in polarised photoemission [155,105]. This band is almost one dimensional as discussed in section 3, so that our (quasi) one-dimensional approximation is a realistic starting point. Solving a two-particle problem with the $\delta$-function potential one obtains a bound state with the binding energy $2\Delta_p = \frac{1}{4}m^*U^2$, and with the radius of the bound state $r = 2/m^*U$. We assume that this radius is less than the inter-carrier distance in cuprates, which puts a constraint on the doping level, $E_F < 2\Delta_p$, where $E_F$ is the polaron Fermi energy, strongly reduced by the electron-phonon interaction. Then real-space bipolarons are formed. If three-dimensional corrections to the energy spectrum of pairs are taken into account, the ground state of the system is the Bose-Einstein condensate (BEC). The chemical potential is pinned below the band edge by about $\Delta_p$ both in the superconducting and normal state [9], so that the normal state single-particle gap is $\Delta_p$. The binding energy $2\Delta_p$ might change due to the same corrections. However, this change does not affect our further results as soon as they are expressed in terms of $\Delta_p$ rather than $U$.

Now we take into account that in the superconducting state $(T < T_c)$ the single-particle excitations interact with the condensate via the same potential $U$. Applying the Bogoliubov approximation [4] we reduce the Hamiltonian, Eq.(206) to a quadratic form as

$$H = \sum_s \int dx \psi_s^\dagger(x) \left( -\frac{1}{2m^*} \frac{d^2}{dx^2} - \mu \right) \psi_s(x) + \int dx [\Delta_c \psi_\uparrow^\dagger(x) \psi_\downarrow^\dagger(x) + H.c.], \quad (207)$$

where the coherent pairing potential, $\Delta_c = -U \langle \psi_\downarrow(x) \psi_\uparrow(x) \rangle$, is proportional to the square root of the condensate density, $\Delta_c = constant \times n_c(T)^{1/2}$. The single-particle excitation energy spectrum $E(k)$ is found using the Bogoliubov transformation as

$$E(k) = \left[ (k^2/2m^* + \Delta_p)^2 + \Delta_c^2 \right]^{1/2}, \quad (208)$$

if one assumes that the condensate density does not depend on position. This spectrum is quite different from the BCS quasiparticles because the chemical potential is negative with respect to the bottom of the single-particle band, $\mu = -\Delta_p$. The single particle gap, $\Delta$, defined as the minimum of $E(k)$, is given by

$$\Delta = \left[ \Delta_p^2 + \Delta_c^2 \right]^{1/2}. \quad (209)$$

It varies with temperature from $\Delta(0) = \left[ \Delta_p^2 + \Delta_c(0)^2 \right]^{1/2}$ at zero temperature down to the temperature independent $\Delta_p$ above $T_c$. The theoretical temperature dependence, Eq.(4) describes well the pioneering experimental observation of the anomalous gap in $YBa_2Cu_3O_{7-\delta}$ in the electron-energy-loss spectra by Demuth et al [156], Fig. 4.10, with $\Delta_c(T)^2 = \Delta_c(0)^2 \times [1 - (T/T_c)^n]$ below $T_c$ and zero above $T_c$, and $n = 4$. We notice that $n = 4$ is the exponent which is expected in the two-fluid model of any superfluid [157].

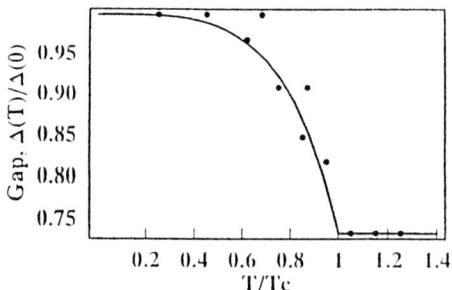

**FIGURE 4.10.** Temperature dependence of the gap, Eq.(209) (line) compared with the experiment [156](dots) for $\Delta_p = 0.7\Delta(0)$. Reprinted from [107], with permission from EDP Sciences.

The normal metal-superconductor (SIN) tunnelling conductance via a dielectric contact, $dI/dV$ is proportional to the density of states, $\rho(E)$ of the spectrum Eq.(208). Taking also into account the scattering of single-particle excitations by a random potential, thermal lattice and spin fluctuations one finds at $T = 0$ (see subsection 4.9)

$$dI/dV = constant \times [\rho\left(\frac{2eV - 2\Delta}{\Gamma_0}\right) + A\rho\left(\frac{-2eV - 2\Delta}{\Gamma_0}\right)], \qquad (210)$$

with

$$\rho(\xi) = \frac{4}{\pi^2} \times \frac{Ai(-2\xi)Ai'(-2\xi) + Bi(-2\xi)Bi'(-2\xi)}{[Ai(-2\xi)^2 + Bi(-2\xi)^2]^2}, \qquad (211)$$

$A$ is the asymmetry coefficient, explained in Ref. [103], $Ai(x), Bi(x)$ the Airy functions, and $\Gamma_0$ is the scattering rate.

We compare the conductance, Eq.(210) with one of the best STM spectra measured in $Ni$-substituted $Bi_2Sr_2CaCu_2O_{8+x}$ single crystals by Hancottee et al [93], Fig. 4.11a. This experiment showed anomalously large $2\Delta/T_c > 12$ with the temperature dependence of the gap similar to that in Fig. 4.10 below $T_c$. The theoretical conductance, Eq.(210) describes well the anomalous $gap/T_c$ ratio, injection/emission assymmetry, zero-bias conductance at zero temperature, and the spectral shape inside and outside the gap region. There is no doubt that the gap,

Fig. 4.11a is s-like.

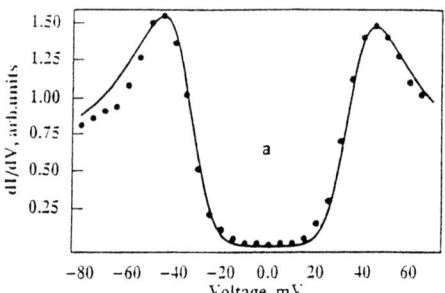
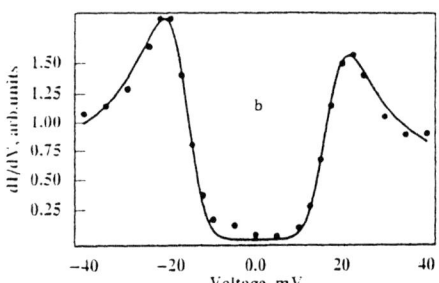

**FIGURE 4.11.** Theoretical tunnelling conductance, Eq.(210) (line) compared with the experimental STM conductance (dots) in (a) Ni-substituted $Bi_2Sr_2CaCu_2O_{8+x}$ [93] with $2\Delta = 90$ meV, $A = 1.05$, $\Gamma_0 = 40$ meV, and (b) in 'pure' Bi2212 [93] with $2\Delta = 43$ meV, $A = 1.2$, and $\Gamma_0 = 18$ meV. Reprinted from [107], with permission from EDP Sciences.

The conductance, Eq.(210) fits also well the conductance curve obtained on 'pure' Bi2212 single crystals, as shown in Fig. 4.11b, while a simple d-wave BCS denstity of states cannot describe the excess spectral weight in the peaks and the shape of the conductance outside the gap region. We notice that the scattering rate, $\Gamma$ is apparently smaller in the 'pure' sample than in the $Ni$- substituted sample, as expected.

A simple theory of the tunnelling into bosonic (bipolaronic) superconductor in the metallic (no-barrier) regime follows from this model. As in the canonical BCS approach applied to the normal metal-superconductor tunnelling by Blonder, Tinkham and Klapwijk [158] and to the normal-superconductor boundary in the intermediate type I state [153], the incoming electron produces only outgoing particles in the superconductor ($x > l$), allowing for a reflected electron and (Andreev) hole in the normal metal ($x < 0$). There is also a buffer layer of the thickness $l$ at the normal metal-superconductor boundary ($x = 0$), where the chemical potential with respect to the bottom of the conduction band changes gradually from a positive large value $\mu$ in the metal to a negative value $-\Delta_p$ in the bosonic superconductor. We approximate this buffer layer by a layer with a constant chemical potential $\mu_b$

($-\Delta_p < \mu_b < \mu$) and with the same strength of the pairing potential $\Delta_c$ as in the bulk superconductor. The Bogoliubov-de Gennes equations may be written as usual [158], with the only difference that the chemical potential with respect to the bottom of the band is a function of the coordinate $x$,

$$E\psi(x) = \begin{pmatrix} -(1/2m)d^2/dx^2 - \mu(x) & \Delta_c \\ \Delta_c & (1/2m)d^2/dx^2 + \mu(x) \end{pmatrix} \psi(x). \quad (212)$$

Thus the two-componet wave function in the normal metal is given by

$$\psi_n(x<0) = \begin{pmatrix} 1 \\ 0 \end{pmatrix} e^{iq^+x} + b \begin{pmatrix} 1 \\ 0 \end{pmatrix} e^{-iq^+x} + a \begin{pmatrix} 0 \\ 1 \end{pmatrix} e^{-iq^-x}, \quad (213)$$

while in the buffer layer it has the form

$$\psi_b(0<x<l) = \alpha \begin{pmatrix} 1 \\ \frac{\Delta_c}{E+\xi} \end{pmatrix} e^{ip^+x} + \beta \begin{pmatrix} 1 \\ \frac{\Delta_c}{E-\xi} \end{pmatrix} e^{-ip^-x} + \gamma \begin{pmatrix} 1 \\ \frac{\Delta_c}{E+\xi} \end{pmatrix} e^{-ip^+x} + \delta \begin{pmatrix} 1 \\ \frac{\Delta_c}{E-\xi} \end{pmatrix} e^{ip^-x}, \quad (214)$$

where the momenta associated with the energy $E$ are $q^{\pm} = [2m(\mu \pm E)]^{1/2}$ and $p^{\pm} = [2m(\mu_b \pm \xi)]^{1/2}$ with $\xi = (E^2 - \Delta_c^2)^{1/2}$. The well-behaved solution in the superconductor with negative chemical potential is given by

$$\psi_s(x>l) = c \begin{pmatrix} 1 \\ \frac{\Delta_c}{E+\xi} \end{pmatrix} e^{ik^+x} + d \begin{pmatrix} 1 \\ \frac{\Delta_c}{E-\xi} \end{pmatrix} e^{ik^-x}, \quad (215)$$

where the momenta associated with the energy $E$ are $k^{\pm} = [2m(-\Delta_p \pm \xi)]^{1/2}$.

The coefficients $a, b, c, d, \alpha, \beta, \gamma, \delta$ are determined from the boundary conditions, which are continuity of $\psi(x)$ and its derivatives at $x=0$ and $x=l$. Applying the boundary conditions, and carrying out an algebraic reduction, we find

$$a = 2\Delta_c q^+ (p^+ f^- g^+ - p^- f^+ g^-)/D, \quad (216)$$

$$b = -1 + 2q^+[(E+\xi)f^+(q^-f^- - p^-g^-) - (E-\xi)f^-(q^-f^+ - p^+g^+)]/D, \quad (217)$$

with

$$D = (E+\xi)(q^+f^+ + p^+g^+)(q^-f^- - p^-g^-) - (E-\xi)(q^+f^- + p^-g^-)(q^-f^+ - p^+g^+), \quad (218)$$

and $f^{\pm} = p^{\pm}\cos(p^{\pm}l) - ik^{\pm}\sin(p^{\pm}l)$, $g^{\pm} = k^{\pm}\cos(p^{\pm}l) - ip^{\pm}\sin(p^{\pm}l)$.

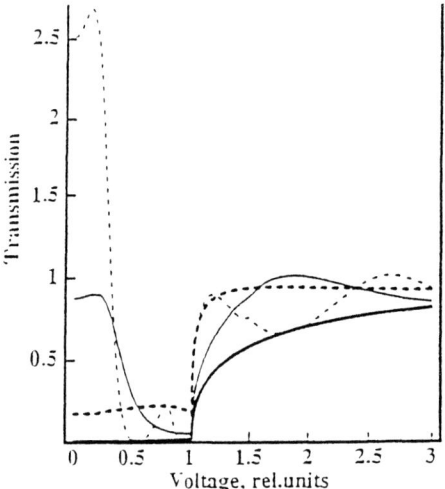

**FIGURE 4.12.** Transmission versus voltage (measured in units of $\Delta_p/e$) for $\Delta_c = 0.2\Delta_p$, $\mu = 10\Delta_p$, $\mu_b = 2\Delta_p$ and $l = 0$ (thick line), $l = 1$ (thick dashed line), $l = 4$ (thin line), and $l = 8$ (thin dashed line) (in units of $1/(2m\Delta_p)^{1/2}$). Reprinted from [107], with permission from EDP Sciences.

The transmisson coefficient for electrical current, $1 + |a|^2 - |b|^2$ is shown in Fig. 4.12 for different values of $l$ when the coherent gap $\Delta_c$ is smaller than the pair-breaking gap $\Delta_p$, and in Fig. 4.13 for the opposite case, $\Delta_p < \Delta_c$. In the first case, Fig. 4.12, we find two distinct energy scales, one is $\Delta_c$ in the subgap region due to electron-hole reflection and the other one is $\Delta$, which is the single-particle band edge. On the other hand, there is only one gap $\Delta_c$, which can be seen in the second case, Fig. 4.13. We notice that the transmission has no subgap structure if the buffer layer is absent ($l = 0$) in both cases. In the extreme case of a wide buffer layer, $l \gg (2m\Delta_p)^{-1/2}$, Fig. 4.12, or $l \gg (2m\Delta_c)^{-1/2}$, Fig. 4.13, there are some oscillations of the transmission due to the bound states inside the buffer layer. It was shown in Ref. [149] that the pair-breaking gap $\Delta_p$ is inverse proportional to the doping level. On the other hand, the coherent gap $\Delta_c$ scales with the condensate density, and therefore with the critical temperature, determined as the

Bose-Einstein condensation temperature of strongly anisotropic 3D bosons.

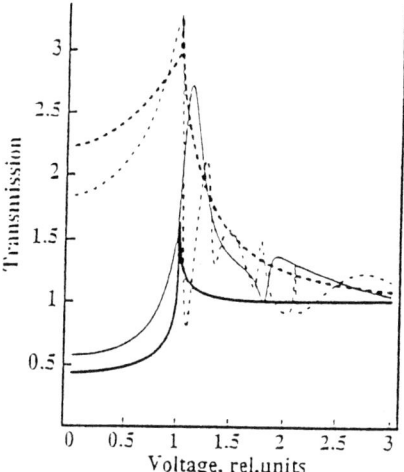

**FIGURE 4.13.** Transmission versus voltage (measured in units of $\Delta_c/e$) for $\Delta_p = 0.2\Delta_c$, $\mu = 10\Delta_c$, $\mu_b = 2\Delta_c$ and $l = 0$ (thick line), $l = 1$ (thick dashed line), $l = 4$ (thin line), and $l = 8$ (thin dashed line) (in units of $1/(2m\Delta_c)^{1/2}$). Reprinted from [107], with permission from EDP Sciences.

Therefore we expect that $\Delta_p \gg \Delta_c$ in the underdoped cuprates, Fig. 4.12, while $\Delta_p \leq \Delta_c$ in the optimally doped cuprates, Fig. 4.13. Thus the model accounts for the two different gaps experimentally observed in Giaver tunnelling and electron-hole reflection in the underdoped cuprates and for a single gap in the optimally doped samples [154]. An oscillating structure, observed in underdoped YBa$_2$Cu$_3$O$_{7-\delta}$ [108], is also found in the theoretical conductance at finite $l$, Fig. 4.12. The transmission, Fig. 4.12,13 is entirely due to the coherent tunnelling into the condensate and (or) into the single-particle band of the bosonic superconductor. There is an incoherent transmission into localised single-particle impurity states and into incoherent ('supracondensate') bound pair states as well, which might explain a significant featureless background in the subgap region [108].

## H  SIS tunnelling

Within the standard approximation [159] the tunnelling current, $I(V)$, between two parts of a superconductor separated by an insulating barrier is proportional to a convolution of the Fourie component of the single-hole Green's function (GF),$G(\mathbf{k},\omega)$, with itself as

$$I(V) \propto \Re \sum_{\mathbf{k},\mathbf{p}} \int_{-\infty}^{\infty} d\omega G(\mathbf{k},\omega) G(\mathbf{p}, e|V| - \omega), \tag{219}$$

where $V$ is the voltage in the junction.

A problem of a hole on a lattice coupled with the bosonic field of lattice vibrations has a solution in terms of the coherent (Glauber) states in the strong-coupling limit, $\lambda > 1$, where the Migdal-Eliashberg theory [13,14] cannot be applied owning to the broken translational symmetry [26]. For any type of electron-phonon interaction conserving the on-site occupation numbers of fermions the $1/\lambda$ perturbation technique yields (at $T = 0$) [71,160]

$$G(\mathbf{k},\omega) = Z \sum_{l=0}^{\infty} \sum_{\mathbf{q}_1,\ldots\mathbf{q}_l} \frac{\prod_{r=1}^{l} |\gamma(\mathbf{q}_r)|^2}{(2N)^l l! (\omega - \sum_{r=1}^{l} \omega(\mathbf{q}_r) - \epsilon(\mathbf{k} + \sum_{r=1}^{l} \mathbf{q}_r) + i\delta)}. \quad (220)$$

Here we generalize Eq.(119) taking into account the phonon frequency dispersion, $\omega_\mathbf{q} \equiv \omega(\mathbf{q})$. The hole energy spectrum, $\epsilon(\mathbf{k})$ is renormalised due to familiar polaronic narrowing of the band, and (in the superconducting state) also due to the interaction with the Bose-Einstein condensate (BEC) of bipolarons [107] as discussed above:

$$\epsilon(\mathbf{k}) = \left[\xi(\mathbf{k})^2 + \Delta_c^2\right]^{1/2}. \quad (221)$$

Here $\xi(\mathbf{k}) = Z'E(\mathbf{k}) - \mu$ is the renormalised polaron band dispersion with the chemical potential $\mu$, $E(\mathbf{k}) = \sum_\mathbf{m} t(\mathbf{m}) exp(-i\mathbf{k} \cdot \mathbf{m})$ is the bare (LDA) disperison in a rigid lattice. GF, Eq.(220) is exact in the extreme strong-coupling regime, $\lambda \to \infty$.

Quite differently from the BCS superconductor the chemical potential $\mu$ is negative in the bipolaronic system, so that the edge of the single-hole band is found above the chemical potential at $-\mu = \Delta_p$. Near the edge the parabolic one-dimensional approximation for the oxygen hole is applied [103], compatible with the ARPES data [105]

$$\epsilon_\mathbf{k} \simeq \frac{k_x^2}{2m^*} + \Delta. \quad (222)$$

Differently from the canonical Migdal-Eliashberg GF there is no damping ('defasing') of low-energy polaronic excitations in Eq.(220) due to the electron-phonon coupling alone (because of the energy conservation, section 3). This coupling leads to the coherent dressing of low-energy carriers by phonons, which is seen in GF as phonon sided bands with $l \geq 1$. On the other hand, the elastic scattering by impurities yields a finite life-time of the Bloch polaronic states. For the sake of analytical transparency we model this scattering as a constant imaginary self-energy, replacing $i\delta$ in Eq.(220) by a finite $i\Gamma/2$. In fact, the 'elastic' self-energy has been found explicitely as a function of energy and momentum (see below). Its particular energy/momentum dependence is essential in the subgap region of tunneling, where it determines the value of the zero-bias conductance. However, it hardly plays any role in the peak region and higher voltages, which are of our prime interest here.

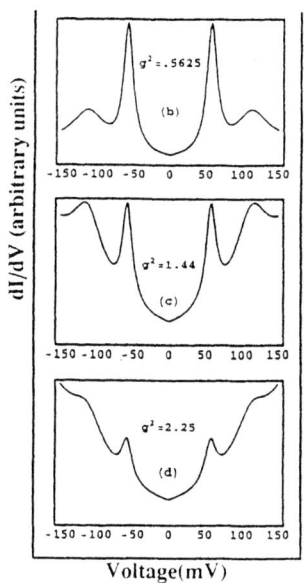

**FIGURE 4.14.** SIS tunnelling conductance in the bipolaronic superconductor for different values of the electron-phonon coupling, $g^2$, and $\Delta = 29$ meV, $\omega_0 = 55$ meV, $\delta\omega = 20$ meV, $\Gamma = 8.5$ meV.

Substiting Eq.(220) into the current, Eq.(219), and performing the intergration with respect to frequency and both momenta, we obtain for the tunnelling conductance, $\sigma(V) = dI/dV$,

$$\sigma(V) \propto \sum_{l,l'=0}^{\infty} \sum_{\mathbf{q},\mathbf{q'}} \frac{\prod_{r=1}^{l} \prod_{r'=1}^{l'} |\gamma(\mathbf{q}_r)|^2 |\gamma(\mathbf{q'}_{r'})|^2}{(2N)^{l+l'} l! l'!} L\left[ e|V| - 2\Delta - \sum_{r=1}^{l} \omega(\mathbf{q}_r) - \sum_{r'=1}^{l'} \omega(\mathbf{q'}_{r'}), \Gamma \right],$$
(223)

where $L[x,\Gamma] = \Gamma/(x^2 + \Gamma^2)$. To perform the remaining integrations and summations, we apply a model analog of the Eliashberg spectral function $\alpha^2 F(\omega)$ by replacing q-sums, $\sum_{\mathbf{q}} |\gamma(\mathbf{q})|^2 A(\omega(\mathbf{q}))/2N$, in Eq.(223) for the integrals $(g^2/\pi) \int d\omega L[\omega - \omega_0, \delta\omega] A(\omega)$ for any arbitrarry function of the phonon frequency $A(\omega(\mathbf{q}))$. In this way we introduce the characteristic frequency $\omega_0$ of phonons strongly coupled with holes, their avarage number $g^2$ in the polaronic cloud, and their dispersion $\delta\omega$.

As soon as $\delta\omega$ is less than $\omega_0$, we can extend the integration over phonon frequencies from $-\infty$ to $\infty$ and obtain

$$\sigma(V) \propto \sum_{l,l'=0}^{\infty} \frac{g^{2(l+l')}}{l!l'!} L\left[e|V| - 2\Delta - (l+l')\omega_0, \Gamma + \delta\omega(l+l')\right]. \tag{224}$$

By replacing the Lorentzian in Eq.(224) with the Fourie integral, we perform the summation over $l$ and $l'$ with the final result for the conductanse as

$$\sigma(V) \propto \int_0^\infty dt \exp\left[2g^2 e^{-\delta\omega t}\cos(\omega_0 t) - \Gamma t\right] \cos\left[2g^2 e^{-\delta\omega t}\sin(\omega_0 t) - (e|V| - 2\Delta)t\right]. \tag{225}$$

From the isotope effect on the carrier mass, phonon densities of states, experimental values of the normal state pseudogap, and the residual resistivity one estimates the coupling strength $g^2$ to be of the order of 1 [87], the characteristic phonon frequency between 20 and 80 meV, the phonon frequency dispersion about a few tens meV, the gap $\Delta$ about 30 meV, and the impurity scattering rate of the order of 10 meV.

SIS conductance, Eq.(225) calculated with the parameters in this range is shown in Fig 4.14,b-d for four different values of the coupling. The conductance shape is remarkably different from the BCS density of states, both s-wave and d-wave. There is no Ohm's law in the normal region, $e|V| > 2\Delta$, the dip/hump features (due to phonon sided bands) are clearly seen already in the first derivative of the current, there is a substantial incoherent spectral weigt beyond the quasiparticle peak for the strong coupling, $g^2 \geq 1$, and there is unusual shape of the quasiparticle peaks. All these features as well as the temperature dependence of the gap are beyond the BSC theory no matter what the symmetry of the gap is. However, they perfectly argee with the experimental SIS tunnelling spectra in cuprates [93-96,98]. In particular, the theory quantitatively describes one of the best tunnelling spectra obtained on $Bi_2Sr_2CaCu_2O_{8+\delta}$ single crystals by the break-junction technique [93], Fig. 4.15. Some excess zero-bias conductance compared with the experiment is due to our approximation of the elastic self-energy. The exact (energy dependent) self-energy provides an excellent agreement in this sub-gap region, as has been shown in Fig. 4.11. A more recent dynamic conductance of Bi-2212 mesas (as shown in Fig.2 of Ref. [98]) is almost identical to our Fig. 4.14b as well.

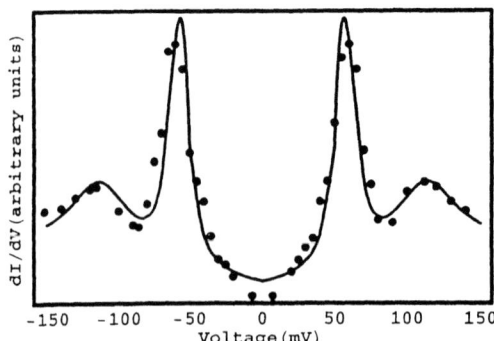

**FIGURE 4.15.** Theoretical conductance of Fig.4.14b (solid line) compared with the tunnelling spetrum obtained on $Bi_2Sr_2CaCu_2O_{8+\delta}$ single crystals by the break-junction technique [93](dots).

The unusual shape of the main peaks (Fig. 4.14b) is a simple consequence of the (quasi) one-dimensional hole density of states near the edge of the oxygen band. The coherent ($l = l' = 0$) contribution to the current with no elastic scattering ($\Gamma = 0$) is given by

$$I_0 \propto \int_\Delta^\infty \frac{d\epsilon}{(\epsilon - \Delta)^{1/2}} \int_\Delta^\infty \frac{d\epsilon'}{(\epsilon' - \Delta)^{1/2}} \delta(\epsilon + \epsilon' - e|V|), \qquad (226)$$

so that the conductance is a $\delta$ function

$$\sigma_0(V) \propto \delta(e|V| - 2\Delta). \qquad (227)$$

Hence, the width of the main peaks in the SIS tunnelling, Fig. 4.14 measures directly the elastic scattering rate.

The disapperence of the quasiparticle sharp peaks above $T_c$ in Bi-cuprates has been also explained in the framework of the bipolaron theory [103,106]. Below $T_c$ bipolaronic BEC provides an effective screening of the long-range (Coulomb) potential of impurities, while above $T_c$ the scattering rate might increase by many times [106]. This sudden increase of *Gamma* in the normal state washes out the sharp peaks from the tunnelling and ARPES spectra.

Finally we would like to comment on a possible role of spin fluctuations in the tunnelling spectra. If they play a role, the peak-dip separation as observed in ARPES and tunnelling should be equal to the resonance peak energy, $E_r$, observed in the spin-flip neutron scattering [161]. However, as discussed recently in the comprehensive review of the experimental constraints on the physics of cuprates [100], the peak-dip separation is nearly independent of $T_c$ or the dopinng level, while $E_r$ is approximately proportional to $T_c$. This controversy as well as the direct comparison of the electron-phonon and spin fluctuation interactions [110] suggest that the origin of the dip/hump features in ARPES and tunnelling is the strong electron-phonon coupling rather than a coupling with damped spin fluctuations.

# I  ARPES

Let us discuss the theory of ARPES in doped charge-transfer Mott insulators based on the bipolaron theory [105], which describes some unusual ARPES features of high-$T_c$ $YBa_2Cu_3O_{7-\delta}$ (Y123), $YBa_2Cu_4O_8$ (Y124) and a few other materials.

ARPES data in cuprates remains highly controversial [162]. One of the surprising features is a *large* Fermi surface claimed to exist in a wide range of doping fitting well the LDA band structures in the earlier studies. This should evolve with doping as $(1 - x)$ in a clear contradiction with low frequency kinetics and thermodynamics (see, for example [163–165,151,152]), which show an evolution proportional to $x$ in a wide range of doping including the overdoped region [151,166] ($x$ is the number of holes introduced by doping). Now it is established, however, that there is a normal state (pseudo)gap in ARPES and tunnelling, existing well above $T_c$ [90,94], so that some segments of a 'large Fermi surface' are actually missing [167]. The temperature and doping dependence of the gap still remain a subject of controversy. While kinetic [151], thermodynamic [152], tunnelling [94] and some ARPES [167] data suggest that the gap opens at any relevant temperature in a wide range of doping, other ARPES studies [89,168,169] claim that it exists only in underdoped samples below a characteristic temperature $T^*$.

Perhaps the most intriguing feature of ARPES is the extremely narrow and intense peak lying below the Fermi energy, which is most clearly seen near the Y and X points in Y124 [170], Y123 [155]. Its angular dependence and spectral shape as well as the origin of the featureless (but dispersive) background remain unclear. Some authors [170] refer to the peak as an extended van Hove singularity (evHs) arising from the plane ($CuO_2$) strongly correlated band. They also implicate the resulting (quasi)1D density of states singularity as a possible origin for the high transition temperature. However, recent polarised ARPES studies of untwinned Y123 crystals of exceptional quality [155] unambiguously refer the peak to as a narrow resonance arising primarily from the quasi-1D $CuO_3$ chains in the buffer layers rather than from the planes. Also, careful analysis of the Eliashberg equations shows that the van Hove singularity can hardly be the origin of high $T_c$, in sharp contrast with a naive weak-coupling estimate. Interestingly, a very similar narrow peak was observed by Park *et al* [171] in high-resolution ARPES near the gap edge of the cubic *semiconductor* FeSi with no Fermi surface at all.

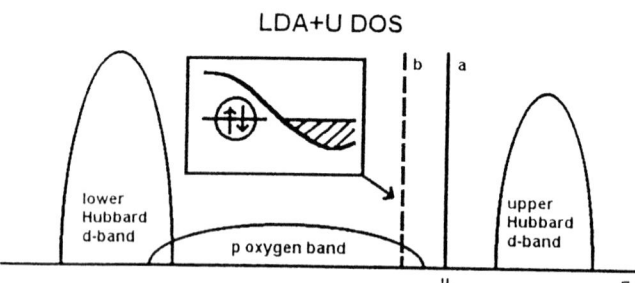

FIGURE 4.16. Schematic LDU+U density of states. The chemical potential is pinned inside the charge transfer gap (a) due to the bipolaron formation in underdoped cuprates. It might enter the oxygen band in overdoped cuprates (b) if the polaron band crosses the bipolaron one (inset).

I take the view that cuprates and many other transition metal compounds are charge transfer Mott-Hubbard insulators at *any* level of doping [172] as established in some 'LDA+U' band structure calculations for cuprates and manganites suggesting the single-particle density of states shown in Fig. 4.16. This means that the first band to be doped is the oxygen band lying within the Hubbard gap, Fig. 4.16.

The oxygen band, which is found inside the Hubbard gap, Fig. 4.16, is completely filled in a parent insulator. As a result, a single photoexited oxygen hole has well defined quasiparticle properties in the absence of the electron-phonon (or spin fluctuation) interaction. Strong coupling with high frequency phonons, unambiguously established for many oxides (subsection 4.1), leads to the high-energy spectral features of an oxygen hole in an energy window about twice the Franck-Condon (polaronic) level shift, $E_p$, and to the band-narrowing effect. On the other hand, the low energy spectral function is influenced by the low frequency thermal lattice, spin and random fluctuations. The latter can be described as 'Gaussian white noise'. The $p$-hole polaron in oxides is almost one-dimensional due to a large difference in the $pp\sigma$ and $pp\pi$ hopping integrals. This allows us to explain the narrow peak in ARPES spectra with the spectral density $A(k, E)$ of a one-dimensional particle in a Gaussian white noise potential [173].

The only role of hole bipolarons in ARPES is to pin the chemical potential inside the charge transfer gap, half the bipolaron binding energy above the oxygen band edge, Fig. 4.16a. In overdoped samples the bipolaron and polaron bands might overlap because the bipolaron binding energy becomes small, so the chemical potential might enter the oxygen band, Fig. 4.16b, and then a Fermi-level crossing might be seen in ARPES. The featureless background is explained as the phonon cloud of a small hole polaron, which spreads over a wide energy interval about $2E_p \simeq 1eV$.

*Polaronic ARPES*

The interaction of the crystal with the electromagnetic field of frequency $\nu$ is described by the Hamiltonian (in the dipole approximation)

$$H_{int} = (8\pi I)^{1/2} sin(\nu t) \sum_{\mathbf{k},\mathbf{k}'} (\mathbf{e} \cdot \mathbf{d}_{\mathbf{k},\mathbf{k}'}) c_{\mathbf{k}}^{\dagger} h_{\mathbf{k}'}^{\dagger} + H.c., \qquad (228)$$

where $I$ is the intensity of the radiation with the polarisation $\mathbf{e}$, $\mathbf{k}$ is the momentum of the final state (i.e. of the photoelectron registered by the detector), $\mathbf{k}'$ is the (quasi)momentum of the hole remaining in the sample after the emission, and $c_{\mathbf{k}}^{\dagger}$ and $h_{\mathbf{k}'}^{\dagger}$ are their creation operators, respectively. For simplicity we suppress the band index in $h_{\mathbf{k}'}^{\dagger}$. Due to the translational symmetry of the Bloch states, $|\mathbf{k}'> \equiv u_{-\mathbf{k}'}(\mathbf{r})exp(-i\mathbf{k}' \cdot \mathbf{r})$, there is a momentum conservation rule in the dipole matrix element,

$$\mathbf{d}_{\mathbf{k},\mathbf{k}'} = \mathbf{d}(\mathbf{k})\delta_{\mathbf{k}+\mathbf{k}',\mathbf{G}} \qquad (229)$$

with

$$\mathbf{d}(\mathbf{k}) = ie(N/v_0)^{1/2} \nabla_{\mathbf{k}} \int_{v_0} e^{-i\mathbf{G}\cdot\mathbf{r}} u_{\mathbf{k}-\mathbf{G}}(\mathbf{r}) d\mathbf{r}, \qquad (230)$$

$v_0$ is the unit cell volume.

The Fermi Golden Rule gives the photocurrent to be

$$I(\mathbf{k}, E) = 4\pi^2 * I|\mathbf{e} \cdot \mathbf{d}(\mathbf{k})|^2 \sum_{i,f} e^{\Omega + \mu N_i - E_i} |<f|h^\dagger_{\mathbf{k}-\mathbf{G}}|i>|^2 \delta(E + E_f - E_i), \qquad (231)$$

where $E$ is the binding energy, $E_{i,f}$ is the energy of the initial and final states of the crystal, and $\Omega, \mu, N_i$ are the thermodynamic and chemical potentials and the number of holes, respectively. By definition the sum in Eq.(4) is $n(E)A(\mathbf{k}-\mathbf{G}, -E)$ where the spectral function $A(\mathbf{k}-\mathbf{G}, E) = (-1/\pi)\Im G^R(\mathbf{k}-\mathbf{G}, E)$ is proportional to the imaginary part of the retarded Green function (GF), and $n(E) = [exp(E/T) + 1]^{-1}$, the Fermi distribution. In the following we consider temperatures well below the experimental energy resolution, so that $n(E) = 1$ if $E$ is negative and zero otherwise, and, for convenience, we put $\mathbf{G} = 0$.

The spectral function depends on essential interactions of a single hole with the rest of the system. The most important interaction in oxides is the Fröhlich long-range electron-phonon interaction of the oxygen hole with the c-axis polarised high-frequency phonons. This interaction results in self-trapped polarons. The Fröhlich interaction is integrated out with the Lang-Firsov canonical transformation, leading to the polaron GF, Eq.(220). As a result we obtain

$$I(\mathbf{k}, E) \sim |d(\mathbf{k})|^2 n(E) Z \delta(E + \xi_{\mathbf{k}}) + I_{incoh}(E), \qquad (232)$$

where $I_{incoh}(E)$ is a structureless function of the binding energy, which spreads from about $-\omega$ down to $-2E_p$. It might be some multiphonon structure in $I_{incoh}(E)$ as observed in the tunnelling, Fig. 4.15.

In the following we concentrate on the angular, spectral and polarisation dependence of the first coherent term, Eq.(232). The present experimental resolution [162] allows probing of the intrinsic damping of the coherent polaron tunnelling. This damping appears due to the random field and low-frequency lattice and spin fluctuations described by the polaron self-energy $\Sigma_p(\mathbf{k}, E)$, so that the coherent part of the spectral function is given by

$$A_p(\mathbf{k}, E) = -(1/\pi) \frac{\Im \Sigma_p(\mathbf{k}, E)}{[E + \Re\Sigma_p(\mathbf{k}, E) - \xi_{\mathbf{k}}]^2 + [\Im\Sigma_p(\mathbf{k}, E)]^2}. \qquad (233)$$

Hence, the theory of the narrow ARPES peak reduces to the determination of the self-energy of the coherent small hole polaron scattered by impurities, low-frequency deformation and spin fluctuations.

*Self-energy of 1D hole in the non-crossing approximation*

Due to energy conservation small polarons exist in the Bloch states at temperatures well below the optical phonon frequency $T << \omega/2$ no matter how strong their interaction with phonons is [66,24,9]. The finite polaron self-energy appears only due to the (quasi)elastic scattering. First we apply the simplest non-crossing (ladder) approximation to derive the analytical results. Within this approximation the self-energy is **k** -independent for a short-range scattering potential like the deformation or a screened impurity potential, so that

$$\Sigma_p(E) \sim \sum_{\mathbf{k}} G_p^R(\mathbf{k}, E), \qquad (234)$$

with $G_p^R(\mathbf{k}, E) = [E - \xi_\mathbf{k} - \Sigma_p(E)]^{-1}$

The oxygen polaron spectrum is parametrised in the tight-binding model as

$$\xi_\mathbf{k}^{x,y} = 2[t\cos(k_{x,y}a) - t'\cos(k_{y,x}a)] - \mu, \qquad (235)$$

If the oxygen hopping integrals in Eq.(235), reduced by the narrowing effect, are positive, the minima of the polaron bands are found at the Brillouin zone boundary in X $(\pi, 0)$ and Y $(0, \pi)$. The wave vectors corresponding to the energy minima belong to the stars with two prongs. Their group has only 1D representations. This means that the spectrum is degenerate with respect to the number of prongs of the star. The spectrum Eq.(235) belongs to the star with two prongs, and, hence it is a two-fold degenerate. The doublet is degenerate if the hole resides on the apical oxygen. In general, the degeneracy can be removed due to the chains in the buffer layers of Y123 and Y124, so that the y-polaronic band corresponding to the tunneling along the chains might be the lowest one.

As mentioned above the oxygen hole is (quasi) one-dimensional due to a large difference between the oxygen hopping integrals for the orbitals elongated parallel to and perpendicular to the oxygen-oxygen hopping $t' \ll t$. This allows us to apply a one-dimensional approximation, reducing Eq.(235) to two 1D parabolic bands near the X and Y points, $\xi_\mathbf{k}^{x,y} = k^2/2m^* - \mu$ with $m^* = 1/2ta^2$ and $k$ taking relative to $(\pi, 0)$ and $(0, \pi)$, respectively. Then, the equation for the self-energy in the non-crossing approximation, Eq.(234) takes the following form

$$\Sigma_p(\epsilon) = -2^{-3/2}[\Sigma_p(\epsilon) - \epsilon]^{-1/2}, \qquad (236)$$

for each doublet component. Here we introduce a dimensionless energy (and self-energy), $\epsilon \equiv (E + \mu)/\Gamma_0$ using $\Gamma_0 = (D^2 m^*)^{1/3}$ as the energy unit. The constant $D$ is the second moment of the Gaussian white noise potential, comprising thermal and random fluctuations as $D = 2(V_0^2 T/M + n_{im}v^2)$, where $V_0$ is the amplitude of the deformation potential, $M$ is the elastic modulus, $n_{im}$ is the impurity density, and $v$ is the coefficient of the $\delta$- function impurity potential. The solution is

$$\Sigma_p(\epsilon) = \frac{\epsilon}{3} - \left(\frac{1 + i3^{1/2}}{2}\right)\left[\frac{1}{16} + \frac{\epsilon^3}{27} + \left(\frac{1}{256} + \frac{\epsilon^3}{216}\right)^{1/2}\right]^{1/3}$$
$$- \left(\frac{1 - i3^{1/2}}{2}\right)\left[\frac{1}{16} + \frac{\epsilon^3}{27} - \left(\frac{1}{256} + \frac{\epsilon^3}{216}\right)^{1/2}\right]^{1/3}. \qquad (237)$$

While the energy resolution in the present ARPES studies is almost perfect [162], the momentum resolution remains finite in most experiments, $\delta > 0.1\pi/a$. Hence we have to integrate the spectral function, Eq.(233), with a Gaussian momentum resolution to obtain the experimental photocurrent,

$$I(\mathbf{k}, E) \sim Z \int_{-\infty}^{\infty} dk' A_p(k', -E) exp[-\frac{(k-k')^2}{\delta^2}]. \qquad (238)$$

The integral is expressed in terms of $\Sigma_p(\epsilon)$, Eq.(237) and the tabulated Error function $w(z)$ as

$$I(\mathbf{k}, E) \sim -\frac{2Z}{\delta}\Im\left(\Sigma_p(\epsilon)[w(z_1) + w(z_2)]\right), \qquad (239)$$

where $z_{1,2} = [\pm k - i/2\Sigma_p(\epsilon)]/\delta$, $w(z) = e^{-z^2} erfc(-iz)$ and $\epsilon = (-E + \mu)/\Gamma_0$. This photocurrent is plotted as dashed lines in Fig. 4.17 for two momenta, $k = 0.04\pi/a$ (almost Y or X points of the Brillouin zone) and $k = 0.3\pi/a$. The chemical potential is placed in the charge transfer gap below the bottom of the hole band, $\mu = -20 meV$, the momentum resolution is taken as $\delta = 0.28\pi/a$ and the damping $\Gamma_0 = 19 meV$.

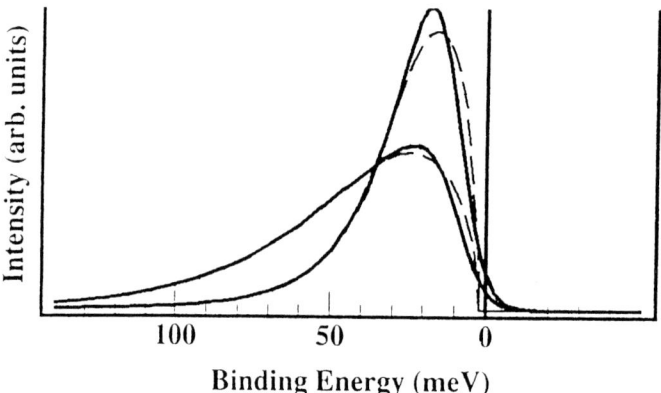

**Figure 4.17.** The polaron spectral function, integrated with the momentum resolution function for two angles, $k = 0.04\pi/a$ (upper curves), and $k = 0.30\pi/a$ with the damping $\Gamma_0 = 19 meV$, the momentum resolution $\delta = 0.28\pi/a$ and the polaron mass $m^* = 9.9 m_e$. The bipolaron binding energy $2|\mu| = 40 meV$. The dashed curves are the spectral density integrated with the momentum resoltion in the non-crossing approximation.

The imaginary part of the self-energy, Eq.(237) disappears below $\epsilon = -3/2^{5/3} \simeq -0.9449$. Hence this approximation gives a well-defined gap rather than a pseudogap. Actually, the non-crossing approximation fails to describe the localised states inside the gap (i.e. the Lifshitz tail of the density of states). One has to go beyond the simple ladder to describe the single-electron tunnelling inside the gap [103] and the ARPES spectra at very small binding energy.

*Exact spectral function of 1D hole*

The exact spectral function for a one dimensional particle in a random Gaussian white noise potential was derived by Halperin [173] and the density of states by Frisch and Lloyd [174]. Halperin derived two pairs of differential equations from whose solutions the spectral function of a 'Schrödinger' particle (i.e. in the effective mass approximation) and of a 'discrete' particle (tight-binding approximation) may be calculated. The QMC polaronic bandwidth is about 100 meV or larger [42], which allows us to apply the 'Schrödinger' particle spectral function, given by [173]

$$A_p(k,\epsilon) = 4 \int_{-\infty}^{\infty} p_0(-z)\Re p_1(z)dz, \qquad (240)$$

where $p_{0,1}(z)$ obey the two differential equations

$$\left[\frac{d^2}{dz^2} + \frac{d}{dz}(z^2 + 2\epsilon)\right] p_0 = 0, \qquad (241)$$

and

$$\left[\frac{d^2}{dz^2} + \frac{d}{dz}(z^2 + 2\epsilon) - z - ik\right] p_1 + p_0 = 0, \qquad (242)$$

with boundary conditions

$$\lim_{z \to \infty} z^{2-n} p_n(z) = \lim_{z \to -\infty} z^{2-n} p_n(z) \qquad (243)$$

where $k$ is measured in units of $k_0 = (D^{1/2}m^*)^{2/3}$. The first equation may be integrated to give

$$p_0(z) = \frac{exp(-z^3/3 - 2z\epsilon) \int_{-\infty}^{z} exp(u^3/3 + 2u\epsilon)du}{\pi^{1/2} \int_0^{\infty} u^{-1/2} exp(-u^3/12 - 2u\epsilon)du}. \qquad (244)$$

The equation for $p_1(z)$ has no known analytic solution, and hence must be solved numerically. There is however an asymptotic expression for $A_p(k,\epsilon)$ in the tail where $|\epsilon| \gg 1$:

$$A_p(k,\epsilon) \sim 2\pi(-2\epsilon)^{\frac{1}{2}} exp[-\frac{4}{3}(-2\epsilon)^{\frac{3}{2}}] cosh^2\left[\frac{\pi k}{(-8\epsilon)^{\frac{1}{2}}}\right]. \qquad (245)$$

In practice, for computational efficiency we use the exact spectral density for $-1.4 \leq \epsilon < 1$, and outside this range we use the asymptotic result for $\epsilon < -1.4$, Eq.(245) and the non-crossing approximation for $\epsilon \geq 1$, where they are almost indistinguishable from the exact result on the scale of the diagrams plotted here.

The result for $A_p(k, -E)$ integrated with the Gaussian momentum resolution is shown in Fig. 4.17 for two values of the momentum (solid lines). Quite differently from the non-crossing approximation the exact spectral function (averaged with the momentum resolution function) has the Lifshitz tail due to the states localised by disorder within the normal state gap. However, besides this tail the non-crossing approximation gives very good agreement, and for binding energy greater than about 30 meV it is practically exact. Owing to the screening of the static (or low-frequency) random potential (bi)polarons can reduce the characteristic length of the tail, $\Gamma_0$, which is familiar to the tendency towards delocalisation of carriers with doping [140] in semiconductors.

The cumulative DOS

$$N_p(\epsilon) = (2\pi)^{-1} \int_{-\infty}^{\epsilon} d\epsilon' \int_{-\infty}^{\infty} dk A_p(k, \epsilon') \tag{246}$$

is expressed analytically [174] in terms of the tabulated Airy functions $Ai(x)$ and $Bi(x)$ as

$$N_p(\epsilon)) = \pi^{-2} \left[ Ai^2(-2\epsilon) + Bi^2(-2\epsilon) \right]^{-1}. \tag{247}$$

The DOS $dN_p(\epsilon)/d\epsilon$ fits well the voltage-current tunnelling characteristics of cuprates, Fig. 4.11.

*Theory of ARPES in Y124 and Y123*

With the polaronic doublet, Eq.(235) placed above the chemical potential we can quantitatively describe high-resolution ARPES in Y123 [155] and Y124 [170]. First we explain the experimentally observed polarisation dependence of the intensity near Y and X [155]. The Bloch periodic function $u_\mathbf{k}(\mathbf{r})$ can be expressed in terms of the Wannier orbitals $w(\mathbf{r})$ as

$$u_\mathbf{k}(\mathbf{r}) = N^{-1/2} \sum_\mathbf{m} e^{i\mathbf{k}\cdot(\mathbf{m}-\mathbf{r})} w(\mathbf{r} - \mathbf{m}). \tag{248}$$

Then the dipole matrix element is given by the derivative of the Fourier component of the atomic (Wannier) orbital, $f_\mathbf{k} \equiv v_0^{-1/2} \int d\mathbf{r} w(\mathbf{r}) exp(i\mathbf{k} \cdot \mathbf{r})$ as

$$\mathbf{d}(\mathbf{k}) = i(\mathbf{e} \cdot \nabla_\mathbf{k}) f_\mathbf{k}. \tag{249}$$

To estimate $f_\mathbf{k}$ we approximate the $x, y$ oxygen orbitals contributing to the x and y polaronic bands, respectively, with $w_x(\mathbf{r}) = (1/8a_0^3\pi)^{1/2}(x/2a_0)exp(-r/2a_0)$ and $w_y(\mathbf{r}) = (1/8a_0^3\pi)^{1/2}(y/2a_0)exp(-r/2a_0)$. As a result we obtain for the x orbital,

$$\frac{\partial f_\mathbf{k}}{\partial k_x} = (8a_0^3\pi/v_0)^{1/2} a_0 \left( [(ka_0)^2 + 1/4]^{-3} - 6(k_x a_0)^2 [(ka_0)^2 + 1/4]^{-4} \right), \tag{250}$$

and

$$\frac{\partial f_\mathbf{k}}{\partial k_{y,z}} = -6(8a_0^3\pi/v_0)^{1/2} a_0^3 k_x k_{y,z} [(ka_0)^2 + 1/4]^{-4}, \tag{251}$$

Here **k** is the photoelectron momentum and $a_0$ is the size of the Wannier function. For the case of y-orbital one should interchange x and y. Near the X and Y points of the Brillouin zone, we have $|k_{y,x}| \ll k$ respectively. Then it follows from Eq.(250) and Eq.(251) that the ARPES peak should be seen at X and almost disappear at Y if the photons are polarised along the x-direction, i.e. **e** $\parallel$ **a**. If the polarisation is along the y-direction (**e** $\parallel$ **b**) the peak appears at Y and almost disappears at X. Precisely this behaviour is observed in ARPES spectra obtained using polarised photons [155]. We also notice a very strong dependence of the dipole matrix element, Eq.(249) on the photon energy, $d \sim \nu^{-3}$ at large $\nu$.

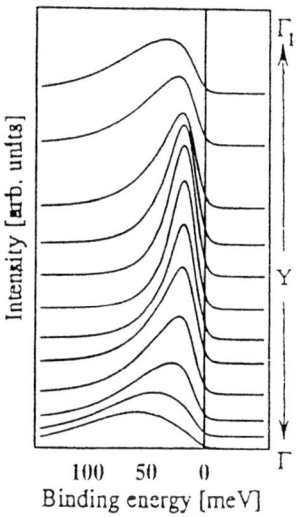

**FIGURE 4.18.** Theoretical ARPES spectra for $Y - \Gamma$ direction. Parameters are those of Fig.4.17. The theory provides a quantitative fit to experiment (Fig. 2a of [170]) in this scanning direction.

The exact 1D polaron spectral function, Eq.(240), integrated with the experimental momentum resolution (shown in Fig. 4.18), provides a quantitative fit to the ARPES spectra in Y124 along the $Y - \Gamma$ direction (Fig. 2(a) of [170].) The angular dispersion is described with the polaron mass $m^* = 9.9 m_e$. The spectral shape is reproduced well with $\Gamma_0 = 19 meV$, in close agreement with the value of this parameter found in tunnelling experiments [103]. That yields an estimate of the polaron scattering rate, which appears to be smaller than the polaron bandwidth (about $100 meV$ or larger), in agreement with the notion that many high-$T_c$ cuprates are in the clean limit. There is also quantitative agreement between theory (Fig. 4.19) and experiment (Fig. 3a of [170]) in the perpendicular direction $Y - S$, in a restricted region of small $k_x$, where almost no dispersion is observed around Y.

However, there is a significant loss of the energy-integrated intensity along both directions, Fig. 4.20, which the theoretical spectral function alone cannot account for. The energy-integrated ARPES spectra obey the sum rule,

$$\int_{-\infty}^{\infty} dE I(\mathbf{k}, E) \sim |d(\mathbf{k})|^2 n_\mathbf{k}, \tag{252}$$

where $n_\mathbf{k} = \langle h_\mathbf{k} h_\mathbf{k}^\dagger \rangle$. If the dipole matrix element is almost k-independent and the chemical potential is pinned well inside the charge-transfer gap, so that $n_\mathbf{k} = 1$ this integral would be k independent as well. This is not the case for Y124, no matter what the scanning direction is, Fig. 4.20. Therefore, we have to conclude that the dipole matrix element strongly depends on k.

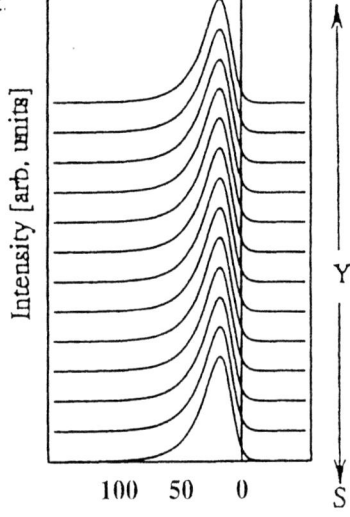

FIGURE 4.19. Theoretical ARPES spectra in Y124 for $Y - S$ direction. The theoretical fit agrees well with experiment (Fig. 3a of [170]) in a restricted range of $k_x$ near the Y-point, and outside this range the theoretical peaks are somewhat higher than in experiment.

The rapid loss of the integrated intensity in the $Y - S$ direction was interpreted by Randeria and Campuzano [168] as a Fermi-surface crossing. While a Fermi-surface crossing is not incompatible with our scenario (see Fig. 4.16 inset), we do not believe that it has really been observed in Y124. The peaks in the $Y - S$ direction are all 15 meV or more below the Fermi level - at a temperature of 1 meV, if the loss of spectral weight were due to a Fermi-surface crossing one would expect the peaks to approach much closer to the Fermi level. Also the experimental spectral shape of the intensity at $\mathbf{k} = \mathbf{k}_F$ is incompatible with any theoretical scenario, including different marginal Fermi-liquid models, Fig. 4.21.

The spectral function on the Fermi surface should be close to a simple Lorentzian,

$$A_p(\mathbf{k}_F, E) \sim \frac{|E|^\beta}{E^2 + constant \times E^{2\beta}}, \tag{253}$$

because the imaginary part of the self-energy behaves as $|E|^\beta$ with $0 \leq \beta \leq 2$. On the contrary, the experimental intensity shows a pronounced minimum at the alleged Fermi-surface, Fig. 4.21. Taking into account the finite temperature and (or) the experimental energy resolution does not help because both of them are well below the characteristic width of the experimental spectral function in Fig. 4.21.

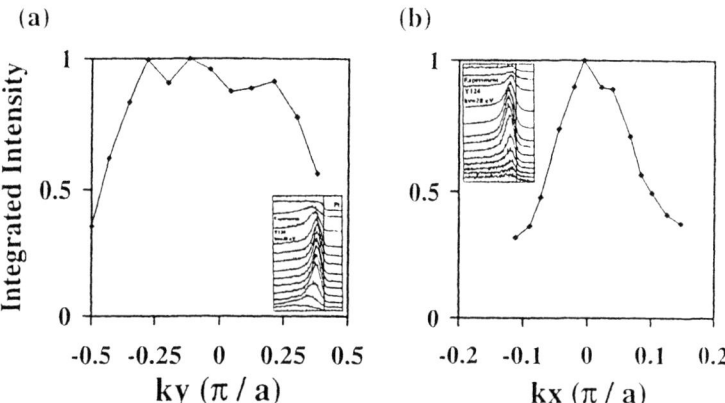

**FIGURE 4.20.** The energy-integrated ARPES intensity in Y124 in the $Y - \Gamma$ (a) and $Y - S$ (b) directions. Momenta are measured relative to the Y-point of the Brillouin zone.

If there is indeed no Fermi-surface crossing in many cuprates, as we argue, why then does some determinations point to a large Fermi surface in cuprates, which is drastically incompatible with their kinetic and thermodynamic properties? We propose that it appears due to the fact that oxygen semiconducting band has its minima at large $k$ inside or even on the boundary of the Brillouin zone. That is why ARPES show intense peaks near large $k$ imitating a large Fermi-surface.

We believe that many cuprates are doped insulators with no Fermi surface at all due to the bipolaron formation. The Fermi-surface crossing, if it were firmly established in the overdoped samples, would correspond to a small Fermi surface of the oxygen band pockets located at finite $k$ like in many ordinary semiconductors, for example, in Ge and Si.

## J  d-wave bipolaronic stripes

The electron-phonon interaction is strong in ionic cuprates and manganites as established both experimentally [87,99,175,176] and theoretically [9,30,110]. The carriers, doped into the Mott insulator, are coupled with the antiferromagnetic background as well. The antiferromagnetic interactions are thought to give rise to spin and charge segregation (stripes) [177,178]. There is growing experimental evidence [179,180] that stripes occur in slightly doped insulators. Their theoretical studies were restricted so far to the repulsive strongly correlated models [177,178], or to an extreme adiabatic limit of the electron-phonon interaction in narrow [181,182] and wide band [183,184] polar semiconductors and polymers. On the other hand there is strong evidence that the nonadiabatic electron-phonon interaction and small polarons are involed in the physics of stripes [175,180]. Also the role of the long-range Coulomb and Fröhlich interactions remains to be properly addressed.

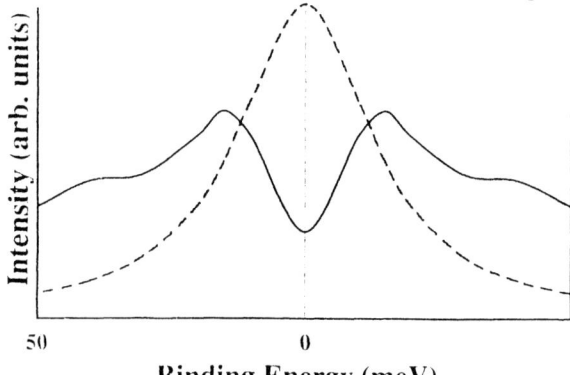

**Binding Energy (meV)**

FIGURE 4.21. The experimental ARPES signal (solid line) on the alleged Fermi-surface does not correspond to a Fermi-liquid spectral function (dashed line). We assume particle-hole symmetry to obtain the spectral function for negative binding energy.

In this final section we prove that the Fröhlich electron-phonon interaction combined with the direct Coulomb repulsion does not lead to charge segregation like strings in doped narrow-band insulators, both in the nonadiabatic and adiabatic regimes [185]. However, this interaction significantly reduces the Coulomb repulsion, which might allow much weaker antiferromagnetic and/or short-range electron-phonon interactions to bound carriers into small bipolarons. Then the d-wave BEC of bipolarons naturally explains the superstripes in cuprates.

As discussed above the extention of the deformation surrounding (Fröhlich) polarons is large, so their deformation fields overlap at finite density. However, taking into account both the long-range attraction of polarons due to the lattice deformations *and* the direct Coulomb repulsion, the net long-range interaction is repulsive [9]. At distances larger than the lattice constant, $|\mathbf{m} - \mathbf{n}| \geq a \equiv 1$, this interaction is significantly reduced to

$$v_{ij} = \frac{e^2}{\epsilon_0 |\mathbf{m} - \mathbf{n}|}. \tag{254}$$

Optical phonons nearly nullify the bare Coulomb repulsion in ionic solids if $\epsilon_0 \gg 1$, which is normally the case in oxides. The kinetic energy term in the exact transformed Hamiltonian, Eq.(86) involves multiphonon events generating a residual *polaron*-phonon interaction [9]. Below we show that in the two opposite limits, the nonadiabatic ($\omega_{\mathbf{q}} \geq t$) and in the extreme adiabatic ($\omega_{\mathbf{q}} \to 0$) regimes, there is no charge segregation or any other instability of the polaronic liquid, but only Wigner crystallization at very low densities.

First we consider the nonadiabtic and intermediate regime. Exact Quantum Monte-Carlo simulations (section 3) showed that the first order $1/\lambda$ perturbation theory is numerically accurate for *any* coupling if the phonon frequency is sufficiently large, $\omega_{\mathbf{q}} > t/2$. The characteristic frequency of phonons strongly coupled with carriers is about $\omega_{\mathbf{q}} = 75$ meV [99] in cuprates, so cuprates are in this regime. Hence, one can replace the hopping operator in Eq.(86) for its phonon average, reducing the problem to narrow-band fermions with the weak repulsive interaction, Eq.(254). Next order corrections in $1/\lambda$ increase the polaron binding energy with little effect on the bandwidth. Because the net long-range repulsion is relatively weak, the relevant dimensionless parameter $r_s = m^* e^2/\epsilon_0 (4\pi n/3)^{1/3}$ is not very large in doped cuprates. Wigner crystallization appears around $r_s \simeq 100$ or larger, which corresponds to the atomic density of polarons $n \leq 10^{-6}$ with $\epsilon_0 = 30$ and the polaronic mass $m^* = 5m_e$ typical for cuprates and manganites. This estimate shows that small polarons in cuprates and manganites are stable in the liquid state at any physically interesting density.

In the opposite adiabatic limit one can apply a discrete version of the continuos nonlinear equation [186] proposed in Ref. [28] for the Holstein (molecular) model of the electron-phonon interaction and extended to the case of the deformation and Fröhlich interactions in Ref. [181,182]. Applying the Hartree approximation for the Coulomb repulsion, the single-particle wave-function, $\psi_{\mathbf{n}}$ (the amplitude of the Wannier state $|\mathbf{n}\rangle$) obeys the following equation

$$-\sum_{\mathbf{m}\neq 0} t(\mathbf{m})[\psi_{\mathbf{n}} - \psi_{\mathbf{n+m}}] - e\phi_{\mathbf{n}}\psi_{\mathbf{n}} = E\psi_{\mathbf{n}}. \tag{255}$$

The potential $\phi_{\mathbf{n},k}$ acting on a fermion $k$ at the site $\mathbf{n}$ is created by the polarization of the lattice $\phi^l_{\mathbf{n},k}$ and by the Coulomb repulsion with the other $M-1$ fermions, $\phi^c_{\mathbf{n},k}$,

$$\phi_{\mathbf{n},k} = \phi^l_{\mathbf{n},k} + \phi^c_{\mathbf{n},k}. \tag{256}$$

Both potentials satisfy the descrete Poisson equation as

$$\kappa \Delta \phi^l_{\mathbf{n},k} = 4\pi e \sum_{p=1}^{M} |\psi_{\mathbf{n},p}|^2, \tag{257}$$

and

$$\epsilon_\infty \Delta \phi^c_{\mathbf{n},k} = -4\pi e \sum_{p=1, p\neq k}^{M} |\psi_{\mathbf{n},p}|^2, \tag{258}$$

with $\Delta\phi_n = \sum_m(\phi_n - \phi_{n+m})$. Differently from Ref. [182] we include the Coulomb interaction in Pekar's functional $J$ [186], describing the total energy, in a selfconsistent manner using the Hartree approximation, so that

$$\begin{aligned} J = & -\sum_{n,p,m\neq 0} \psi^*_{n,p} t(\mathbf{m})[\psi_{n,p} - \psi_{n+m,p}] \\ & -\frac{2\pi e^2}{\kappa} \sum_{n,p,m,q} |\psi_{n,p}|^2 \Delta^{-1} |\psi_{m,q}|^2 \\ & +\frac{2\pi e^2}{\epsilon_\infty} \sum_{n,p,m,q\neq p} |\psi_{n,p}|^2 \Delta^{-1} |\psi_{m,q}|^2. \end{aligned} \qquad (259)$$

If we assume, following Ref. [181] that the single-particle function of a fermion trapped in a string of the length $N$ is a simple exponent, $\psi_n = N^{-1/2}\exp(ikn)$ with the periodic boundary conditions, then the functional $J$ is expressed as $J = T + U$, where $T = -2t(N-1)\sin(\pi M/N)/[N\sin(\pi/N)]$ is the kinetic energy, proportional to $t$, and

$$U = -\frac{e^2}{\kappa} M^2 I_N + \frac{e^2}{\epsilon_\infty} M(M-1) I_N, \qquad (260)$$

corresponds to the polarisation and the Coulomb energies. Here the integral $I_N$ is given by

$$\begin{aligned} I_N = & \frac{\pi}{(2\pi)^3} \int_{-\pi}^{\pi} dx \int_{-\pi}^{\pi} dy \int_{-\pi}^{\pi} dz \frac{\sin(Nx/2)^2}{N^2 \sin(x/2)^2} \\ & \times (3 - \cos x - \cos y - \cos z)^{-1}. \end{aligned} \qquad (261)$$

It has the following asymptotics:

$$I_N = \frac{1.31 + \ln N}{N}. \qquad (262)$$

If we split the first (attractive) term in Eq.(260) into two parts by replacing $M^2$ for $M + M(M-1)$, then it becomes clear that the net interaction between polarons remains repulsive in the adiabatic regime as well because $\kappa > \epsilon_\infty$. Hence, there are no strings or stripes within the Hartree approximation for the Coulomb interaction. Strong correlations do not change this conclusion. Indeed, if we take the Coulomb energy of spinless one-dimensional fermions comprising both Hartree and exchange terms as [187]

$$E_C = \frac{e^2 M(M-1)}{N\epsilon_\infty}[0.916 + \ln M], \qquad (263)$$

the polarisation and Coulomb energy per particle becomes (for large $M \gg 1$)

$$U/M = \frac{e^2 M}{N\epsilon_\infty}[0.916 + \ln M - \alpha(1.31 + \ln N)], \qquad (264)$$

where $\alpha = 1 - \epsilon_\infty/\epsilon_0 < 1$. Minimising this energy with respect to the length of the string $N$ we find

$$N = M^{1/\alpha} \exp(-0.31 + 0.916/\alpha), \qquad (265)$$

and

$$(U/M)_{min} = -\frac{e^2}{\kappa} M^{1-1/\alpha} \exp(0.31 - 0.916/\alpha). \qquad (266)$$

One can see that the potential energy per particle increases with the number of particles. Hence, the energy of $M$ well separated polarons is lower than the energy of polarons trapped in a string no matter correlated or not. The opposite conclusion of Ref. [182] originates in an incorrect approximation of the integral $I_N \propto N^{0.15}/N$. The correct asymptotic result is $I_N = \ln(N)/N$.

The Fröhlich interaction is, of course, not the only electron-phonon interaction in ionic solids. As discussed in Ref. [9], any short range electron-phonon interaction, like, for example, the Jahn-Teller distortion can overcome the residual weak repulsion of Fröhlich polarons to form small bipolarons. At large distances small nonadibatic bipolarons weakly repel each other due to the long-range Coulomb interaction like polarons. Hence, they form a liquid state [9], or bipolaronic-polaronic crystal-like structures [56] depending on their effective mass and density. The fact, that the Fröhlich interaction almost nullifies the Coulomb repulsion in oxides justifies the use of the Holstein-Hubbard model [30,31]. The ground state of the 1D Holstein-Hubbard model is a liquid of intersite bipolarons with a significantly reduced mass as shown recently [188]. The bound states of three or more polarons are not stable in this model, thus ruling out phase separation. However, the situation might be different if the antiferromagnetic interaction with the parent Mott insulator or a short (but finite)-range electron-phonon interaction are strong enough.

Due to long-range nature of the Coulomb repulsion the length of a string should be finite (see, also Ref. [180,181]). One can readily estimate its length by the use of Eq.(88) for any type of the short-range electron-phonon interaction. If, for example, we take dispersive phonons, $\omega_\mathbf{q} = \omega + \delta\omega(\cos q_x + \cos q_y + \cos q_z)$ with a $q$-independent matrix element $\gamma(\mathbf{q}) = \gamma$, we obtain a short-range polaron-polaron attraction as

$$v_{att}(\mathbf{n} - \mathbf{m}) = -E_{att}(\delta\omega/\omega)\delta_{|\mathbf{n}-\mathbf{m}|,1}, \qquad (267)$$

where $E_{att} = \gamma^2 \omega/2$. Taking into account the long-range repulsion as well, Eq.(254), the potential energy of the string with $M = N$ polarons becomes

$$U = \frac{e^2}{\epsilon_0} N^2 I_N - \frac{N E_{att} \delta\omega}{\omega}. \qquad (268)$$

Minimization of this energy yields the length of the string as

$$N = \exp\left(\frac{\epsilon_0 E_{att}\delta\omega}{e^2\omega} - 2.31\right). \tag{269}$$

Actually, this expression provides a fair estimate of the string length for any kind of attraction (not only generated by phonon dispersion), but also for the antiferromagnetic exchange and/or Jahn-Teller type of interactions. There is no need to go into details of these interactions to apply Eq.(269). Due to the numerical coefficient in the exponent in Eq.(269) one can expect only short strings (if any) with the realistic values of $E_{att}$ (about a hundred millivolts), and static dielectric constant $\epsilon_0 \leq 100$.

We conclude that there are no strings in ionic narrow band doped insulators with the Fröhlich interaction alone. Depending on their density and mass polarons remains in the liquid state or Wigner crystal. On the other hand the short-range electron-phonon and/or antiferromagnetic interactions might provide a liquid bipolaronic state and/or charge segregation (strings of a finite length) because the long-range Fröhlich interaction significantly reduces the Coulomb repulsion in highly polarizable ionic insulators. However, with the typical values of $\epsilon_0 = 30$, $a = 3.8\text{Å}$ and $E_{att} = 2J = 0.3$ eV of the $t - J$ model one obtains $N < 2$ from Eq.(269). Hence the antiferromagnetic spin fluctuations are not strong enough to segregate charges into strings of any length, at least not in all high-$T_c$ cuprates. Lattice/spin polarons form a liquid state in most of them.

If (bi)polaronic carriers in many cuprates are in the liquid state, one can pose a key question how stripes can be seen at all. I suggest that the bipolaron liquid might be striped owing to the bipolaron energy band dispersion, Eq.(149,150), rather than to any particular interaction. If bipolaron bands have their minima at finite **k** (including the Brillouin zone boundaries), then the condensate wave function is modulated in real space Fig. 4.5. As a result, the hole density, which is about twice of the condensate density at low temperatures, is striped, with the characteristic period of stripes determined by the inverse band-minima wave vectors. Our interpretation of stripes is consistent with the recent inelastic neutron scattering in $YBa_2Cu_3O_{7-\delta}$, where the incommensurate peaks in neutron scattering have been only observed in the *superconducting* state [189]. The vanishing at $T_c$ of the incommensurate peaks can actually be anticipated over a wide part of the phase diagram and in other neutron data [190]. It is inconsistent with the usual stripe picture where a characteristic distance needs to be observed in the normal state as well. On the contrary, with the d-wave *striped* Bose-Einstein condensate the incommensurate neutron peaks disappear at $T_c$ or slightly above $T_c$ (due to superconducting fluctuations), as observed.

## V  CONCLUSIONS

We have developed the multi-polaron theory of strongly coupled electrons and phonons. Extending the BCS theory to the strong-coupling regime, we have shown

that low energy physics of the multi-polaron system in doped ionic insulators (strongly-correlated or not) is that of charged bosons. We have discussed a few recent applications of the bipolaron theory to cuprates, in particular the parameter-free expression for the superconducting critical temperature, upper crtitical field, symmetry of the order parameter, London penetration depth, vortex structure, tunnelling and Andreev reflection, normal and superconducting gaps, angle-resolved photoemission, and stripes. The bipolaron theory describes remarkably well these and many other experimental observations in cuprates, for example
- oxygen isotope effect on the carrier mass [87],
- uniform magnetic susceptibility [150] and NMR [101],
- different normal state gaps in charge and spin channels [191],
- 'ab' and 'c' axis normal state transport [192,149,193,194],
- large anisotropy and its doping dependence [165].

We believe that strong evidence is accumulating for a novel state of electronic matter in the layered cuprates; this is the charged Bose-liquid of bipolarons.

# REFERENCES

1. J.G. Bednorz and K.A. Müller, Z.Phys. B**64**, 189 (1986).
2. J. Bardeen, L.N. Cooper and J.R. Schrieffer, Phys.Rev. **108**, 1175 (1957).
3. F. London, Phys. Rev. **54**, 947 (1938).
4. N.N. Bogoliubov, Izv. Academy of Sciences (USSR) **11**, 77 (1947).
5. L.D. Landau, J. Phys. (USSR) **11**, 91 (1947)
6. R.A. Ogg Jr, Phys. Rev. **69**, 243 (1946).
7. M.R. Schafroth, Phys. Rev. **100**, 463 (1955).
8. H. Fröhlich, Phys. Rev. **79**, 845 (1950).
9. A.S. Alexandrov and N.F. Mott, Rep. Prog. Phys. **57** 1197; '*High Temperature Superconductors and Other Superfluids*' (Taylor and Francis, London) (1994); '*Polarons and Bipolarons*', (World Scientific, Singapore) (1995).
10. A.S. Alexandrov, in '*Models and Phenomenology for Conventional and High-temperature Superconductivity*' (Course CXXXVI of the Intenational School of Physics 'Enrico Fermi'), eds. G. Iadonisi, J.R. Schrieffer and M.L. Chiofalo, (IOS Press, Amsterdam), p. 309 (1998).
11. A.S. Alexandrov, Zh.Fiz.Khim. **57**, 273 (1983) (Russ.J.Phys.Chem.**57**, 167 (1983)).
12. J.G. Bednorz and K.A. Müller, Angew.Chem.Int.Ed.Engl. **27**, 735 (1988).
13. A.B. Migdal, Zh. Eksp. Teor. Fiz. **34**, 1438 (1958) (Sov. Phys. JETP **7**, 996 (1958)).
14. G.M. Eliashberg, Zh. Eksp. Teor. Fiz. **38**, 966 (1960); ibid **39**, 1437 (1960) (Sov. Phys. JETP **11**, 696 (1960; **12**, 1000 (1960)).
15. L.N. Cooper, Phys. Rev. **104**, 1189 (1957).
16. T. Matsubara, Prog. Theor. Phys. **14**, 351 (1955).
17. L.P. Gor'kov, Zh. Eksp. Teor. Fiz. **34**, 735 (1958) (Soviet Phys. JETP **7**, 505 (1958)).
18. Y. Nambu, Phys. Rev. **117**, 648 (1960).
19. V.V. Tolmachev, in '*A new Method in the Theory of Superconductivity*', ed. N.N. Bogoliubov et al, Consultant Bereau, New York (1959).
20. W.J. McMillan, Phys. Rev. B**167**, 331 (1968).
21. B.M. Klein, L.L. Boyer, D.A. Papaconstantopoulos, Phys. Rev. B **18**, 6411 (1978).
22. B.T. Geilikman, Usp. Fiz. Nauk **115**, 403 (1975) (Sov. Phys. Usp. **18**, 190 (1975)).
23. D.J. Scalapino, in '*Superconductivity*', ed. R.D. Parks, Marcel Deccer, p. 449 (1969).
24. I.G. Lang and Yu.A. Firsov, Zh.Eksp.Teor.Fiz. **43**, 1843 (1962) ( Sov.Phys.JETP **16**, 1301 (1963)).
25. A.S. Alexandrov, Phys. Rev. B**46**, 2838 (1992).
26. A.S. Alexandrov and E.A. Mazur, Zh. Eksp. Teor. Fiz. **96**, 1773 (1989).
27. L. D. Landau, J. Phys. (USSR) **3**, 664 (1933).
28. V.V. Kabanov and O.Yu. Mashtakov, Phys. Rev. B**47**, 6060 (1993).
29. A.S. Alexandrov, V.V. Kabanov, and D.K. Ray, Phys. Rev. B **49**, 9915 (1994).
30. A.R. Bishop and M. Salkola , *in: 'Polarons and Bipolarons in High-$T_c$ Superconductors and Related Materials'*, eds E.K.H. Salje, A.S. Alexandrov and W.Y. Liang, Cambridge University Press, Cambridge, 353 (1995).
31. H. Fehske et al, Phys. Rev. B**51**, 16582 (1995).
32. F. Marsiglio, Physica C **244**, 21 (1995).

33. Y. Takada and T. Higuchi, Phys. Rev. B **52**, 12720 (1995).
34. J. T. Devreese, in *Encyclopedia of Applied Physics*, vol. 14, p. 383, VCH Publishers (1996).
35. H. Fehske, J. Loos, and G. Wellein, Z. Phys. B**104**, 619 (1997).
36. T. Hotta and Y. Takada, Phys. Rev. B **56**, 13 916 (1997).
37. A.H. Romero, D.W. Brown and K. Lindenberg, J. Chem. Phys. **109**, 6504 (1998).
38. A. La Magna and R. Pucci, Phys. Rev. B **53**, 8449 (1996).
39. P. Benedetti and R. Zeyher, Phys. Rev. B**58**, 14320 (1998).
40. T. Frank and M. Wagner, Phys. Rev. B **60**, 3252 (1999).
41. J. Bonca, S.A. Trugman, and I. Batistic, Phys. Rev. B **60**, 1633 (1999).
42. A.S. Alexandrov and P.E. Kornilovich, Phys. Rev. Lett. **82**, 807 (1999).
43. L. Proville and S. Aubry, S. Eur. Phys. J. B **11**, 41 (1999).
44. A. S. Alexandrov, Phys. Rev. B **61**, 12315 (2000).
45. A.A. Abrikosov, Phys. Rev. B **56**, 446 (1997).
46. O. V. Danylenko and O.V. Dolgov, cond-mat/0007189, to appear in Phys. Rev. B (2001).
47. A.S. Aleksandrov (Alexandrov), V.N. Grebenev, and E.A. Mazur, Pis'ma Zh. Eksp. Teor. Fiz. **45**, 357 (1987) (JETP Lett. **45**, 455 (1987)).
48. P.B. Allen and B. Mitrovic, in *'Solid State Physics'*, ed. H. Ehrenreich, F. Seitz, and D. Turnbull (Acadenic, New York (1982)), v.37, p.1.
49. C. Grimaldi *et al*, Phys. Rev. Lett. **75**, 1158 (1995).
50. T. Holstein, Ann.Phys. **8**, 325; ibid 343 (1959).
51. In the momentum representation the electron-phonon interaction is $H_{e-ph} = (2N)^{-1/2} \sum_{\mathbf{q},\mathbf{k}} \gamma \omega c_{\mathbf{k}}^{\dagger} c_{\mathbf{k}-\mathbf{q}} (d_{\mathbf{q}}^{\dagger} + d_{\mathbf{q}})$ with $\gamma^2 = 2\lambda z t/\omega$, $c_{\mathbf{k}}, d_{\mathbf{q}}$ the electron and phonon operators, respectively, and $z$ the lattice coordination number.
52. E.I. Rashba, Opt. Spectr. **2**, 75 (1957), see also in *Excitons*, eds. E.I. Rashba and D.M. Struge (Nauka, Moscow (1985)).
53. P.W. Anderson, Phys. Rev. Lett. **34**, 953 (1975).
54. A.S. Alexandrov and V.V. Kabanov, Phys. Rev. B **54**, 3655 (1996).
55. A.S. Alexandrov and J. Ranninger, Phys. Rev. B **23**, 1796 (1981).
56. S. Aubry, in: *'Polarons and Bipolarons in High-$T_c$ Superconductors and Related Materials'*, eds E.K.H. Salje, A.S. Alexandrov and W.Y. Liang, Cambridge University Press, Cambridge, 271 (1995).
57. A.S. Alexandrov, Phys. Rev. B **46**, 14932(1992).
58. G. M. Zhao, M. B. Hunt, H. Keller, & K. A. Müller, Nature (London) **385**, 236-238 (1997).
59. A.S. Alexandrov, Guo-meng Zhao, H. Keller, B. Lorenz, Y. S. Wang, and C. W. Chu, cond-mat/0011436.
60. T. Hotta and Y. Takada, Technical Rep. ISSP, Ser. A, No 3093, January (1996).
61. H. Hiramoto and Y. Toyozawa, J. Phys. Soc. Jpn. **54**, 245 (1985).
62. S.V. Tjablikov, Zh.Eksp.Teor.Fiz. **23**, 381 (1952).
63. J. Yamashita and T. Kurosawa, J.Phys.Chem.Solids **5**, 34 (1958).
64. G.L. Sewell, Phil.Mag. **3**, 1361 (1958).
65. E.K. Kudinov and Yu.A. Firsov, Fiz. Tverd. Tela (Leningrad), **7**, 546 (1965) (Soviet Physics- Solid State **7**, 435 (1965).

66. J.Appel, in *Solid State Physics*, eds. F. Seitz, D. Turnbull and H. Ehrenreich, Academic Press **21** (1968).
67. Yu.A. Firsov (ed), Polarons, Nauka (Moscow) (1975).
68. H. Böttger and V.V. Bryksin, *'Hopping Conduction in Solids'*, Academie-Verlag Berlin (1985).
69. G.D. Mahan, *'Many Particle Physics'*, Plenum Press, (1990).
70. A.S. Aleksandrov (Alexandrov), Pis'ma Zh. Eksp.Teor.Fiz. (Prilozh.) **46**, 128 (1987) (JETP Lett Suppl. **46**, 107 (1987)).
71. A.S. Alexandrov, in *'Models and Phenomenology for Conventional and High-temperature Superconductivity'* (Course CXXXVI of the Intenational School of Physics 'Enrico Fermi'), eds. G. Iadonisi, J.R. Schrieffer and M.L. Chiofalo, (IOS Press, Amsterdam), p. 309 (1998).
72. A.S. Alexandrov and A.M. Bratkovsky, Phys. Rev. Lett. **82**, 141 (1999).
73. A.S. Alexandrov, Phys. Rev. B**53**, 2863 (1996).
74. D.M. Eagles, Phys. Rev. **181**, 1278 (1969).
75. A.A. Gogolin, Phys.Status Solidi B**109**, 95 (1982).
76. G.J. Kaye, Phys. Rev. B **57**, 8759 (1998).
77. L.D. Landau and E.M. Lifshitz, *Quantum Mechanics*, Third Edition, Pergamon press, (1991).
78. C.R.A. Catlow, M.S. Islam and X. Zhang, J. Phys.: Condens. Matter **10**, L49 (1998).
79. L.L. Foldy, Phys.Rev. **124**, 649(1961).
80. A.L. Fetter, Ann. Phys. (N.Y.) **64**,1 (1971).
81. R.F. Bishop, J.Low Temp.Phys. **15**, 601(1974).
82. S.R. Hore and N.E. Frankel, Phys. Rev. B **12**, 2619(1975); ibid **14**, 1952(1976).
83. A.S. Alexandrov, Doctoral thesis MIFI (Moscow) (1984); Phys. Rev. B**48**, 10571 (1993).
84. A.S. Alexandrov, W.H. Beere, and V.V. Kabanov, Phys. Rev. B**54**, 15363 (1996).
85. A.S. Alexandrov and W.H. Beere, Phys. Rev. B**51**, 5887 (1995).
86. A.S. Alexandrov, W.H. Beere, and V.V. Kabanov, Phys. Rev. B**54**, 15363 (1996).
87. G. Zhao, M.B. Hunt,H. Keller, and K.A. Müller, Nature **385**, 236 (1997).
88. A.S. Alexandrov and A.M. Bratkovsky, J. Phys.: Condens. Matt. **11**, L531 (1999).
89. H. Ding *et al*, Nature (London) **382**, 51 (1996).
90. Z.-X. Shen and J.R. Schrieffer, Phys. Rev. Lett. **78**, 1771 (1997) and references therein.
91. N.L. Saini *et al*, Phys. Rev. Lett. **79**, 3467 (1997).
92. Yu.G. Ponomarev *et al*, Solid St. Commun. **111**, 315 (1999).
93. H. Hancotte *et al*, Phys. Rev. B**55**, R3410 (1997).
94. Ch. Renner *et al*, Phys. Rev. Lett. **80**, 149 (1998).
95. Y. DeWilde *et al*, Phys. Rev. Lett. **80**, 153 (1998).
96. A. Mourachkine, Europhys. Lett. **49**, 86 (2000).
97. P. Müller *et al*, unpublished.
98. V.M. Krasnov *et al*, Phys. Rev. Lett. **84**, 5860 (2000).
99. T. Timusk *et al*, in *Anharmonic Properties of High-$T_c$ Cuprates*, eds. D. Mihailović *et al*, (World Scientific, Singapore, 1995), p.171.

100. Guo-meng Zhao, Proceedings of the International Summer School on Strongly Correlated Systems (Debrecen, Hungary, September 2000)), to appear in Phil. Mag (2001).
101. A.S. Alexandrov, Physica C **182**, 327 (1991).
102. A.S. Alexandrov and D.K. Ray, Phil. Mag. Lett. **63**, 295, (1991).
103. A.S. Alexandrov, Physica C (Amsterdam) **305**, 46 (1998).
104. A.S. Alexandrov and C. Sricheewin, cond-mat/0102284.
105. A.S. Alexandrov and C.J. Dent, Phys. Rev. B **60**, 15414 (1999).
106. A.S. Alexandrov and C.J. Dent, cond-mat/0012234.
107. A.S. Alexandrov and A.F. Andreev, Europhys. Lett. **54**, 373 (2001).
108. Y. Yagil et al, Physica C (Amsterdam) **250**, 59 (1995).
109. A.S. Alexandrov and P.P. Edwards, Physica C **331**, 97 (2000).
110. A.S. Alexandrov and A.M. Bratkovsky, J. Phys.: Condens. Matter **11**, L531 (1999).
111. A.S. Alexandrov, Phys. Rev. Lett. **82**, 2620 (1999).
112. A.S. Alexandrov and V.V. Kabanov, Phys. Rev. B **59** (1999).
113. O.V. Dolgov, D.A. Kirzhnits, and E.G. Maximov, Rev. Mod. Phys. **53**, 81 (1981).
114. P.B. Allen and R.C. Dynes, Phys. Rev. B**12**, 905 (1975).
115. B.J. Alder and D.S. Peters, Europhys. Lett. **10**, 1 (1989).
116. V.L. Pokrovsky, Pis'ma Zh. Teor. Fiz. **47**, 539 (1988).
117. V.J. Emery and S.A. Kivelson, Nature, **374**, 434 (1995).
118. A. Junod, in *Studies of High Temperature Superconductors*, Vol. 19, ed. A. Narlikar (Nova Science, Commack, NY) (1996) p.1.
119. A.S. Alexandrov et al, Phys. Rev. Lett. **79**, 1551 (1997).
120. V.N. Zavaritsky, M. Springford and A.S. Alexandrov, EuroPhys. Lett. **51**, 334 (2000).
121. Y. Ando et al, Phys. Rev. Lett. **77**, 2065 (1996).
122. A.P. Mackenzie et al, Phys. Rev. Lett. **71**, 1238 (1993).
123. M.S. Osofsky et al, Phys. Rev. Lett. **71**, 2315 (1993); **72** 3292 (1994).
124. A.S. Alexandrov et al, Phys. Rev. Lett. **76**, 983 (1996).
125. D.D. Lawrie et al, J. Low Temp. Phys. **107**, 491 (1997).
126. G. S. Boebinger, private communication (1998).
127. V.F. Gantmakher et al, Zh. Eksp. Teor. Fiz. **115**, 268 (1999) ( JETP **88**, 148 (1999)).
128. Y.F. Yan et al, Phys. Rev. B**52**, R571 (1995).
129. N.E. Hussey et al, Phys. Rev. B**58**, R611 (1998).
130. V.B. Geshkenbein, L.B. Ioffe, and A.J. Millis, Phys. Rev. Lett. **80**, 5778 (1998).
131. H.H. Wen et al, Phys. Rev. Lett. **82**, 410 (1999).
132. A. Junod et al, Physica C **294**, 115 (1998).
133. C.J. Dent, A.S. Alexandrov and V.V. Kabanov, Physica C **341-348**, 153 (2000).
134. J. Annett, N. Goldenfeld and A.J. Legget, in : D.M. Ginsberg (ed), Physical Properties of High Temperature Superconductors, vol. 5, World Scientific, Singapore, 375 (1996).
135. D.A. Wollman et al, Phys. Rev. Lett. **71**, 2134 (1993); C.C. Tsuei et al, Phys. Rev. Lett. **73**, 593 (1994); J.R. Kirtley et al, Nature **373**, 225 (1995); C.C. Tsuei et al, Science **272**, 329 (1996).

136. D.A. Bonn et al, Phys. Rev. B **50**, 4051 (1994).
137. for a review see T. Xiang, C. Panagopoulos and J. R. Cooper, Int. J. Mod. Phys. B **12**, 1007 (1998).
138. H. Walter et al, Phys. Rev. Lett. **80**, 3598 (1998).
139. A.S. Alexandrov and R.T. Giles, Physica C **325**, 35 (1999).
140. N.F. Mott and E. Davis, in 'Electronic processes in non-crystalline materials', 2nd edn. Oxford University Press, Oxford (1979).
141. A.S. Alexandrov, Phys. Rev. B **60**, 14573 (1999).
142. V.L. Ginsburg and L.D. Landau, Zh. Eksp. Teor. Fiz. **20**, 1064 (1950).
143. E.P. Gross, Nuovo Cimento **20**, 454 (1961).
144. L.P. Pitaevskii, Zh. Eksp. Teor. Fiz. **40**, 646 (1961) ( Soviet Phys. JETP **13**, 451 (1961)).
145. F. London, 'Superfluids', vol. I,II, (Wiley, New York) (1950).
146. W.F. Vinen, in *'Superconductivity'* vol. II, ed R.D. Parks (Marcel Deccer, New York), p.1167 (1969).
147. D. Saint-James, G. Sarma, and E.J. Thomas, 'Type II Superconductivity' (Pergamon Press, Oxford) (1968).
148. L.P. Gor'kov, Zh. Eksp. Teor. Fiz. **36**, 1918 (1959) ( Soviet Phys. JETP **9**, 1364 (1959)).
149. A.S. Alexandrov, V.V. Kabanov and N.F. Mott, Phys. Rev. Lett. **77**, 4796 (1996).
150. K.A. Müller et al, J.Phys.: Condens. Matter **10**, L291 (1998).
151. B. Batlogg et al, Physica C (Amsterdam) **135-140**, 130 (1994);
152. J.W. Loram et al, Physica C (Amsterdam), **235**, 134 (1994).
153. A.F. Andreev, Zh. Eksp. Teor. Fiz. **46**, 1823 (1964) [Sov. Phys. JETP **19**, 1228 (1964)].
154. G. Deutscher, Nature **397**, 410 (1999).
155. M. C. Schabel et al, Phys. Rev. B **57**, 6090 (1998).
156. J.E. Demuth et al, Phys. Rev. Lett. **64**, 603 (1990)
157. M. Tinkham, *Introduction to Superconductivity*, McGraw-Hill, Inc. (Singapore) (1996), p.20.
158. G.E. Blonder, M. Tinkham, and T.M. Klapwijk, Phys. Rev. B**25**, 4515 (1982).
159. G.D. Mahan, *Many Particle Physics* (Plenum, New York, 1990), p. 793.
160. A.S. Alexandrov and C. Sricheewin, Europhys. Lett., **51**, 188 (2000).
161. A. Abanov, and A.V.Chubukov, Phys. Rev. Lett. **83**, 1652 (1999).
162. for a review see Zhi-xun Shen, in 'Models and Phenomenology for Conventional and High-temperature Superconductivity' (Course CXXXVI of the Intenational School of Physics 'Enrico Fermi'), eds. G. Iadonisi, J.R. Schrieffer and M.L. Chiofalo, IOS Press (Amsterdam), p. 141 (1998).
163. D.C. Johnston, Phys. Rev. Lett **62**, 957 (1989).
164. D. Mihailovi'c et al, J. Superconductivity, **10**, 337 (1997).
165. J. Hofer et al, Physica C **297**, 103 (1998).
166. C. Kendziora et al, Phys. Rev. Lett. **79**, 4935 (1997).
167. N.L. Saini et al, Phys. Rev. Lett. **82**, 2619 (1999)
168. M. Randeria and J.-C. Campuzano, in 'Models and Phenomenology for Conventional and High-temperature Superconductivity' (Course CXXXVI of the Intena-

tional School of Physics 'Enrico Fermi'), eds. G. Iadonisi, J.R. Schrieffer and M.L. Chiofalo, IOS Press (Amsterdam), p. 115 (1998).
169. J. Mesot et al, Phys. Rev. Lett. **82**, 2619 (1999).
170. K. Gofron et al, Phys. Rev. Lett. **73**, 3302 (1994).
171. C.-H. Park et al, Phys. Rev. B **52**, R16981 (1995).
172. A.S. Alexandrov, Philos. Trans. R. Soc. London, Ser. A**356**, 197 (1998).
173. B.I. Halperin, Phys. Rev. **139**, A104 (1965).
174. H.L. Frisch and S.P. Lloyd, Phys. Rev. **120**, 1175 (1960).
175. A. Lanzara, et al, Journ. Phys.: Condens. Mat. **11**, L541 (1999)
176. D.R. Temprano et al, Phys. Rev. Lett. **84**, 1982 (2000).
177. J. Zaanen and O. Gunnarsson, Phys. Rev. B **40**, 7391 (1989).
178. V. J. Emery et al, Phys. Rev. B **56**, 6120 (1997) and refrences therein.
179. J.M. Tranquada et al, Nature **375**, 561 (1996).
180. A. Bianconi, J. Phys. IV France, **9**, 325 (1999) and references therein.
181. F.V. Kusmartsev, J. Phys. IV France, **9**, 321 (1999).
182. F.V. Kusmartsev, Phys. Rev. Lett., **84**, 530 (2000).
183. L.N. Grigorov, Makromol. Chem., Macromol. Symp. **37**, 159 (1990).
184. L.N. Grigorov et al, Makromol. Chem., Macromol. Symp. **37**, 177 (1990).
185. A.S. Alexandrov and V.V. Kabanov, JETP Lett. **72**, 569 (2000).
186. S.I. Pekar, Zh. Eksp. Teor. Fiz. **16**, 335 (1946).
187. This expression differs from Ref. [181,182] by the numerical coefficients.
188. J. Bonca et al, Phys. Rev. Lett., **84**, (2000).
189. P. Bourges et al, Science **288**, 1234 (2000); P. Bourges et al, cond-mat/0006085.
190. H.A. Mook et al, Nature **395**, 580 (1998); M. Arai et al, Phys. Rev. Lett. **83**, 608 (1999); P. Dai et al, Phys. Rev. Lett. **80**, 1738 (1998).
191. D. Mihailovic, Phys. Rev. B**60**, R6995 (1999).
192. A.S. Alexandrov, A.M. Bratkovsky and N.F. Mott, Phys. Rev. Lett. **72**, 1734 (1994).
193. W.M. Chen, J.P. Franck, and J. Jung, Physica C **341**, 1875 (2000).
194. V.N. Zverev and D.V. Shovkun, JETP Lett. **72**, 73 (2000).

# Transport Properties in Superconducting Layered Systems

## Luigi Maritato

*INFM-Unità di Ricerca di Salerno, Salerno, Italy*
*Dipartimento di Fisica, Università di Cagliari, Cagliari, Italy*

**Abstract.** The principles of the type II superconductor behavior are briefly and simply introduced and are applied to particular cases, such as the magnetic field dependence of pinning mechanisms in $Bi_2Sr_2CaCu_2O_x$ thin films, the melting of the vortex lattice in Nb/CuMn multilayers and the transport properties of High Temperature Superconductor based superlattices. The common behaviors of these three systems are pointed out, in view of a complete and deeper comprehension of the superconducting phenomena observed in the new family of the oxide superconductors.

## I. INTRODUCTION

After the discovery of the High Temperature Superconductors (HTS) [1], all showing a layered structure with superconducting planes alternating non superconducting blocks [2], the study of the superconducting phenomena in systems with reduced dimensionality, has gained new interest and strength.
In particular, due to the difficulty of using HTS in many practical applications, because of their poor performances in carrying high currents in the presence of externally applied magnetic field, the study of the transport mechanisms and properties of layered systems has been the focus of an intense research activity both from the point of view of the theoretical analyses and the experimental measurements [3].
In this respect, artificial layered systems, based on HTS and conventional superconductors, have proven to be an ideal issue of study, because of the possibility to suitably select the materials involved and to change, under control, many parameters which seem to be important in determining the observed behaviors.
In particular, conventional superconductor based layered systems, viewed as models for HTS, have been intensively studied to better understand their transport properties, with the aim of a better comprehension of what is also seen in HTS.
On the other hand, HTS based artificial superlattices have been a completely new field of investigation, which has allowed to point out the importance of mechanisms such as the charge transfer, in determining the observed properties of these new compounds.
Moreover, playing with the fabrication of artificial HTS based superlattices, revealed surprising aspects, allowing the synthesis of new materials, with properties that, in principle, can be designed in order to satisfy specific applicative needs [4].

In the following, after having introduced some fundamental notions about the type II superconductors transport properties, we will firstly focus our attention on the correct use of these notions in analyzing the behaviors of HTS. In particular, we will see that a complete characterization of the dissipative behavior of Bi based HTS, rely on the simultaneous measurements of their properties changing the temperature, the external magnetic field and the applied current [5].

Then we will discuss the characterization of Nb/CuMn multilayers, where the non superconducting material is a well known metallic spin glass. We will see that the behavior of these multilayers can be very useful to better understand many aspects also observed in HTS [6-8].

Finally, we will describe the properties of artificial HTS based superlattices, in order to get a deeper look to the microscopic mechanisms which are the base of the superconductivity phenomena in HTS [9,10].

## II. TRANSPORT PROPERTIES OF TYPE II SUPERCONDUCTORS

Due to the presence of a regular array of vortices, type II ideal superconductors, in external magnetic fields higher than a critical value, are in principle dissipative systems even below the critical temperature $T_c$. In fact, this type of superconductors are able to completely screen external magnetic fields H only up to a value $H_{c1}$, called the lower critical magnetic field. When $H > H_{c1}$, magnetic flux starts to penetrate into the superconducting system, the presence of surfaces separating superconducting and normal zones being energetically favored [11]. To minimize the energy then, each normal zone will carry a magnetic flux as small as possible. The smallest achievable value of magnetic flux in quantum mechanics, is called the flux quantum $\phi_o$ and is equal to $2.07 \times 10^{-7}$ gausscm$^2$. An ideal type II superconductor in the presence of external magnetic fields higher than $H_{c1}$, therefore, will be threaded by circular zones carrying a magnetic flux $\phi_o$, surrounded by superconducting zones where screening currents are flowing. These structures are called Abrikosov vortices and the inner part is generally addressed as the core of the vortices or the fluxon. The penetration of the magnetic flux into the type II superconductor continues up to a critical value of the external magnetic field, $H_{c2}$, called upper critical magnetic field, at which the system becomes normal. The state of type II superconductors presenting vortices is called mixed state, to distinguish from the Meissner state in which complete expulsion of the magnetic flux is reached. In the Magnetic Field- Temperature H-T plane, the phase diagram for a type II superconductor is shown in Figure 1.

Simple calculations give explicit relations for $H_{c1}$ and $H_{c2}$. In particular, $H_{c1} = 4\pi\varepsilon_1/\phi_o$, is related to the change in the Helmholtz free energy per unit length $\varepsilon_1$, associated to the penetration of the first vortex, while $H_{c2} = \phi_o/2\pi\xi^2$, is connected with the value of the Ginzburg-Landau coherence length $\xi$, which in turn, gives an estimate of the radius of the vortex core. Typical values of $H_{c1}$, in the case of conventional superconductors, are in the range of few hundreds of Oersted while $H_{c2}$ can vary from $10^{-1}$ to 10-20 Tesla.

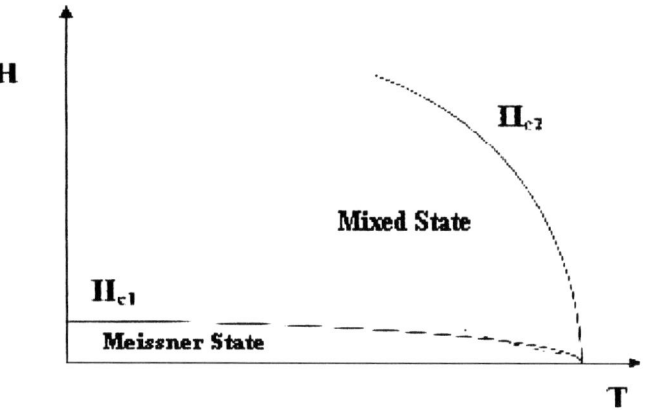

**FIGURE 1.** H-T phase diagram for a type II superconductor.

Due to the presence of the screening currents, the interaction between vortices is repulsive. In an ideal type II superconductor, the vortices will arrange themselves in an ordered array resulting in a configuration with the lowest possible energy. For a fixed density of vortices, the lattice minimizing repulsive energy is the triangular one.

When a current density **J** acts on a vortex, a force **F**, given by

$$F = J \times \phi_0/c$$

where $\phi_0$ is a vector with the modulus equal to the flux quantum and the direction of the magnetic flux crossing the vortex, will tend to accelerate the fluxon. In the case of a vortex lattice, the force $F_s$ acting on a single vortex and due to the presence of all the other vortices, will be

$$F_s = J_s \times \phi_0/c$$

where now $J_s$ is the current density in the position of the considered vortex, given by the contribution of the screening currents of all the other fluxons in the lattice. Due to the symmetry of the lattice, this force $F_s$ will be zero. The picture changes when an external current density $J_{ext}$ is present. In this case the force density $F$ acting on the vortex will be

$$F = J_{ext} \times B/c$$

and the vortex will move, producing an electric field $E = B \times v/c$, with **v** the vortex velocity parallel to $J_{ext}$, dissipating energy at a rate of $E*J_{ext}$. Therefore, an ideal type II superconductor, in the presence of external fields higher than $H_{c1}$, will show electric dissipation with electrical resistivity different from zero, due to the motion of all the vortices in the lattice.

The actual situation in the case of a real superconductor is strongly different, even disregarding the problem of the edges of the finite system. In fact, due to local changes in microscopic parameters such as the coherence length $\xi$ and the energy gap $\Delta$, vortices find more convenient to be located in places where the free energy difference between the normal and the superconducting state is lower. These places act, therefore, as pinning centers for the vortices and to produce their motion a certain critical value of the force density $F_c$ will have to be overcome. As a consequence, only at current density values higher than a critical threshold $J_c$ defined from the relation $F_c = J_c \times B/c$, the superconducting system will show dissipation, while at $J < J_c$ it will present, in principle, zero electrical resistance. Obviously the actual situation can be much more complicated, with each pinning center having associated a different $F_c$ and, consequently, a different $J_c$. In fact, the value of the free energy difference between the normal and the superconducting state in a particular position depend upon the local values taken by several microscopic parameters (the penetration depth $\lambda$, $\xi$, $\Delta$) and moreover, the force density is strictly related to the spatial variation of the free energy. Finally, even with pinning centers in the superconducting system, one should not forget the presence of thermal effects, which can provide the energy needed to overcome the pinning barrier potential.

As a result of the simultaneous presence of all these phenomena, the H-T phase diagram of a real type II superconductor can be much more complicated than that shown in figure 1, presenting different zones with different transport properties and dissipative behaviors.

From the point of view of applications, it is worthwhile to point out that for type II superconductors $J_c$ is not an intrinsic property of the material, but can be varied, in principle, introducing suitable pinning centers in the system.

## II. A. Flux Flow Regime

In the case of weak pinning centers in the system, the transport properties of a type II superconductor can be described in terms of the so-called flux flow regime. In fact, if the critical density force $F_c$ associated to each pinning center can be overcome applying small external currents $J > J_c$, then all the vortices will be free to move simultaneously. On the other hand, the presence of the underlying pinning centers will result in a viscous drag, and the motion of vortices will produce a dissipation related to a resistivity $\rho_{ff}$ of the same order of the normal state value $\rho_n$. If we assume that a viscous force density $F_L$ directly proportional to its velocity $v_L$

$$F_L = -\eta\, v_L$$

acts on each single vortex, with $\eta$ coefficient of viscosity, then in stationary conditions one should have

$$J\phi_0/c = \eta v_L$$

In the case of vortices moving with a velocity $v_L$ in the presence of a magnetic field B perpendicular to the velocity direction, the electric field is given by $E=Bv_L/c$ and the flux flow resistivity $\rho_{ff}$ will be

$$\rho_{ff}=E/J=B\phi_o/\eta c^2$$

In the limit of $\eta$ independent from B, the flux flow resistivity $\rho_{ff}$ is therefore directly proportional to B. On the other hand, an explicit expression for $\eta$ can be obtained assuming a specific model to describe the vortices in the system. In particular, Bardeen and Stephen developed a simple description in which a vortex is thought as made of a normal cylinder of radius $\xi$ surrounded by superconducting material where the screening currents flow. The dissipation associated to the motion of the vortices is then related to the ordinary resistive processes in this normal cylindrical core. The Bardeen-Stephen expression for $\eta$ is the following

$$\eta=\phi_o^2/2\pi\xi^2c^2\rho_n=\phi_o H_{c2}/\rho_n c^2$$

where $\rho_n$ is the resistivity in the normal state. Using the expression found for $\rho_{ff}$ it is then immediate to get

$$\rho_{ff}/\rho_n=B/H_{c2}$$

with $\rho_{ff}$ smoothly joining $\rho_n$ at $H_{c2}$. Moreover, the last expression directly shows that the value of $\rho_{ff}$ is of the same order of magnitude of $\rho_n$. Therefore, in the presence of weak pinning centers, a type II superconductor can present a substantial dissipative behavior.

## II. B. Flux Creep Regime

On the other hand, even in the case of strong pinning centers, at finite temperature, thermal energy may allow vortices to jump from one pinning center to another. This vortex motion does not give any contribution to dissipation if no external force is applied to the system. In the presence of a driving force such as an externally applied current, this thermally activated vortex motion will produce a longitudinal resistive voltage, proportional to the average velocity of the fluxon motion. This effect, generally called thermally activated flux creep, also leads to small changes in the trapped magnetic field in the system, therefore determining not only the presence of dissipation at temperatures below $T_c$, but also the disappearance of a perfect diamagnetic state.

Flux creep phenomena can be described, using the Anderson-Kim model, in terms of vortex bundles jumping between adjacent pinning centers. The jump rate R is given by

$$R=\omega_o e^{-U_o/kT}$$

where $\omega_o$ is a characteristic frequency of vortex vibration and $U_o$ is the pinning barrier energy. The presence of an external force introduces an asymmetry in the spatial dependence of the energy barrier, rendering easier a jump in the direction of the applied excitation. If $\Delta F$ is the work done by the driving force in going over the barrier, then the net jumping rate in the direction of the force will be

$$R=\omega_o e^{-U_o/kT}(e^{\Delta F/kT}-e^{-\Delta F/kT})$$

and a net vortex creep velocity $v_c$ will be present in the system, given by

$$v_c=2v_o e^{-U_o/kT}\sinh(\Delta F/kT)$$

with $v_o=\omega_o L$, where L is a typical length associated to the creep phenomena. Assuming $U_o$ of the order of the condensation superconducting energy, $v_c$ will always be unosservably small, for any physical value given to $\omega_o$ and at any temperature below $T_c$. The only possible case in which $v_c$ can be appreciably high, is obtained when $\Delta F \gg kT$. In this case, in fact, we will have

$$2\sinh\Delta F/kT \cong e^{\Delta F/kT}$$

and the vortex creep motion will produce, in the presence of a magnetic field B, an electric field $E=Bv_c$ equal to

$$E=Bv_o e^{-U_o/kT} e^{\Delta F/kT}$$

On the other hand, if $N_p$ is the volume density of pinning centers in the system, it is [12]

$$N_p\Delta F=JBL$$

and if $J_o$ represents the critical current density in the absence of creep, then

$$J_o BL=U_o N_p$$

and we can immediately write

$$E=Bv_o e^{(\Delta F/kT-U_o/kT)}=Bvoe^{U_o(J-J_o)/J_o kT}$$

which gives

$$J=J_o[1+(kT/U_o)\ln(E/Bv_o)]$$

In this case, therefore, vortex creep will produce exponential voltage-current characteristics, and as long as J<<J$_o$, typical electric fields will be again of the order of $e^{-U_o/kT}$.

If we are instead in the limit ΔF<<kT, then the E-J relation will be given by

$$E=(2v_oBJU_o/J_okT)e^{-U_o/kT}$$

and we will have now linear voltage-current characteristics, strongly dependent on the temperature. It is worth to point out that even though the electrical resistivity could seem to be linearly dependent on the magnetic field B as for the flux flow value $\rho_{ff}$, this is not the case for the flux creep, because of the strong U$_o$ dependence on B.

The observation that, for a type II real superconductor the presence of weak or strong pinning centers, does not result in non dissipative electrical conduction, can appear discouraging in view of practical applications. Fortunately, the dissipation associated with the flux creep effects, for example, is however very low and almost negligible for many practical situations, especially in the case of low transition temperature systems [11]. For HTS, the higher obtainable values of temperature in the superconducting state render flux creep effects much more easy to be seen and much more important in determining the electrical transport properties. Moreover, in this class of material the actual dimensionality of the system and the energy distribution of the pinning centers, play a very important role in the superconducting behavior. As a result, the H-T phase diagram of HTS is generally much more complicated than that shown in figure 1 and its interpretation is still an open question of study for many research groups in the world.

In the next sections, we will discuss some points about the superconducting properties of HTS in the light of the observations made so far, and we will see how the study of artificial layered systems can improve our understanding of the physics involved.

We will start by considering the role played by the magnetic field dependence of pinning in the transport properties of Bi based HTS, and then we will analyze the H-T phase diagram of conventional superconductor based multilayers, used as model for HTS systems. Finally, we will describe the superconducting properties of HTS artificially obtained superlattices which have allowed to get a deeper understanding of the microscopic mechanisms upon which the behavior of these materials is based.

## III. MAGNETIC FIELD DEPENDENCE OF PINNING MECHANISMS IN BSCCO THIN FILMS

As we have seen, in the case of flux creep regime the resistivity can be described in the limit of small imposed stresses (due both to the external magnetic field and to the bias current) by [12,13]

$$\rho=\rho_oe^{-U/kT}$$

where $\rho_o$ is a coefficient of the order of the normal state resistivity and U is the pinning barrier energy which depends on the temperature T, the external magnetic field B and the bias current I. In figure 2a and 2b the ρ(T) curves plotted in Arrhenius fashion, in the case of two $Bi_2Sr_2CaCu_2O_{8+x}$ thin films obtained by Molecular Beam Epitaxy (MBE) deposition [5], are shown for different applied perpendicular magnetic fields. The slope of the lnρ versus $T^{-1}$ curves, is related to the pinning barrier U by

$$U_{eff} = -kd(\ln\rho)/dT^{-1} = U(T) - TdU/dT$$

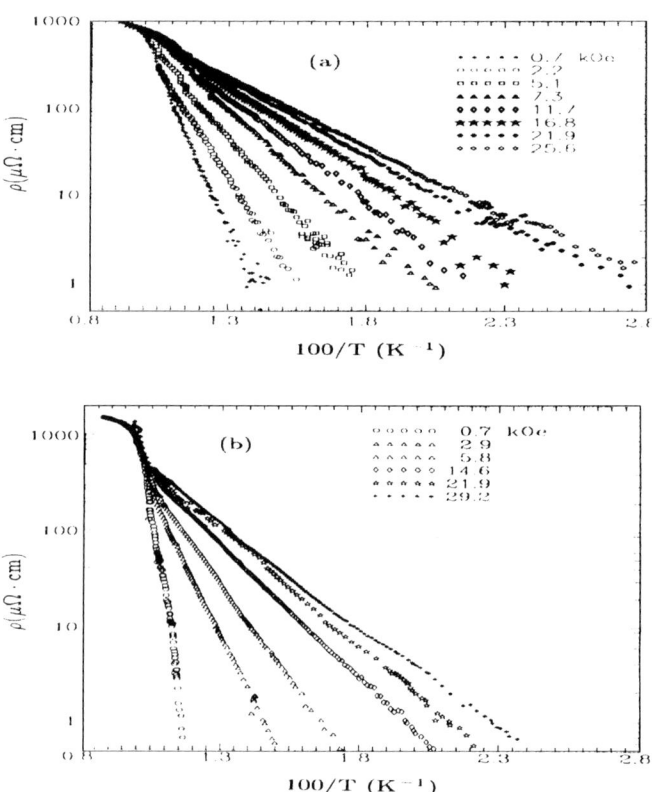

**FIGURE 2.** Arrhenius plot of ρ(T) for two BSCCO patterned films (a) and (b), for different values of the perpendicular applied magnetic field [5]. Reprinted from [5], Copyright 1995, with permission from Elsevier Science.

To analyze the data in figure 2, it is necessary to take into account the temperature dependence of the pinning potential. Tinkham [14] proposed the following behavior for U(t),

$$U(t) = U(0)(1-t^2)^{2-n}(1-t^4)^{n/2}$$

where $t=T/T_c$ is the so called reduced temperature and the exponent n could be equal to 1,2 or 3 [14-16]. Substituting this expression for U(t) we can write $U_{eff}$ as

$$U_{eff}=U(t)[1+(2-n)2t^2/(1-t^2) + 2nt^2/(1-t^4)]=U(t)\beta(t,n)$$

The slope of the Arrhenius plot gives therefore the $U_{eff}$ which is enhanced by $\beta(t,n)$ with respect to the real value U(t). The U(H) dependence obtained according to the Tinkham approach, for the data in figure 2b, is presented in figure 3 for different values of n. The U values were calculated in the limit of zero temperature. The character of the U(H) dependencies does not depend upon the choice of n, and can be described by a law of the kind $U(H) \cong H^{-\alpha}$, with $\alpha$ almost equal to 0.5 independently of the value of n ($\alpha$=0.43) [5]. This behavior for U(H) is predicted considering as the mechanism of vortex depinning the plastic deformation of the vortex lattice in an anisotropic superconductor. In this case the energy of plastic deformation is written as [17]

$$U_{pl}(T,H)=2\varepsilon_v a_o \cong \phi_o^2 a_o \lambda^{-2} \cong (T_c-T)H^{-0.5}$$

where $\varepsilon_v$ is the vortex energy per unit length in the $CuO_2$ planes and $a_o$ is the intervortex lattice spacing. On the other hand, it is known that the shape of the pinning potential and the pinning strength distribution influence the U(H) curve. As pointed out by Inui et al. [18], a cross over from individually pinned vortices to a regime of elastic vortex interaction is expected in BSCCO at fields of the order of 1 Tesla. At small magnetic field the intervortex forces are negligibly small and one should take into account the motion of individual vortices in the field of unperturbed pinning centers. For an exponential type of pinning energy distribution U(H) is then given by

$$U(H)=U_{av}+\sigma \ln(H_o/H)$$

where $U_{av}$ is some average activation energy corresponding to $H_o$, the cross over field between the single vortex and the collective behavior.
For large magnetic fields, the motion of vortices in the pinning center field has to be taken into account and the role of the elastic forces is then important. It can be shown, using a harmonical one-dimensional model in the limit $\delta \ll U_{av}K^2$, where $\delta$ is the elasticity modulus of the vortex lattice [18] and K is related to the period of the potential, that the amplitude of the potential barrier is

$$U(H)=U_{av}-\delta(\pi/K)^2$$

Therefore, in terms of this model, we should expect a logarithmic behavior of U(H) for $H<H_o$, and a linear decrease for $H>H_o$. On the other hand, if one looks only to the U(T) and U(H) dependencies, this cross over it is not shown and, as pointed out previously, regarding for example U(H), everything seems to be explained assuming a U(H) law of the type $H^{-\alpha}$ in all the investigated magnetic field range [5].

The situation completely changes if we also analyze the current dependence of U. As shown in figure 4a and 4b, the U(J) behavior strongly depends upon the external magnetic field.

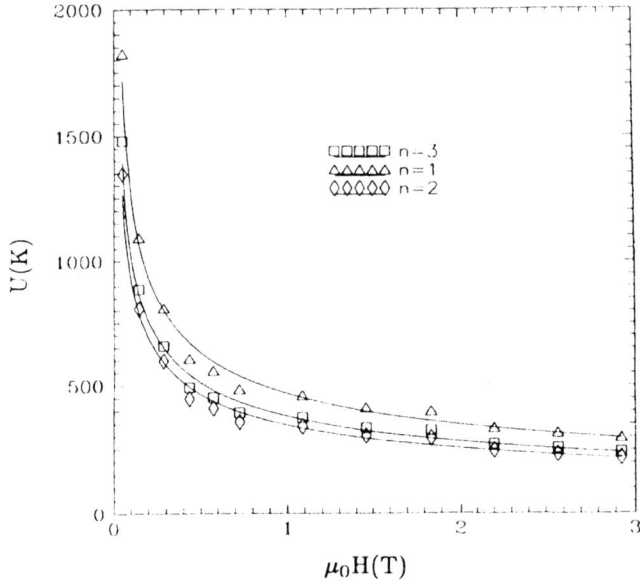

**FIGURE 3.** The U(H) dependence for the sample in figure 2b with the zero temperature value for the pinning potential calculated with different values of n, see text. The solid lines correspond to the behaviors outlined in the text [5]. Reprinted from [5], Copyright 1995, with permission from Elsevier Science.

In fact, at low magnetic field, figure 4a, U(J) can be well fitted by the expression

$$U(j) = U(0)[(1-j^2)^{1/2} - j\cos^{-1}j] \cong U(0)(1-j)^{1.5}$$

with $j=J/J_o$, which follows from the assumption of a washboard pinning potential spatial shape [19-21], while at large magnetic fields the U(J) curve is well described using a logarithmic dependence for U(J). The squares and the triangles in figure 4a and 4b are related to different procedures to extract the U value in the zero temperature limit. In particular, the squares are the values obtained using the Tinkham model with n=1, while the triangles are the values deduced by the plastic deformation model [17]. As it is clear from the figures, the type of behavior does not depend upon the used method.

Assuming two different kinds of U(H) dependencies in the case of large and small magnetic fields, we can analyze the U(H) curves in the light of the Inui et al. model [18]. In figure 5, we show the data obtained for the two samples investigated, along with the best fits calculated from the logarithmic behavior at small magnetic fields (exponential distribution) and from the linear dependence at large magnetic fields. The agreement with the experimental data is good and the cross over field is 0.7 Tesla. The collective pinning theory predicts a cross over from single vortex regime to collective behavior at fields of the order of $H_{sb}=5(J_o/J_{GL})H_{c2}$, where $J_{GL}$ is the

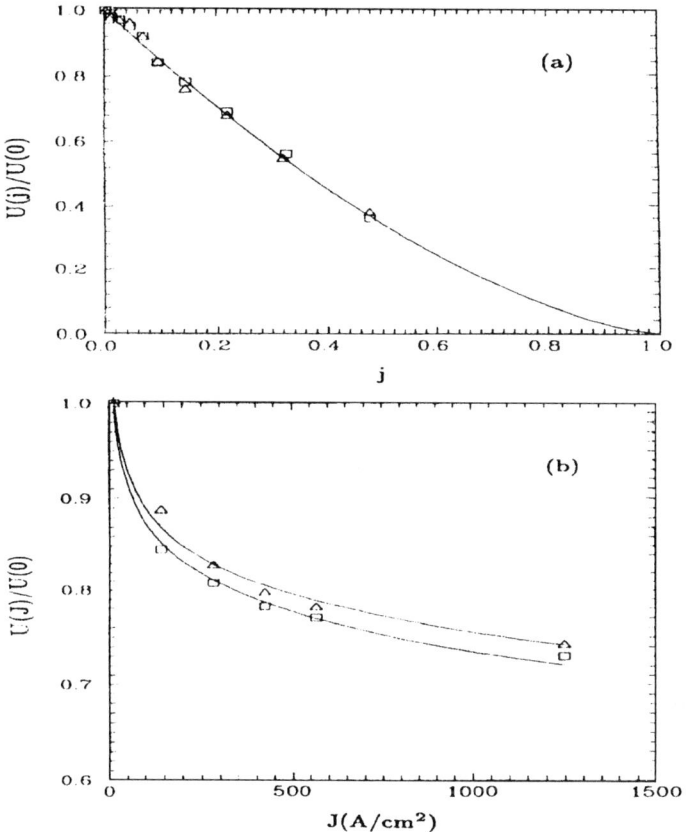

**FIGURE 4.** The U(H) dependence for two samples (squares and triangles) as obtained in Tinkham model with n=1. The solid lines are the best fits discussed in the text [5]. Reprinted from [5], Copyright 1995, with permission from Elsevier Science.

Ginzburg Landau depairing current [3,22,23]. In our case, $H_{sb}$ should be in the range 0.6-1.1 T, in reasonable agreement with the obtained value for $H_o$. It is then clear that, to obtain an exhaustive understanding of the pinning properties of a type II superconductor, all the three dependencies U(H), U(J) and U(T) have to be simultaneously analyzed. Only in this way we have been able to point out a cross over from a certain behavior at low magnetic fields and another at high fields, which was completely hidden by looking only at the U(T) and U(H) dependencies.

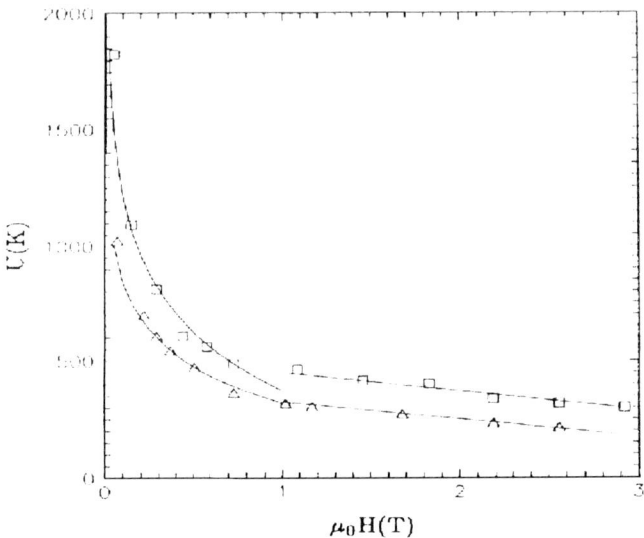

**FIGURE 5.** The U(H) dependences as obtained in Tinkham model with n=1. Triangles correspond to the sample in figure 2a squares to that in figure 2b. The solid lines are the best fits obtained as discussed in the text [5]. Reprinted from [5], Copyright 1995, with permission from Elsevier Science.

On the other hand, the study of the HTS pinning properties is complicated also by the difficulty to independently control many parameters which seem to play a role in their behaviors. A way to solve this problem and to go deeper in the comprehension of the dissipative phenomena in type II superconductors, is to study artificial layered conventional superconducting systems, which can be viewed as models for HTS, where it is easier to change several parameters under control. In the following section we will analyze the transport properties of superconducting (Nb)/magnetic (CuMn) multilayers, and we will see how many interesting observations can be done which are also extremely important to better understand HTS.

# IV. H-T PHASE DIAGRAM OF CONVENTIONAL SUPERCONDUCTING/MAGNETIC ARTIFICIAL MULTILAYERS

As we have seen so far, the transport properties of a type II superconductor can be influenced by many different parameters. In the case of HTS, this influence is amplified by the layered nature of these compounds (i.e. by the reduced dimensionality of the systems), and by the very small values (only few angstroms) of the superconducting coherence lengths. Unfortunately, for HTS, this causes a dramatic reduction of their ability to be used in practical applications, because of the appearance of the so called irreversibility line in the H-T phase diagram below the $H_{c2}$

curve, which separates the superconducting state into two zones, one where the dissipative effects are still important and the other where the true non dissipative superconducting state is present [3].

The location of the irreversibility line in the H-T phase diagram is very important for applications, the closer it is to the $H_{c2}$ curve the larger it is the range of temperatures and fields where the superconductor can be used in practical non dissipative devices. The understanding of the basic mechanisms determining in HTS the appearance of such line is therefore central in trying to improve their performances. On the other hand, the detailed study of this effect in HTS is complicated by the simultaneous presence of different mechanisms which all seem to contribute, with similar energies, to this phenomenon. As an example, the nature of the irreversibility line has been attributed to depinning of the vortices (i.e. the escape of vortices from the pinning barrier when its height starts to be comparable to the thermal energy), to melting of the vortex lattice or to a glassy transition of the entangled vortices [3]. The amount of disorder present in the system seems to play an important role in determining which of these mechanisms is the most important, so that, even looking at the same compound, one can observe different situations.

A way to solve this problem can be the use of model systems, based on conventional superconductors, which can help at least in separating the different contributions, determining their typical characteristics. From this point of view, multilayers of conventional superconductor/magnetic materials are ideal candidates to be used as model of HTS in studying vortex dynamics. In fact, they can be realized in such a way to show variable anisotropies in order to analyze the importance of this parameter on the location of the irreversibility line, while presenting distances between adjacent superconducting layers of few angstroms (due to the strong pair breaking effect of magnetism). Moreover, the interplay between superconductivity and magnetism seems to be important in many of the HTS peculiar properties [24,25].

In Nb/CuMn (superconducting/spin glass) multilayers obtained by sputtering and presenting different layer thicknesses of CuMn, an irreversibility line has been observed below $H_{c2}$, and its nature has been related to vortex melting mainly due to quantum fluctuations [6].

In figure 6, the Arrhenius plots of the transition curves of two Nb/CuMn multilayers, one with 250 Å of Nb and 4 Å of CuMn (a), and the other with the same thickness of Nb and 6 Å of CuMn (b) are shown for different perpendicularly applied magnetic fields. A kink is always present in the curves of the plots, with position depending upon the external field. This kink signals a change in the vortex dynamics of the system, with a cross over between two different activation energies. Plotting the positions of these kinks in the H-T plane, they locate a line below the $H_{c2}$ curve, which can be drawn back to an irreversibility line.

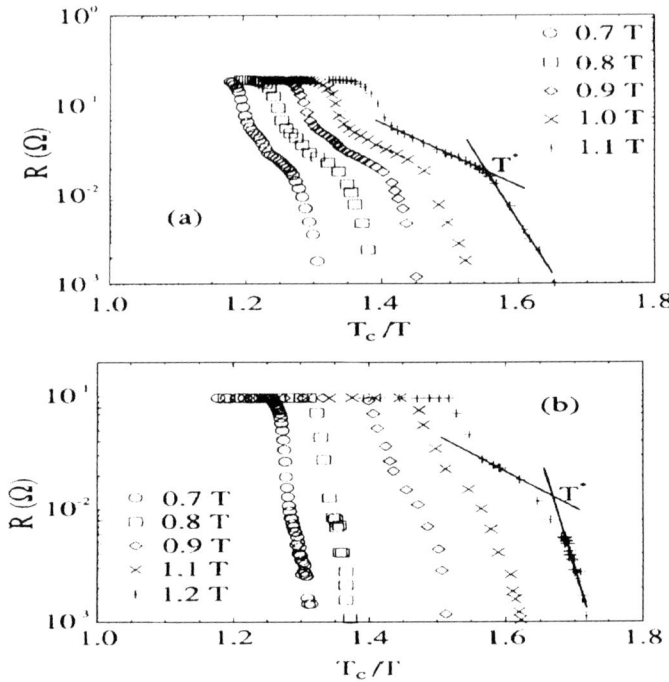

**FIGURE 6.** ln R vs. $T_c/T$ curves for two Nb/CuMn multilayers with the same nominal thickness of the Nb layers (250 Å) and CuMn thickness of 4 Å (a) and 6 Å (b) [6].

Similar results have been also observed on Nb/CuMn multilayered samples where a regular square array of antidots was obtained by lift off procedure [7]. In the case of these samples, the irreversibility line location has been measured by three different methods (Arrhenius plot, disappearance of hysteresis in the I-V characteristics, change in the curvature of the LnI-LnV characteristics), figure 7, and all the experimental points fall on the same curve locating the line which separates dissipative behavior from superconducting state.

The value of the matching field in the antidotted samples was always very small, of the order of 5-10 Oe, and the distance d between adjacent antidots was generally such that $d >> \xi(T) \cong 100$ Å. Therefore, vortex lattice was present in the zones among antidots in the superconducting state, and the pinning of these vortices was mostly determined by the intrinsic properties of the material. In a Nb based system we do not expect low values of the ratio U/kT, and in the case of Nb/CuMn antidotted multilayers this expectation has been confirmed by direct measurements of the critical current density [7].

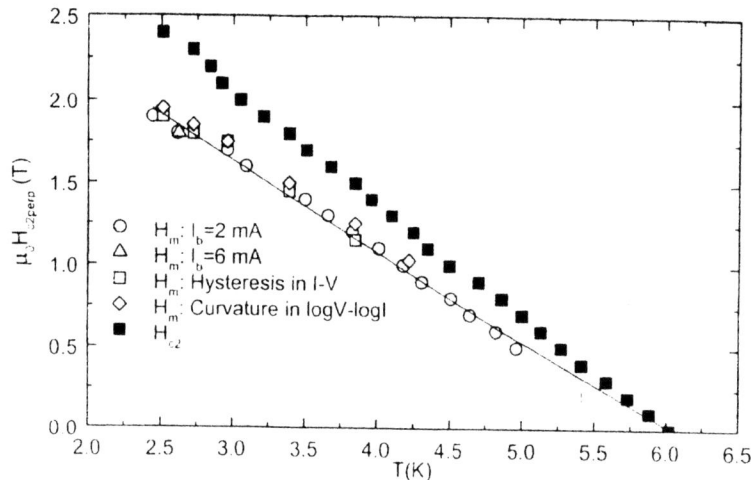

**FIGURE 7.** H-T phase diagram for an antidotted multilayer of Nb/CuMn with 250 Å of Nb and 12 Å of CuMn. The antidots have circular shape with diameter of 1 µm and distance between adjacent dots of 1.6 µm. Solid squares correspond to $H_{c2}$, open squares correspond to the disappearance of nhysteris in the I-V curves, diamonds to the change of the curvature in the logV-logI curves, circles (up triangles) to the kinks in the Arrhenius plots of the transition curves using a bias current of 2 (6) mA. The solid line is the best fit discussed in the text [7].

As a consequence, in these samples, flux creep effects should not be important at least in the range of few microvolts, and depinning cannot play any important role in determining the presence of the observed irreversibility line.

On the other hand, the values of the anisotropy parameter $\gamma$ measured on the multilayers were always lower than 16 and both the angular and the temperature behavior of $H_{c2}$ indicate that we were in the presence of strongly coupled superconducting systems [26,27].

The strongly coupled nature of the vortices in adjacent layers was also assured by the fact that any externally applied field was much smaller than the critical decoupling value $H_{cd} \cong 4\phi_0/\gamma^2 s^2 \cong 10^3$ T, where s is the distance between adjacent superconducting layers [28]. Therefore, entaglement or twisting of the vortices is very difficult to happen and cannot be used to explain the nature of the irreversibility line.

The other possible mechanism to explain the nature of the observed irreversibility lines, is the vortex melting. Before starting to analyze the data in the light of this explanation, the dimensionality of the vortex lattice has to be determined. A vortex lattice, for example, can behave bidimensionally when its shear modulus is much smaller than the tilt modulus. Thermal fluctuations can cause tilt deformation of the vortices. The critical thickness value $d_{cr}$ beyond which this effect becomes relevant is given by [29]

$$d_{cr} \cong 4.4\xi_{par}/[h(1-h)]^{1/2}$$

where $\xi_{par}$ is the Ginzburg Landau superconducting coherence length in the plane of the film and $h=H/H_{c2}$. In multilayers this value of $d_{cr}$ is reduced by a factor $\gamma^2$ [28] and for the Nb/CuMn antidotted samples typical values were around 2-300 Å while the total thickness of the multilayers was always in the range 1500-1700 Å. This means that the vortex lattice in these multilayers was always in a strong three dimensional regime. In a three dimensional superconducting system, the temperature $T_m$ at which the vortex lattice goes from a solid to a liquid phase because of thermal fluctuations is given by [30]

$$c_L^4 \cong 3G_i h\, t_m^2 / \pi^2 (1-h)^3 (1-t_m)$$

where $c_L$ is the Lindemann number, $G_i$ is the Ginzburg number [7] and $t_m=T_m/T_c$ is the reduced melting temperature. Melting typically occurs for values of $c_L$ in the range 0.1-0.3 [31]. The fit to the experimental data for whole and antidotted Nb/CuMn multilayers was obtained with the last formula only using values of $c_L$ unphysically low, of the order of $10^{-4}$, rendering unreasonably to relate the observed irreversibility lines to three dimensional pure thermal melting.

On the other hand, as first pointed out by Blatter and Ivlev [31], at low temperatures, in moderately anisotropic superconductors the contribution of quantum fluctuations to the melting can be relevant. In this case, the total averaged fluctuation displacement of the vortex $<u>^2$ is the sum of two terms, $<u>_{th}^2$, due to thermal fluctuations, and $<u>_q^2$ related to quantum fluctuations. The amplitude of $<u>_q^2$ depends on the ratio $Q^*/(G_i)^{1/2}$, where $Q^*=2\pi e^2 \rho_n/h^*s$ with $h^*$ the Planck constant and $e$ the elementary charge. When $Q^*/(G_i)^{1/2} >> 1$, the contribution of quantum fluctuations to the melting is crucial. A critical parameter in the value of $Q^*$ is the interlayer spacing s, which, in the case of the Nb/CuMn multilayers studied, was never larger than 12 Å. For the Nb/CuMn samples investigated the values of $Q^*/(G_i)^{1/2}$ were always higher than 30 and the quantum contribution to melting should be taken into account [32]. In this case, the melting line in the H-T phase diagram is given by [33]

$$h_m = 4\Theta^2 / [1+[1+4Q\Theta/t]^{1/2}]^2$$

where $h_m = H_m/H_{c2}$, $t=T/T_c$, $\Theta=\pi c_L^2(t^{-1}-1)/G_i^{1/2}$ and $Q=Q^*\Omega\tau/\pi G_i^{1/2}$, with $\Omega$ a cut off frequency of the order of the Debye frequency and $\tau$ an effective electronic relaxation time [33]. In spite of the many quantities appearing in the previous formula, the only one that can be used as free fit parameter is the Lindemann number $c_L$. In fact, all the other terms in the expression are known or can be independently measured for Nb/CuMn multilayers [6,7]. For example, in the case of the data in figure 7, the solid line has been obtained with $c_L=0.16$. In the case of the data in figure 8, measured on Nb/CuMn multilayers without antidots, the solid lines, calculated from the previous formula, have been obtained using $c_L$ values in the range 0.1-0.2 [6].

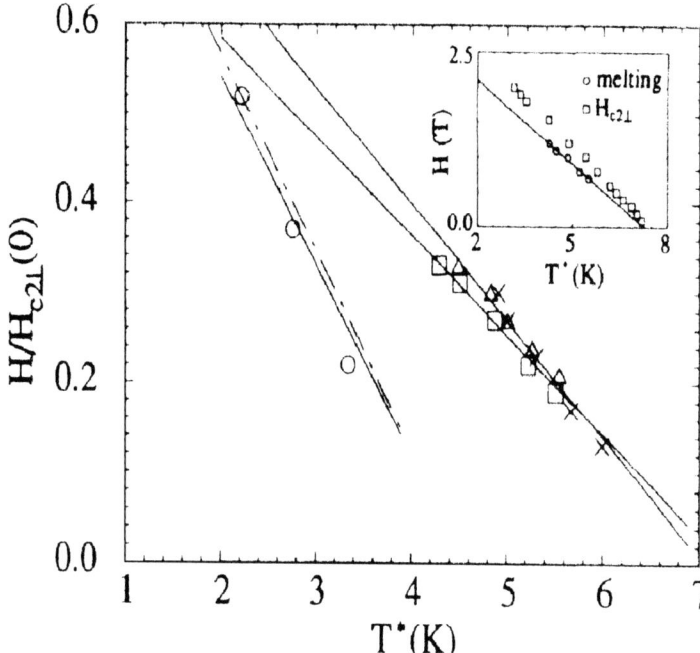

**FIGURE 8.** Melting lines for Nb/CuMn multilayers without antidots and having all the same Nb thickness (250 Å) and CuMn layers with 14% of Mn and thicknesses of 4 Å (Δ), 6 Å (X), 16 Å (O), or 6% of Mn and thickness of 6 Å (squares). In the inset of the figure is shown the H-T phase diagram for the last (squares) multilayer [6].

From the study on the transport properties of Nb/CuMn multilayers with and without a regular array of antidots, it is then possible to gain interesting information about the nature of the irreversibility line in layered superconducting systems where the distances between superconducting planes is only of few angstroms. A fundamental result is that quantum fluctuations can play a key role in the position of such a line. Various experiments on HTS had already addressed the possible influence of quantum fluctuations [33,34], and it is very interesting to have found the discussed results in a system used as model for HTS although based on conventional superconductors. Another important aspect that has been possible to point out studying Nb/CuMn multilayers, is the influence of the anisotropy in the position of the irreversibility line in the H-T plane. In the case of these samples, in fact, is simple to change values of the anisotropy parameter γ by changing the relative thicknesses or the amount of Mn in the CuMn layers. For the antidotted Nb/CuMn multilayers [7], the distance from the $H_{c2}$ curve increases monotonously with the value of γ.

A final interesting result regards the influence on the superconducting state of the interplay with the magnetic order present in the samples. A striking consequence of this interplay is the presence of a non trivial ground state in the layered system for certain suitable values of the relative thicknesses, the so called π phase, where the sign of the superconducting parameter changes going from one layer to the next. This phase determines a non monotonic behavior of the critical temperature $T_c$ versus the CuMn thickness layer $d_{CuMn}$ behavior [8,35]. In figure 9, it is shown the $T_c$ vs. $d_{CuMn}$ behavior, for a series of Nb/CuMn multilayers with the same Nb thickness and different CuMn layer thicknesses. The non monotonic character of the curve is clear and can be described in terms of the presence of the π phase (the solid line in figure 9) [36].

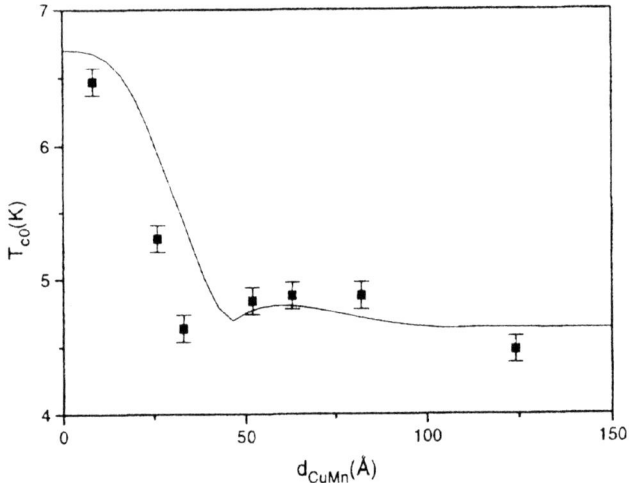

**FIGURE 9.** $T_c$ versus $d_{CuMn}$ curve for a series of Nb/CuMn multilayers having all the same Nb thickness (250 Å) and with the 1.3% of Mn in the CuMn layers. The solid line is the best fit obtained as discussed in the text [8].

For HTS, many typical properties have been explained assuming a d-type symmetry of the superconducting parameter [37,38]. This kind of symmetry is justified in terms of magnetic effects at work in these compounds. On the other hand, the d-type symmetry is also very similar to the π phase observed in Nb/CuMn multilayers and in general in Superconducting/Magnetic layered systems [39]. In fact, in the case of the π phase the sign of the order parameter changes by going from one Nb s-type layer to the next, moving along the direction perpendicular to the plane of the film, while for the d-wave in HTS the change in the sign of the order parameter happens when going from one lobe to the adjacent in the plane where the superconducting phenomena are confined (the $CuO_2$ planes) [37,38]. This similarity opens the possibility to many experiments in the case of the conventional multilayers which, following the example of those already performed in HTS, can clarify the presence and the nature of the π phase, allowing its better understanding especially in view of a better comprehension of the basic phenomena which determine the appearing of the d-type symmetry in HTS.

# V. HTS ARTIFICIAL SUPERLATTICES

In the last years, the developments in the vacuum techniques (Molecular Beam Epitaxy (MBE), Sputtering and Laser Ablation) used for the fabrication of layered systems with epitaxial structural properties using perovskitic materials, have allowed to realize artificial superlattices completely made of HTS [4]. In particular, the fabrication of artificial layered systems in which the single layers can be changed at wish [40], gives the opportunity of studying the microscopic phenomena upon which the HTS is based, and to look for new systems with improved properties. As an example, the charge transfer mechanism from carrier reservoir blocks and active planes, which is generally assumed to explain many of the observed properties of HTS [41], can be directly tested realizing suitable HTS artificial superlattices.

At the base of this theoretical interpretation there is the observation that the unit cell of almost all the HTS can be thought as made of two different kinds of blocks, a charge reservoir one and an active one, where the Cooper pairs are confined and generally containing the $CuO_2$ planes [2]. The electronic properties of these blocks and their relative thicknesses play a critical role in determining the optimal number of carriers in the $CuO_2$ planes and therefore the optimization of the critical temperature and of the superconducting performances. An obvious way to check this mechanism is to realize artificial superlattices in which one can freely choose the constituent layers in order to study the behavior of the superconducting properties. These analyses have also a second important interesting aspect related to the search of new artificial materials with improved properties respect to specific applicative needs.

Many of the studies in this field have dealt with the realization of artificial superlattices based on compounds of the so called Infinite Layer family [42]. In fact, practically all the HTS compounds, have a structure that can be thought of alternate Infinite Layer planes intercalated with atoms of suitable species. In particular, superlattices of $BaCuO_2/CaCuO_2$ and $BaCuO_2/SrCuO_2$ obtained by Laser Ablation have extensively studied and have shown surprising superconducting properties [43]. For examples, $BaCuO_2/CaCuO_2$ superlattices with two unit cells of $BaCuO_2$ and two or three unit cells of $CaCuO_2$ have shown critical temperature as high as 80 K [43], despite the fact that the constituent materials are insulators. Moreover, the behavior of the critical temperature versus the number of $CaCuO_2$ unit cells in superlattices with two unit cells of $BaCuO_2$, has strongly indicated, see figure 10, that a charge transfer mechanism can be used to explain the experimental data. In fact, in these superlattices the curve of $T_c$ versus the number of $CaCuO_2$ unit cells showed a maximum similarly to what observed in the dependence of $T_c$ upon the carrier concentration in many HTS [44].

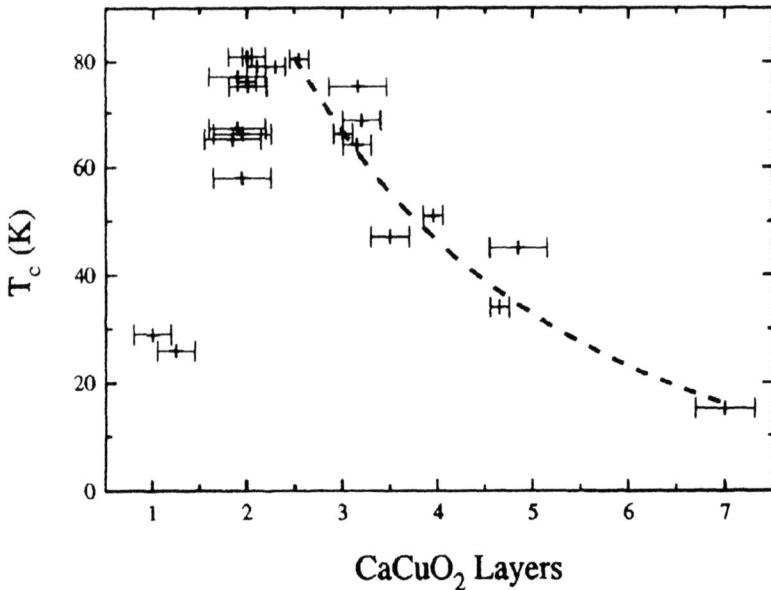

**FIGURE 10.** $T_c$ versus the number of $CaCuO_2$ unit cells in $BaCuO_2/CaCuO_2$ superlattices in which the number of $BaCuO_2$ unit cells is kept fixed to two [43].

Similar results have been obtained also in superlattices of $Bi_2Sr_2CuO_x/CaCuO_2$ realized by MBE technique [9], based on a superconducting material as the $Bi_2Sr_2CuO_x$ and on Infinite Layer such as the $CaCuO_2$. In this case, using the properties of the BSCCO compounds, the role of the charge reservoir blocks is played by the $Bi_2Sr_2CuO_x$ layers, deposited in such a way to be in the overdoped regime. Again, the curve of $T_c$ versus the number of $CaCuO_2$ unit cells in the system, figure 11, showed a defined maximum at $T_c$=41 K, which can be related to the achievement of the optimal doping on the $CuO_2$ planes.

For $Bi_2Sr_2CuO_x/CaCuO_2$ superlattices, transport property measurements have also clearly shown that the $CuO_2$ planes responsible for the superconducting effects, are those contained in the $CaCuO_2$ layers [10]. In fact, the irreversibility lines measured on these superlattices both in the perpendicular and in the parallel direction respect to the plane of the film, figure 12, have given values of the anisotropy parameter smaller than those observed in the case of single $Bi_2Sr_2CuO_x$ films.

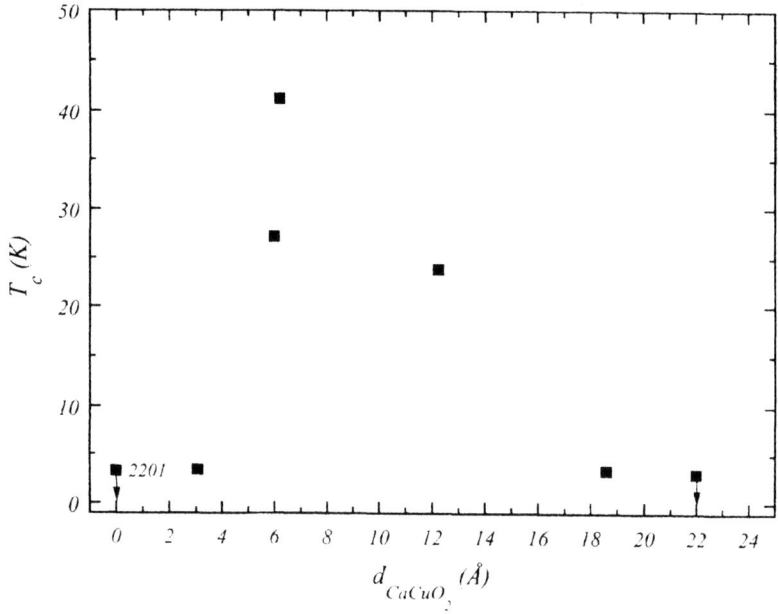

**FIGURE 11.** $T_c$ versus the number of $CaCuO_2$ unit cells in $Bi_2Sr_2CuO_x/CaCuO_2$ superlattices in which the number of $Bi_2Sr_2CuO_x$ unit cells was kept fixed to two [9]. Reprinted from [9], Copyright 1999, with permission from Elsevier Science.

If the superconducting planes in the system were those belonging to the BSCCO layers, their distance in the superlattices should increase, resulting in an increased anisotropy parameter value. On the other hand, the distance between superconducting planes in the $CaCuO_2$ cells is reduced with respect to that in BSCCO layers (3 Å against 12 Å) and this should give better coupling and reduced values of $\gamma$.

Other important observations can be made measuring the transport properties of $Bi_2Sr_2CuO_y/(Ca_xSr_{1-x})CuO_2$ superlattices, in which Ca is partially substituted with Sr. Beside the structural differences between the two atomic species, the conducting properties of the Infinite Layer block should change adding Sr in the structure, rendering less insulating the character of these spacers. This, in turns, should influence the superconducting properties of the overall system. In figure 12, the transition curves measured on three different $Bi_2Sr_2CuO_y/(Ca_xSr_{1-x})CuO_2$ superlattices (x=1, 0.5, 0) are shown. It is interesting to point out that the critical temperature of the system changes abruptly, reaching zero in the case of the sample with x=0.

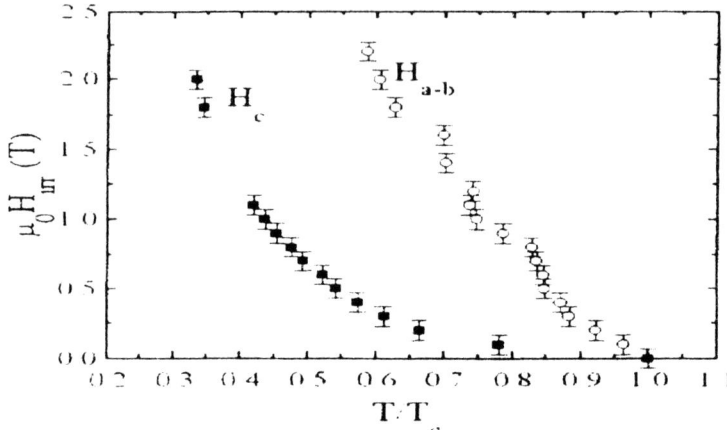

**FIGURE 12.** Irreversibility lines in the case of parallel and perpendicular externally applied magnetic field measured on a $Bi_2Sr_2CuO_x/CaCuO_2$ superlattice with one unit cell of $Bi_2Sr_2CuO_x$ and four unit cells of $CaCuO_2$ [10]. Reprinted from [10], Copyright 2000, with permission from Elsevier Science.

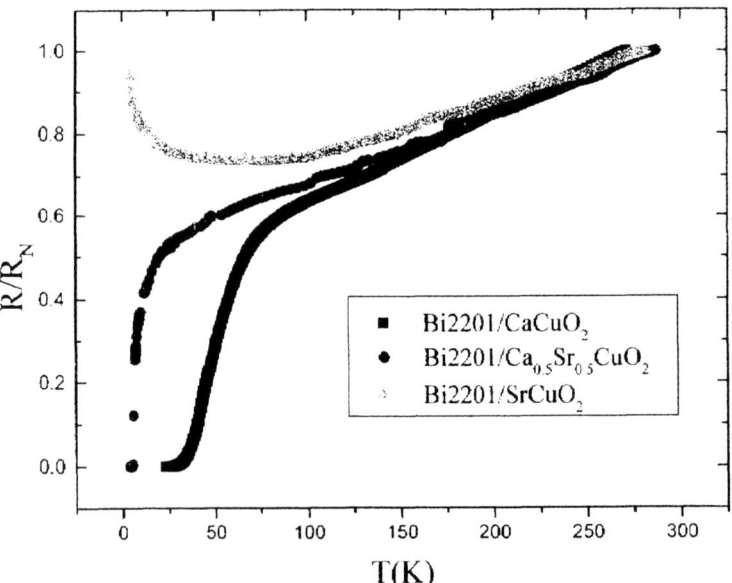

**FIGURE 13.** Transition curves for $Bi_2Sr_2CuO_y/(Ca_xSr_{1-x})CuO_2$ superlattices with x=1, 0.5, 0, and one unit cell of $Bi_2Sr_2CuO_y$ and two unit cells of $(Ca_xSr_{1-x})CuO_2$

This can be a strong indication that the transport properties of the charge reservoir block are playing a central role in the overall behavior.

## VI. CONCLUSIONS

The study of the transport properties of superconducting layered systems is central to obtain a deeper understanding of many phenomena related both to the basic physics and to the possible applications of HTS. In particular, artificial multilayers based on conventional superconductors intercalated by magnetic materials can be used as model for HTS, in studying vortex dynamics in layered system with variable anisotropy and small distance between superconducting planes. We have seen that the role of quantum fluctuations in determining the location of the irreversibility, melting related, line, can be clarified by looking to the H-T phase diagram of Nb/CuMn multilayers, and that the interplay between superconductivity and magnetism in this system can produce effects similar to those observed in HTS.

In HTS based artificial superlattices, the analysis of the transport properties allows to emphasize the role played by the charge transfer mechanism and to look for new materials with improved properties in view of specific applications.

Finally, the study of the pinning properties of naturally layered system, such as the BSCCO compounds, has pointed out the importance of considering simultaneously the dependence of the pinning potential upon Temperature, Magnetic Field and Current for a correct interpretation of the pinning mechanisms at work in these systems.

## VII. REFERENCES

1. Bednorz, J. G., and Muller, K. A., *Z. Phys B* **64**, 189, (1986).
2. Cava, R. J., *Science* **247**, 656, (1990).
3. Blatter, G., et al., *Rev. Mod. Phys.* **66**, 1125, (1994).
4. See, e.g., *Superlattices II: Native and Artificial*, edited by I.Bozovic and D.Pavuna, *Proc. SPIE* **3480**, (1998).
5. Attanasio, C., et al., *Physica C* **225**, 239, (1995).
6. Attanasio, C., et al., *Phys. Rev. B* **53**, 1087, (1996).
7. Attanasio, C., et al., *Phys. Rev. B* **62**, 14461, (2000).
8. Mercaldo, L. V., et al., *Phys. Rev. B* **53**, 4040, (1996).
9. Salvato, M., et al., *Physica C* **316**, 215, (1999).
10. Salvato, M., et al., *Physica C* **341-348**, 1903, (2000).
11. See, e.g., Tinkham, M., *Introduction to Superconductivity*, (McGraw-Hill, New York, 1996)
12. Dew-Hughes, D., *Cryogenics* **28**, 674, (1988).
13. Anderson, P. W., and Kim, Y. B., *Rev. Mod. Phys.* **36**, 39, (1964).
14. Tinkham, M., *Phys. Rev. Lett.* **61**, 1658, (1988).
15. Yeshurun, Y., and Malozemoff, A. P., *Phys. Rev. Lett.* **59**, 2202, (1988).
16. Ferrari, M. J., et al., *Phys. Rev. Lett.* **64**, 72, (1990)
17. Geshkenbein, V., et al, *Physica C* **162-164**, 239, (1989).
18. Inui, M., Littlewood, P. B., and Coppersmith, S. N., *Phys. Rev. Lett.* **63**, 2421, (1989).
19. Beasley, M. R., Labush, R., and Webb, W. W., *Phys. Rev.* **181**, 682, (1969).
20. Matsushita, T., and Otabe, E. S., *Jpn. J. Appl. Phys.* **31**, L33, (1992).
21. Likharev, K. K., *Dynamics of JosephsonJunctions and Circuits*, (Gordon and Breach, 1986), chap. 3.1.
22. Feigel'man, M. V., and Vinokur, V. M., *Phys. Rev. B* **41**, 8986, (1990).

23. Cooper, S. L., and Gray, K. E., in *Physical Properties of High Temperature Suprerconductors* Vol. IV, 63, edited by Ginzberg, D. M., (World Scientific, 1994).
24. Millis, A. J., and Monien, H., *Phys. Rev. Lett.* **70**, 2810, (1993).
25. Katano, S., et al., *Phys. Rev. B* **41**, 2009, (1990).
26. Attanasio, C., et al., *J. Appl. Phys.* **77**, 2081, (1995).
27. Attanasio, C., et al., *Phys. Rev. B* **57**, 6056, (1998).
28. Glazman, L. I., and Koshelev, A. E., *Phys. Rev. B* **43**, 2835, (1991).
29. Vinokur, V. H., Kes, P. H., and Koshelev, A. E., *Physica C* **169**, 29, (1990).
30. Houghton, A., Pelcovits, R.A., and Sudbo, A., Phys. Rev. B **40**, 6763, (1989).
31. Blatter, G., and Ivlev, B. I., *Phys. Rev. Lett.* **70**, 2621, (1993).
32. Blatter, G., et al., Phys. Rev. B **50**, 13013, (1994).
33. Blatter, G., and Ivlev, B. I., *J. Low Temp. Phys.* **95**, 365, (1994).
34. Schilling, A. et al., *Physica B* **194-196**, 1555, (1994).
35. Attanasio, C., et al., *Phys. Rev. B* **57**, 14411, (1998).
36. Radovic, Z., et al., *Phys. Rev. B* **44**, 759, (1991).
37. Wollman, D. A., et al., *Phys. Rev. Lett.* **71**, 2134, (1993).
38. Tsuei, C. C., et al., *Phys. Rev. Lett.* **73**, 593, (1994).
39. Jang, J. S., et al., *Phys. Rev. Lett.* **74**, 314, (1995).
40. Bozovic, I., et al., *J. Supercond.* **5**, 19, (1992).
41. Hybertsen, M. S., and Mattheiss, L. F., *Phys. Rev. Lett.* **60**, 1661, (1998).
42. Norton, D. P., et al., *Appl. Phys. Lett.* **65**, 2869 (1994).
43. Arciprete, F., et al., *Appl. Phys. Lett.* **71**, 959, (1997).
44. Retoux, R., et al., *Phys. Rev. B* **41**, 193, (1990).

# Spin-fluctuation Pairing in Strongly Correlated Systems

## Nikolai M. Plakida

*Joint Institute for Nuclear Research, 141980 Dubna, Russia*

**Abstract.** A microscopical theory of spin-fluctuation superconducting pairing in systems with strong electron correlations as high-temperature superconductors is formulated. Basic models with strong electron correlations: $p$-$d$ model, Hubbard and $t$-$J$ models are considered. The Dyson equations for the matrix Green functions in the Nambu notation for the Hubbard operators are derived. The equations are solved in the self-consistent Born approximation and quasi-particle spectra and superconducting $T_c$ are calculated. By comparison the results for the Hubbard and $t$-$J$ models it is shown that the antiferromagnetic superexchange interaction acting in a broad energy region with negligible retardation effects gives the major contribution to the superconducting pairing. We argue that spin-fluctuations which have been observed in many experiments in cuprates are the driving force for superconducting pairing in high-temperature superconductors.

# I  INTRODUCTION

## A  Phenomenological Approach

One of the universal and unique features of copper-oxide compounds is the antiferromagnetic (AFM) ordering of copper spins in the $CuO_2$ planes. In stoichiometric compounds, the copper ions are in the state $Cu^{2+}$ and have one hole with spin $S = 1/2$ in the $3d$ shell. A strong superexchange interaction (via oxygen ions) of the order of 1500 K between hole spins at copper sites gives rise to a three-dimensional long-range AFM order with relatively high Néel temperatures $T_N = 300 - 500$ K. Although the long range order disappears in the metallic and superconducting phases, strong dynamical spin fluctuations with a wide spectrum of excitations are observed even at temperatures above 100 K (see, e.g., [1] and recent results [2]- [4]). This fact has led to a number of hypotheses on possible electron pairing in copper-oxide compounds via magnetic degrees of freedom. The AFM spin fluctuations are also vitally important in explaining many of the anomalous properties of high-temperature superconductors in the normal phase (see, e.g., [1]).

Recent experimental evidences in favor of the $d$-wave superconducting pairing in high-$T_c$ cuprates strongly support the spin-fluctuation pairing mechanism for high-temperature superconductivity (see, for example, [5]- [7]).

Earlier the magnetic mechanism of pairing was proposed as an explanation for superconductivity in systems with heavy fermions [8]- [10] where $d$-wave pairing is observed. This was then investigated with regard to high temperature superconductors [11]. It was shown that, under the exchange of AFM paramagnons, an attraction appears in the $d$-channel and acts most effectively near the AFM instability [6]. In a more rigorous treatment, a self-consistent account of all three types of instability connected with the formation of spin or charge density waves and with the transition into the superconducting state is required [12]. The spin-fluctuation pairing mechanism on the basis of phenomenological approach was considered by several groups (see, e.g., [13]- [25]).

The retarded interaction between the quasi-particles on the two-dimensional square lattice under the exchange of the AFM paramagnons was most extensively studied by Pines et al. [16]- [19]. They considered a model interaction between two quasi-particles (QP) with spins $\sigma_1, \sigma_2$ mediated by spin fluctuation exchange

$$V_{mag}^{eff}(\mathbf{q}, \omega) = g^2 \chi(\mathbf{q}, \omega) \sigma_1 \sigma_2, \tag{1}$$

where the dynamical spin susceptibility was described by the phenomenological model

$$\chi(\mathbf{q}, \omega) = \chi_{QP}(\mathbf{q}, \omega) + \chi_{AF}(\mathbf{q}, \omega).$$

The QP term

$$\chi_{QP}(\mathbf{q}, \omega) = \bar{\chi}_0 \frac{1}{1 - i\omega\pi/\hbar\Gamma}$$

is determined by two parameters, namely, the static susceptibility $\bar{\chi}_0$ and the specific electronic energy $\hbar\Gamma/\pi$ which is of order of the Fermi energy $E_F$. The term due to AFM fluctuations

$$\chi_{AF}(\mathbf{q}, \omega) = \chi_Q \frac{1}{1 + \xi^2(\mathbf{Q} - \mathbf{q})^2 - i\omega/\omega_{sf}} \tag{2}$$

is described by the static susceptibility $\chi_Q$ for wave the vector $\mathbf{Q}$ and the typical energy of AFM fluctuation $\hbar\omega_{sf}$. According to [26] these parameters are related to the QP parameters by the relations $\chi_Q = \bar{\chi}_0(\xi/\xi_0)^2$, $\omega_{sf} = (\hbar\Gamma/\pi)(\xi_0/\xi)^z$ where $\xi(T)$ is the AFM correlation length and $z = 1$ or 2 [27]. It is assumed that $(\xi/\xi_0)^2 \gg 1$, and therefore $\chi_Q \gg \bar{\chi}_0$ and $\Gamma \gg \omega_{sf}$. The authors numerically solved the Eliashberg equations in $(\mathbf{q}, \omega)$ space and observed a superconductivity in the $d$-wave channel. The results of their calculations can be represented in the form of a formula of the BCS type

$$T_c = \alpha(\Gamma(T_c)/\pi^2)\exp[-1/\lambda(T_c)] \tag{3}$$

where $\lambda(T_c) = \eta g_{eff}^2(T_c)\chi_0(T_c)N(0)$ is the effective coupling constant ($\eta \sim 1$ and $\alpha \sim 1$ are constants) and $\Gamma(T_c)/\pi^2$ is the characteristic spin fluctuation energy.

The formula (3) gives $T_c$ which are experimentally observed for LSCO and YBCO compounds even for a sufficiently weak coupling, $\lambda \simeq 0.33 - 0.48$ due to the large value of the characteristic electron Fermi energy, $\Gamma(T_c) \simeq 0.4$ eV.

Studies of the temperature dependence of the gap $\Delta(T)$ show that near $T_c$ the gap grows rapidly when the temperature decreases and reaches maximum value $2\Delta(0)/kT_c = 6 - 8$ which agrees with experiments. In these calculations, the high $T_c$ and the $d$-wave pairing are conditioned by the quasi-two-dimensional character of the electron spectrum and by a strongly anisotropic interaction due to the AFM spin fluctuations. Strong enhancement of the spin susceptibility (2) near $\mathbf{Q} = (\pi, \pi)$ is capable of bringing about high $T_c$ in spite of strong pair-braking effects due to spin scattering.

Above the paramagnetic ground state was considered. For the AFM ground state, transverse fluctuations of the AFM order parameter ensure that an effective attraction is more favorable for the manifestation of a magnetic pairing mechanism. In this connection, we consider the concept of the spin bag proposed by Schrieffer et al. [13]. Although in copper-oxide compounds superconductivity arises for hole concentration $n_h \geq 0.05$, where long-range AFM order is already absent, due to the small superconducting correlation length $\xi_0$ we can use the theory in the region $\xi_N > \xi_0$ (where $\xi_N$ is an AFM correlation length).

The theory of Schrieffer et al. [13] is based on the assumption of a local depression by a hole of the AFM gap in the electron spectrum. As a result, a magnetic polaron – a spin bag which moves together with the cloud of spin deformation – appears. In this case, two holes, i. e., polarons, attract each other due to the overlap of their regions of deformation of the AFM order. A more detailed analysis shows that the longitudinal spin fluctuations lead to a singlet pairing with the maximum contribution coming from the $d$-wave channel. The transverse fluctuations here suppress the effective attraction.

In the paramagnetic phase, the contribution of spin fluctuations near the AFM wave vector, $\mathbf{q} = \mathbf{Q}$, leads to the appearance of a pseudo-gap in the electron spectrum near half-filling [14]. The additional exchange of the AFM spin fluctuations for two quasi-particles – spin bags – decreases their energy, similar to the case of long-range AFM order, and results in their mutual attraction. In this case, the effective interaction potential $V(\mathbf{k}\mathbf{k}')$ is attractive in the range of small momentum transfer $\mathbf{q} = \mathbf{k} - \mathbf{k}',$ and of a repulsive character for large $\mathbf{q} \simeq \mathbf{Q}$. Therefore, in this case, the symmetry of the superconducting order parameter, $\Delta(\mathbf{k})$, and the value of $T_c$ strongly depend on the form of the Fermi surface for quasi-particles.

Thus, the pairing mechanism due to the two-magnon scattering processes proposed by Kampf and Schrieffer [14] is of a universal character but requires complicated self-consistent calculations of the quasi-particle spectrum and spin fluctuations in order to yield quantitative results. These calculations were performed under the assumption of weak or intermediate couplings and within the framework of the one-band Hubbard model (for $U \leq 4t$). A generalization of the theory for the case of strong coupling and for the multi-band $p$-$d$ model is important to compare its conclusions with experiments in the copper-oxide superconductors.

# B  Microscopical Theory

As was first pointed out by Anderson [28], AFM exchange in copper oxides is of the superexchange origin and can be properly described only in the framework of models with strong electron correlations like the Hubbard model or the $t - J$ model [29,30]. The one-band Hubbard model reads [31]:

$$H = -t \sum_{\langle ij \rangle \sigma} (a^+_{i\sigma} a_{j\sigma} + \text{H.c.}) + U \sum_i n_{i\uparrow} n_{i\downarrow} \qquad (4)$$

where the first term is the kinetic energy of electrons with an effective transfer integral $t$ for the pair of nearest neighbor sites, $\langle ij \rangle$, $i > j$ , and $U$ is the Coulomb single-site energy. For a large enough $U > U_c \simeq W$ ( $W$ is the band width for free electrons, $W = 8t$ for a two-dimensional square lattice) the model (4) describes the Hubbard-Mott insulator at half filling, $n = 1$. In the strong correlation limit, $U \gg t$, when only singly occupied sites are taken into account since doubly occupied site costs a large additional energy $U$ the model (4) can be reduced to the $t - J$ model (see Sec. II B):

$$H_{t-J} = -t \sum_{\langle ij \rangle, \sigma} (\tilde{a}^+_{i\sigma} \tilde{a}_{j\sigma} + \text{H.c.}) + J \sum_{\langle ij \rangle} (\mathbf{S}_i \mathbf{S}_j - \frac{1}{4} n_i n_j) \qquad (5)$$

Here the projected electron operators $\tilde{a}^+_{i\sigma} = a^+_{i\sigma}(1 - n_{i-\sigma})$ act in the subspace without double occupancy and $n_i = n_{i\uparrow} + n_{i\downarrow}$ is the number operator for electrons. The spin operators are $S^\alpha_i = (1/2) \sum_{s,s'} \tilde{a}^+_{is} \sigma^\alpha_{s,s'} \tilde{a}_{is'}$. The second term describes the spin-1/2 Heisenberg antiferromagnet with the exchange energy $J = 4t^2/U$ for the nearest neighbors.

So starting from the original model (4) for fermions with the Coulomb repulsion $U$ we have arrived to the $t - J$ model for electron-like particles but with non-fermionic commutation relations with an additional exchange interaction acting between them. While the latter dynamical type interaction is well known in the many-body theory the kinematic interaction caused by the non-fermionic commutation relations presents great obstacles for employing the conventional diagram technique. Presently there are no rigorous methods to study models with strong electron correlations (4), (5) for dimensions $d \geq 2$. The recently developed dynamical mean field theory (for review see [32,33]) which is exact in the infinite dimensions is unable to treat non-local correlations as e.g., exchange interaction, and to study the non-local $d$-wave type superconducting pairing. To overcome this deficiency a cluster dynamical mean field theory was elaborated [34], [35].

## *Weak Correlation Limit*

In the weak correlation limit, $U \ll t$, a perturbation approach for the Hubbard model (4) can be applied. In a number of studies (see [36]-[39] and references therein) a self-consistent set of equations in the fluctuation-exchange approximation (FLEX) [40] for the one-electron Green functions and the spin and charge susceptibility were numerically solved. The FLEX approximation was also applied for the 3-band $p-d$ model with Coulomb repulsion $U$ on copper sites [41] and for electronically doped cuprates [42]. Within this approach a qualitative description of the copper-oxide materials was obtained at moderate and large doping regimes at intermediate values of the Coulomb repulsion, $U \leq W = 4t$. However, starting from the Fermi liquid picture, the theory fails to describe the underdoped regime with strong correlations and large values of the Coulomb repulsion, $U \geq W$, close to the insulating (and AFM) state. In that region higher order corrections to the effective interactions mediated by spin fluctuations are important [43].

Let us briefly discuss these studies of the strong coupling equations for the Hubbard model in the weak correlation limit [36]- [39]. In the framework of FLEX, a self-consistent system of equations for the diagonal and off-diagonal one-electron Green functions was written as [37]:

$$G(p,\omega_n) = \frac{i\omega_n Z(p,\omega_n) + X(p,\omega_n)}{[i\omega_n Z(p,\omega_n)]^2 - X^2(p,\omega_n) - \phi^2(p,\omega_n)}, \tag{6}$$

$$F(p,\omega_n) = \frac{\phi(p,\omega_n)}{[i\omega_n Z(p,\omega_n)]^2 - X^2(p,\omega_n) - \phi^2(p,\omega_n)} \tag{7}$$

with $X(p,\omega_n) = \epsilon_p - \mu + \xi(p,\omega_n)$. The renormalization parameter $Z(p,\omega_n)$, the energy shift $\xi(p,\omega_n)$, and the gap parameter $\phi(p,\omega_n)$ obey the standard Eliashberg-type equations with interactions mediated by spin and charge fluctuations:

$$V_s(q,\omega_n) = \frac{3}{2}U^2 \frac{\chi_0^s}{1 - U\chi_0^s} - \frac{1}{2}U^2 \chi_0^s, \tag{8}$$

$$V_c(q,\omega_n) = \frac{1}{2}U^2 \frac{\chi_0^c}{1 + U\chi_0^c} - \frac{1}{2}U^2 \chi_0^c. \tag{9}$$

The irreducible spin, $\chi_0^s(q,\omega_n)$, and charge, $\chi_0^c(q,\omega_n)$, susceptibility are calculated self-consistently by using the Green functions (6), (7):

$$\chi_0^{s,c}(q,\omega_n) = -T \sum_{k,m} [G(k+q,\omega_n+\omega_m)G(k,\omega_m)$$

$$\pm F(k+q,\omega_n+\omega_m)F(k,\omega_m)]. \tag{10}$$

where $+(-)$ stands for for $\chi_0^s(\chi_0^c)$. Numerical solutions of the self-consistent set of equations for intermediate values of the Coulomb repulsion, $U \leq 4t$, for the full momentum- and frequency dependent renormalization parameter $Z(p,\omega)$, energy shift $\xi(p,\omega)$, and gap function $\Delta(p,\omega) = \phi(p,\omega)/Z(p,\omega)$ have been obtained. The gap function has been found to have $d_{(x^2-y^2)}$ symmetry. In the vicinity of AFM instability near half filling a superconducting temperature $T_c$ reaches the values of order $0.02t \simeq 60K$. Quasiparticle and spin fluctuation spectra in the normal and superconducting state, calculated also on the real frequency axis by Dahm and Tewordt [39], reveal strong dependence of the spectra on temperature and strong feedback effects for the effective interactions arising from the spin and charge susceptibility as the superconducting gap opens. These direct numerical solutions of the strong coupling Eliashberg equations for the Hubbard model in the weak correlation limit unambiguously confirm a possibility of $d$-wave pairing mediated by spin fluctuation exchange. However, superconductivity exists only in a narrow range of $U$, very close to the AFM instability in this model. Higher order corrections to the effective interactions (8), (9) may be also important [43].

Thus, the electronic interaction mediated by AFM spin fluctuations can lead to a superconducting pairing. The magnetic pairing mechanism is most effectively manifested in the $d$-channel where strong local Coulomb repulsion is suppressed. In spite of the pair-breaking effects due to spin scattering which considerably decreases $T_c$ strong enhancement of spin susceptibility near AFM instability results in $d$-wave pairing.

An exceptionally strong suppression of $T_c$ in all cuprates by nonmagnetic Zn impurities which are also destroy the local magnetic order and in this way suppress the effective spin-fluctuation pairing provides a "smoking gun" for the magnetic mechanism [17].

## *Strong Correlation Limit*

*Finite cluster calculations.* To deal with the strong correlation limit for the Hubbard model and the $t - J$ model a number of numerical methods for finite clusters has been developed (for a review see [44,45], and a recent publication [46]). These studies show strong AFM correlations which lead to the formation of the $d_{x^2-y^2}$ pairing correlations (see, for example, [6]). However, the finite cluster calculations due to known limitations can give only restricted information. For instance, as was shown recently by applying the constrained-path Monte Carlo method [47] to the two-dimensional Hubbard model, small lattice sizes and weak interactions show $d_{x^2-y^2}$ pairing correlations while with increasing lattice size or interaction they vanish. Close results was also obtained for the $t$-$J$ model [48]. So to prove the superconducting pairing in the strong correlation limit an analytical treatment is highly demanded.

*Slave-boson representation.* To take into account the kinematic interaction caused by the projected character of the electron operators in the $t - J$ model (5), different types of slave-boson (-fermion) technique were proposed (see [49] - [51] and references therein). In the simplest version of the slave-boson theory the projected electron operators $\tilde{a}_{i\sigma}^+$ in (5) are replaced by a product of fermion (spinon) $f_{i\sigma}^+$ and boson (holon) $b_i$ operators: $\tilde{a}_{i\sigma}^+ = f_{i\sigma}^+ b_i$. However, to reduce the enlarged Hilbert space of the spinon-holon particles to the physical one of the projected electron operators one has to introduce a local constraint

$$q_i = b_i^+ b_i + \sum_\sigma f_{i\sigma}^+ f_{i\sigma} = 1. \qquad (11)$$

In the MFA the local constraint are substituted by a global one, $q = \langle q_i \rangle = 1$ that reduces the problem to free spinons and bosons in the mean-field. The decoupling of spin and charge degrees of freedom in the slave-boson theory results also in quite a different description of the superconducting phase transition by two order parameters, a bose-condensate expectation value $\langle b_i \rangle$, and a spinon pairing function, $\langle f_{i\sigma}^+ f_{i,-\sigma}^+ \rangle$, instead of one order parameter for the original model, $\langle \tilde{a}_{i\sigma}^+ \tilde{a}_{i,-\sigma}^+ \rangle$. The results of such enlargement of the Hilbert space is difficult to evaluate. To treat the constraint in a systematic way a large-$N$ expansion was proposed [52,50] with $N/2$ being a number of states (orbitals) at a lattice cite. In that approach the local constraint (11) are relaxed to a much weaker one

$$q_i = b_i^+ b_i + \sum_\sigma f_{i\sigma}^+ f_{i\sigma} = N/2. \qquad (12)$$

By using the $1/N$ expansion [51], the $d$-wave superconducting instability induced by the superexchange interaction was obtained in the $t$-$J$ model close to half filling. However, the kinematic interaction in the large-$N$ limit becomes weak and it is difficult to extrapolate the obtained results to real systems at $N = 2$ with strong kinematic restriction.

*Diagram technique.* A rigorous method to treat the unconventional commutation relations for the projected electron operators in the $t$-$J$ model (5) is based on the Hubbard operator technique [53] since in this representation the local constraint (11) are rigorously implemented by the Hubbard operator algebra. A superconducting pairing due to the kinematic interaction in the Hubbard model in the limit of strong electron correlations ($U \to \infty$) was first obtained by Zaitsev and Ivanov [54] who studied the two-particle vertex equation by applying a diagram technique for Hubbard operators [55]. However, they studied only the lowest order diagrams which are equivalent to the MFA for a superconducting order parameter and obtained only the $s$-wave pairing.

Systematic investigation of the $t$-$J$ model within the diagram technique was performed by Izymov et al. [56]. In the framework of this approach spin fluctuations and superconducting pairing in the $t$-$J$ model in the limit of small $J$ were studied in [57], [58]. The first order diagrams for the self-energy reproduced the results of the MFA in [59], [60]. In calculations of the second order diagrams only the exchange interaction $J$ was taken into account while the corresponding contributions due to the kinematic interaction of the order $t^2$ was disregarded. As a result, estimations in the weak coupling limit for the Eliashberg equation for a three-dimensional model near AFM instability resulted in quite a low superconducting $T_c$.

*Equation of motion method.* The equation of motion method for the thermodynamic Green functions [61] was used in several studies. The method appears to be much more simple then the diagram technique while it accurately reproduces the results of the latter within the second order of an effective interaction for the self-energy. We will discuss applications of this method to the $t$-$J$ and Hubbard models in Sec. III and Sec. IV in details while here point out only the main publications concerning the method. At first superconducting pairing was considered for the Hubbard model in [62], [63] by applying a decoupling procedure within the equation of motion method for the Green functions. The $s$-wave pairing due to the kinematic interaction proposed in [54] was obtained in the mean-field type approximation. However, as was shown later [59], [60] the $s$-wave pairing in the limit of strong correlations violates an exact requirement of no single-site pairs and should be disregarded. In that papers the BCS type theory for the $t$-$J$ model was developed within the formally exact projection technique [64] for the Green functions in terms of the Hubbard operators. It was proved that the $d$-wave superconducting pairing mediated by the exchange interaction $J$ is stable and has high $T_c \simeq 0.1t$ for $J \simeq 0.4$.

The theory of one-electron spectra and superconducting pairing beyond MFA was formulated for the $t$-$J$ model by Plakida and Oudovenko [65] by applying the projection technique for the Green functions in terms of the Hubbard operators. A formally exact representation for the Dyson equation with the self-energy as the many-particle Green functions was derived. To close the system of equations the non-crossing approximation for the matrix self-energy operator was employed. A numerical solution of the self-consistent system of equations revealed narrow QP peaks for the one-electron spectral density near the Fermi surface (FS) and a broad incoherent band below the FS. The incoherent part of the spectra resulted in the nonzero occupation numbers $N(\mathbf{k})$ throughout the Brillouin zone and only a small decrease in the occupation numbers $N(\mathbf{k})$ at FS crossing which is characteristic for electron systems with strong correlations. By using the obtained one-electron spectral density a direct numerical solution of the linearized gap equation was performed which showed the $d$-wave like symmetry for the gap function with high $T_c$ at optimal doping. These results supported the previous calculations for the $t$-$J$ model in the polaron representation [66].

The equation of motion method for the Green functions in terms of the projected electron operators and spin operators was also used in [67] to study the electron spectra in the normal state at zero temperature for the $t$-$J$ model. The projection technique for the Green functions in terms of the Hubbard operators was employed in [68] to consider the superconducting pairing in the $t$-$J$ model in MFA with electron-phonon interaction. A generalization of this method for the two-band $p-d$ Hubbard model was considered in [69] for a normal state and in [70], [71] for a superconducting state.

*Hubbard model in mean-field approximation.* Mean-field type approximations of the strong correlation limit $U > W$ of the original Hubbard model (4), within the projection technique for GF, were reported in a number of papers. Roth method as a decoupling scheme for calculations of higher order correlation functions was used in [72] where the $d$-wave superconducting pairing was obtained. To escape uncontrollable decoupling, Mancini et al. [73] imposed kinematic restrictions to the Hubbard operators (the Pauli principle) to obtain superconducting solutions of the $s$- and $d$-type symmetry. Later on, Stanescu et al. [74] have showed that the superconducting pairing for conventional electron (hole) pairs in MFA is absent and they have proposed a pairing mechanism for the composite excitations $\eta_{ij,\sigma} = c_{i\sigma} n_{js}$ (with $(i,j)$ being the nearest neighbors). The use of the Roth decoupling procedure for the calculation of the anomalous correlation function $\langle c_{i\sigma} c_{i-\sigma} n_{js} \rangle$ and of the Pauli principle for the calculation of higher order normal correlation functions were found to emerge in high $T_c$ superconductivity in the $d$-wave channel for the composite excitations [74]. However, the onset temperature following from these calculations strongly depends on the scheme of approximation [72], [73] or on the particular solution [74]. In the limit of large $U$, extremely small $T_c$ values are obtained as a consequence of the fact that the anomalous correlations are induced by pairing at the same lattice site, hence they should disappear in the limit $U \to \infty$. In addition, in the MFA, the self-energy operator caused by kinematic and exchange interactions was ignored, in spite of the fact that it results in finite life-time effects and gives a substantial contribution to the renormalization of QP spectrum in the normal state. Finally, as shown for the $t-J$ model, [65] the self-energy of the anomalous GF is also responsible for the non-local spin-fluctuation $d$-wave superconducting pairing. A detailed discussions of the superconducting $d$-wave pairing is given in Sect.IV.

*Low electron density.* To go beyond the MFA the $t$-matrix approach was used in [75] to study the limit of a low electron density in the $t$-$J$ model. Various forms of electron pairing at low temperatures including the $d$-wave instability were observed at large values of $J/t > 1$. By combining a generalized Lanczos scheme with the variation Monte Carlo method, a low hole density region in the $t$-$J$ model was studied and a finite $d$-wave long-range superconducting order was observed below the phase-separation region [76].

*Baym-Kadanoff Green function.* Another method is based on the Baym-Kadanoff variation technique for the Green functions in terms of the Hubbard operators [77]. The method was used in [78,79] in the limit of large pseudo-spin degrees of freedom $N$ to consider superconducting pairing in the $t$-$J$ model. It was observed that in the lowest order of $1/N$ there is a strong compensation of different contributions to the pairing interaction and for $J = 0$ the superconducting $T_c$ is extremely small. For a finite $J$ the $d$-wave superconducting instability mediated by the exchange and spin- and charge-fluctuations was obtained below $T_c \simeq 0.01t$. In [79] it was also proved that the results in the Hubbard operator technique differs from that one in the slave-boson representation even in the same order of $1/N$ expansion due to different Hilbert spaces used in two cases. However, in the limit of large $N$ the kinematic interaction, as in the slave-boson method, is suppressed and this approach, being rigorous in the limit $N \to \infty$, is difficult to extrapolate to real spin systems with $N = 2$.

*Spin polaron model.* Important analytical results for the $t$-$J$ model were obtained in the limit of small concentrations of holes when one can consider the motion of holes on the AFM background within the spin polaron model [80], [81]. A number of studies of this model (see [82]- [86] and references therein) predicts that a doped hole dressed by strong AFM spin fluctuations can propagate coherently as a QP, spin polaron, in a narrow band of order $J$ even for a finite hole doping [84], [85]. It is quite natural to suggest that the same spin fluctuations could mediate a superconducting pairing of the spin polarons.

This problem was treated in the framework of the weak coupling BCS formalism. Simple phenomenological models of QP with numerically evaluated spectrum and effective pairing interaction in the atomic limit [87] or mediated by AFM magnon exchange [88] were studied. By applying the rigid band approximation high superconducting transition temperature was obtained for the $d$-wave pairing. However, since the pairing spin-fluctuation energy $J$ is of the same order as the QP band width the weak coupling BCS equation is inadequate to treat the problem. Also the rigid band approximation for QP fails to describe a strong doping dependence of the QP spectrum [84,85].

A self-consistent numerical treatment of the strong coupling equations for spin polarons in the $t$-$J$ model was given in [66]. It was observed a strong renormalization of the QP hole spectrum due to spin-fluctuations and $d$- wave pairing with maximum $T_c \simeq 0.01t$ at optimal concentration of doped holes $\delta \simeq 0.2$ (see Sect.III A). In [89] the authors cannot obtain superconducting instability within the spin polaron model. They have to introduce an additional electron-phonon coupling to obtain superconducting $d$-wave pairing. However, a two-sublattice representation used in the spin polaron model can be rigorously proved only for a small doping within the long-range AFM state. At a moderate doping one has to consider a paramagnetic (spin-rotational invariant) state in the $t$-$J$ model. In that case the Hubbard operator representation for the $t$-$J$ model is the most convenient (see Sec. III B).

## II MODELS WITH STRONG CORRELATIONS

### A Two-Band Singlet-Hole Model

An important contribution to the understanding of the low-energy electronic spectrum of copper-oxides has been done by Zhang and Rice [90] who pointed out the remarkable role of singlet formation for doped oxygen holes due to strong Coulomb correlations. Starting from the original $p$-$d$ model proposed by Emery [91] and Varma et al. [92] they have derived an effective one-band $t$-$J$ model for the copper oxide plane. The appearance of singlet quasiparticle states inside the $p$-$d$ gap was proved by different methods based on exact diagonalization [93], cluster calculations [94,95], projection technique [96,97] and other calculations. It should be noted that the commonly used local density approximation [98] cannot describe such a singlet band formation due to insufficient treatment of electronic correlations.

The original Zhang-Rice procedure has been considerably improved in terms of the so-called cell-perturbation method [99–101]. It results in simple analytical formulas to reduce the $p$-$d$ model to an effective singlet-triplet model. Applying the equation of motion method for GF it has been found, however, that the coupling between singlet and triplet band is very small [100]. So we are left with a one-band $t$-$J$ like model [102].

The reduction to the $t$-$J$ model has some disadvantages, however, which may be listed as follows. First of all, it neglects completely the charge fluctuations between the singlet and the one-hole $d$-like states. As will be seen in the following, such a hybridization modifies the spectrum in a considerable way and cannot be neglected. Secondly, the $t$-$J$ model does not reproduce in a correct way the spectral weight transfer which occurs with doping [103,104]. More important difficulties may arise if one would like to calculate the pairing induced by the exchange interaction. The $J$-term in the Hamiltonian describes this effect only in a static and instantaneous way. That is quite similar to the BCS Hamiltonian for the electron-phonon mechanism. It does not deal with the dynamical effects of the exchange interaction. Below we consider the reduction of the original two-band $p$-$d$ model to an effective Hubbard model for one-hole $d$-like states and two-hole singlet states. In distinction to the conventional model one ends up with an asymmetric model with different hopping integrals for the upper Hubbard (singlet) subband and the lower Hubbard ($d$-like hole) subband. That may give a good starting point to study the differences between electron and hole doping. The use of such a Hamiltonian has also some practical reasons since several techniques are easier to apply for a Hubbard model than for a $t$-$J$ model. A two-band spectrum for $d$-like holes and Zhang-Rice-singlets in the normal state and superconducting pairing will be considered in the Sec. IV.

We consider the original $p$-$d$ model [91,92] in the limit of strong correlations at the copper sites, $U_d \to \infty$. By taking into account only the most important terms it can be written in a simple form:

$$H = \epsilon_d \sum_{i,\sigma} \tilde{d}^+_{i\sigma} \tilde{d}_{i\sigma} + \epsilon_p \sum_{m,\sigma} p^+_{m\sigma} p_{m\sigma} + t \sum_{i,m,\sigma} S_{im} \left( \tilde{d}^+_{i\sigma} p_{m\sigma} + \text{H.c.} \right), \qquad (13)$$

where $\tilde{d}^+_{i\sigma} = d^+_{i\sigma}(1 - n_{i\bar{\sigma}})$ denotes the creation of a hole on a copper site $i$ provided there is no other hole with spin $\bar{\sigma} = -\sigma$, $p^+_{m\sigma}$ creates a hole on an oxygen site $m$ and $S_{im} = \pm 1$ depending on the position of the site $m$ in the unit cell $i$ in agreement with [90]. The hopping $p$-$d$ integral $t$ and the difference between the hole energy levels for oxygen and copper, $\Delta = \epsilon_p - \epsilon_d$, are the only two parameters in the model (13).

To derive the singlet band it is reasonable to simplify this Hamiltonian (13) further, following mainly [99,100]. Let us summarize the main steps: introducing the symmetric combination of oxygen operators $p^{(s)}_{i\sigma}$ in the unit cell $i$ according to [90] we can define the orthogonal Wannier states $c_{i\sigma}$ by the equation

$$p_{i\sigma}^{(s)} = \frac{1}{2} \sum_m S_{im} p_{m\sigma} = \sum_j \nu_{ij}\, c_{j\sigma} \,. \tag{14}$$

The overlapping parameters

$$\nu_{jl} = \frac{1}{N} \sum_k \sqrt{1 - \frac{1}{2}(\cos k_x + \cos k_y)}\; e^{ik(j-l)}$$

decrease rapidly, but non-exponentially with the distance $(j-l)$: $\nu_0 = \nu_{jj} \simeq 0.96$, $\nu_1 = \nu_{j\, j\pm a_{x/y}} \simeq -0.14$, and $\nu_2 = \nu_{j\, j\pm a_x \pm a_y} \simeq -0.02$. (see [99]). Taking into account in the following inter-cell perturbation theory *all* the Wannier coefficients $\nu_{jl}$ we would obtain an artificial sharp cusp in the dispersion curve at the $\Gamma$-point ($k = (0,0)$) [100]. That is a known artefact of the Wannier representation for non-isolated bands [105]. Therefore, we consider in the present calculation the Wannier coefficients $\nu_0$, $\nu_1$ and $\nu_2$, only. Using the orthogonal Wannier states $c_{i\sigma}$ in (14) we can write the Hamiltonian in the form:

$$H = \sum_{i\sigma} \{\epsilon_d\, \tilde{d}_{i\sigma}^+ \tilde{d}_{i\sigma} + \epsilon_p\, c_{i\sigma}^+ c_{i\sigma} + V_0\, (\tilde{d}_{i\sigma}^+ c_{i\sigma} + \text{H.c.})\}$$
$$+ \sum_{i\neq j\sigma} V_{ij}\, \{\tilde{d}_{i\sigma}^+ c_{j\sigma} + \text{H.c.}\}\,, \tag{15}$$

where $V_{ij} = 2t\, \nu_{ij}$ and $V_0 = 2t\, \nu_0$. Since $|V_{ij}| \ll V_0$ one can consider the last inter-cell term in (15) as a small perturbation to the intra-cell part given by the first term in (15).

As was shown in [99,100] the first intra-cell part can be diagonalized within one unit cell. That gives for the lowest one-hole $d$-type state

$$|D_\sigma\rangle = \cos\theta_1\, d_\sigma^+ |0\rangle - \sin\theta_1\, c_\sigma^+ |0\rangle\,, \tag{16}$$

and the two-hole state with the lowest energy is the singlet state

$$|\psi\rangle = \cos\theta_2\, \frac{1}{\sqrt{2}}\, (d_\uparrow^+ c_\downarrow^+ - d_\downarrow^+ c_\uparrow^+)\, |0\rangle - \sin\theta_2\, c_\uparrow^+ c_\downarrow^+ |0\rangle\,, \tag{17}$$

where the vacuum state $|0\rangle$ has no holes and $\tan 2\theta_1 = 2V_0/\Delta$, $\tan 2\theta_2 = 2\sqrt{2} V_0/\Delta$. The corresponding one-hole $E_D$ and two-hole energies $E_\psi$ are given by:

$$E_D = \frac{1}{2}(\epsilon_d + \epsilon_p) - \frac{1}{2}\sqrt{\Delta^2 + 4V_0^2}\,, \tag{18}$$

$$E_\psi = \frac{1}{2}(\epsilon_d + 3\epsilon_p) - \frac{1}{2}\sqrt{\Delta^2 + 8V_0^2}\,. \tag{19}$$

Another one-hole $p$-type state has higher energy then the $d$-type state (16) and can be neglected in the subspace of one-hole states. The singlet states (17) are the lowest among the two-hole states and have to be filled first with doping. At small doping we can also neglect the triplet states with the energy $E_\tau = (\epsilon_d + \epsilon_p)$ since the mixing between singlet and triplet bands is rather small [100].

By introducing the Hubbard operators in the subspace of the one-hole states $|D_\sigma\rangle$ (16) and the singlet states $|\psi\rangle$ (17)

$$X_i^{\sigma\sigma} = |D_{i\sigma}\rangle\langle D_{i\sigma}|, \quad X_i^{\sigma 0} = |D_{i\sigma}\rangle\langle 0|, \qquad (20)$$

$$X_i^{22} = |\psi_i\rangle\langle\psi_i|, \quad X_i^{20} = |\psi_i\rangle\langle 0|, \quad X_i^{2\sigma} = |\psi_i\rangle\langle D_{i\sigma}|, \qquad (21)$$

we can write the intra-cell part of the effective Hamiltonian in the form

$$H_0 = E_D \sum_{i\sigma} X_i^{\sigma\sigma} + E_\psi \sum_i X_i^{22}. \qquad (22)$$

By projecting the original $p$- and $d$-operators in the inter-cell part of the Hamiltonian (15) onto the subspace of one- and two-hole states (16), (17)

$$c_\sigma^+ = 2\sigma A_c X^{2\bar\sigma} - \sin\theta_1 X^{\sigma 0} \qquad \tilde{d}_\sigma^+ = 2\sigma A_d X^{2\bar\sigma} + \cos\theta_1 X^{\sigma 0} \qquad (23)$$

where $2\sigma = \pm 1$ we can write the inter-cell term in (15) in the form

$$H_{int} = \sum_{i\neq j\sigma}\{t_{ij}^\psi X_i^{2\bar\sigma} X_j^{\sigma 2} + t_{ij}^D X_i^{\sigma 0} X_j^{0\sigma} + 2\sigma t_{ij}^{\psi D}(X_i^{2\bar\sigma} X_j^{0\sigma} + X_i^{\sigma 0} X_j^{\bar\sigma 2})\}. \qquad (24)$$

The effective hopping parameters are given by

$$\begin{aligned}&t_{ij}^\psi = V_{ij} K_{\psi\psi}, && K_{\psi\psi} = 2 A_d A_c, \\ &t_{ij}^D = V_{ij} K_{DD}, && K_{DD} = -2\sin\theta_1 \cos\theta_1, \\ &t_{ij}^{\psi D} = V_{ij} K_{\psi D}, && K_{\psi D} = A_c \cos\theta_1 - A_d \sin\theta_1,\end{aligned} \qquad (25)$$

with the coefficients

$$A_d = -\frac{1}{\sqrt{2}}\sin\theta_1 \cos\theta_2, \quad A_c = \sin\theta_1 \sin\theta_2 + \frac{1}{\sqrt{2}}\cos\theta_1 \cos\theta_2. \qquad (26)$$

Therefore, the total Hamiltonian of the two-band model for $d$-like holes and singlets takes the form:

$$H = H_0 + H_{int} - \mu N, \qquad (27)$$

where we have introduced the chemical potential $\mu$ and the number operator

$$N = \sum_i N_i = 2\sum_i X_i^{22} + \sum_{i,\sigma} X_i^{\sigma\sigma}. \qquad (28)$$

It is easy to prove that the number operator (28) acting in the subspace of one- and two-hole states (16) and (17) satisfies the necessary condition $[N, H_0 + H_{int}] = 0$.

This condition is not satisfied for the number operator for the original $p$- and $d$-holes in (13) written in terms of the Hubbard operators given by (23) since the higher energy one– and two–hole states were ignored in the model Hamiltonian (27).

To prove the importance of the hybridization term in (24) between $D$-holes and singlets we estimate the hopping parameters (25) for the case of strong intra–cell coupling: $2t = \Delta = \epsilon_p - \epsilon_d$. Direct calculation in (25) gives us

$$K_{\psi\psi} \simeq -0.477, \quad K_{DD} \simeq -0.887, \quad K_{\psi D} \simeq 0.834 \,. \tag{29}$$

For these parameters we can also estimate renormalization of the single-site energies (18), (19):

$$E_D \simeq \epsilon_d - 0.58\Delta, \quad E_\psi \simeq \epsilon_p + \epsilon_d - 0.95\Delta, \quad E_\psi - 2E_D = \tilde{\Delta} \simeq \Delta + 0.2\Delta \,. \tag{30}$$

These estimations shows that the $\psi$-$D$ hybridization is rather strong being much larger than the singlet–triplet coupling $K_{\psi\tau}$ considered in [100]: $K_{\psi D} >> | K_{\psi\tau} | \simeq 0.08$. In the limit of small $p$-$d$ hybridization, $t/\Delta \to 0$, while all the coefficients $K_{\psi\psi}$, $K_{DD}$ and $K_{\psi\tau}$ tend to zero, $K_{\psi D}$ has a finite value, $K_{\psi D} \to 1/\sqrt{2}$. Therefore, in this limit the effective hopping parameter $t_{ij}^{\psi D}$ vanishes linearly with $V_{ij} \propto t$ while all the others, $t_{ij}^{\psi}, t_{ij}^{D}$, are proportional to $(t^2/\Delta)$. As a result, the inter–band $\psi$-$D$ hybridization gives a rather strong renormalization of the singlet band dispersion being of the same order of $(t^2/\Delta)$ as the original one $t_{ij}^{\psi}$.

The two–band Hubbard–like model (27) in comparison with the original $p$-$d$ model (13) takes into account the formation of a new singlet band for doped $p$–holes due to strong Coulomb correlations on copper sites. The appearance of the singlet band due to many–body correlations was proved by different methods (see [93]- [97]) while it cannot be obtained in the framework of standard band–structure calculations based on the local–density approximation [98]. On the other hand, the one–band $t$-$J$ model for singlets [90] considers only the one–hole $d$–like band in a static way by an effective exchange $J$-term and it neglects charge–carrier fluctuations that prevents a proper study of charge transport in the $CuO_2$ planes. In general, the two–band ($p$-$d$) model (27) can be considered as the standard Hubbard model with one–hole and two–hole (lower and upper) subbands but with asymmetric hopping parameters (25) and the single–site correlation energy given by the charge–transfer gap $\Delta = \epsilon_p - \epsilon_d$ instead of Coulomb repulsion $U$. Therefore we can apply to this model well–developed methods in the theory of the standard Hubbard model as will be shown below in Sec. IV.

## B  Reduced t-J Model

In this Section we show how the asymmetric Hubbard model can be reduced to the one subband $t$-$J$ model. We write the former as follows

$$H = H_0 + H_1 + V = E_1 \sum_{i,\sigma} X_i^{\sigma\sigma} + E_2 \sum_i X_i^{22}$$
$$+ \sum_{i \neq j, \sigma} \{t_{ij}^{11} X_i^{\sigma 0} X_j^{0\sigma} + t_{ij}^{22} X_i^{2\sigma} X_j^{\sigma 2} + 2\sigma t_{ij}^{12}(X_i^{2\bar{\sigma}} X_j^{0\sigma} + \text{H.c.})\}, \qquad (31)$$

where $X_i^{nm} = |in\rangle\langle im|$ are the Hubbard operators for the four states $n, m = |0\rangle, |\sigma\rangle, |2\rangle = |\uparrow\downarrow\rangle$, $\sigma = \pm 1/2$, $\bar{\sigma} = -\sigma$. The energy parameters are given by $E_1 = E_D - \mu$ and $E_2 = E_\psi - 2\mu$, and the renormalized charge transfer energy $\tilde{\Delta} = E_\psi - 2E_D$ (see Eqs. (18), (19), (30)).

The Hubbard operators obey the multiplication rules on the same lattice site:

$$X_i^{\alpha\nu} X_i^{\mu\beta} = \delta_{\nu,\mu} X_i^{\alpha\beta}, \qquad (32)$$

which results in the commutatuion relations:

$$\left[ X_i^{\alpha\mu}, X_j^{\nu\beta} \right]_\pm = \delta_{ij} \left( \delta_{\nu,\mu} X_i^{\alpha\beta} \pm \delta_{\beta\alpha} X_i^{\nu\mu} \right). \qquad (33)$$

The plus sign refers to the Fermi-like Hubbard operators (as, e. g., $X_i^{0\sigma}$) while the minus sign stands if one or both operators are Bose-like as the spin or number operators. The constraint of no double occupancy at any lattice site $i$ rigorously preserves by the completeness relation

$$X_i^{00} + X_i^{\sigma\sigma} + X_i^{\bar{\sigma}\bar{\sigma}} + X_i^{22} = 1. \qquad (34)$$

Let us assume now that the hybridization term $V = T_{21} + T_{12}$ is a small perturbation in comparison with the energy $E_2 - 2E_1 = \tilde{\Delta}$. In that case we can exclude the hybridization by the Schrieffer-Wolf transformation [106]:

$$\tilde{H} = e^S H e^{-S} \simeq H_0 + H_1 + V + [S, H_0] + [S, V] + \frac{1}{2}[S, [S, H_0]] + \ldots \qquad (35)$$

where we took into account only the lowest order terms in respect to $t_{ij}^{\alpha\beta}/\Delta$. In this approximation $S$ is defined by the equation

$$V + [S, H_0] = 0, \qquad (36)$$

which has the solution in the form $S = \lambda(T_{21} - T_{12}) = -S^+$, with the parameter $\lambda$ given by the equation:

$$T_{21} = \lambda[H_0, T_{21}]. \qquad (37)$$

By using the condition (36), we obtain the equation for the reduced Hamiltonian:

$$\tilde{H} \simeq H_0 + H_1 + \frac{1}{2}[S, V] = H_0 + H_1 + H_J.$$

For the effective exchange interaction we get

$$H_J = \frac{1}{2}[S, V] = \frac{\lambda}{2}[T_{21} - T_{12}, T_{21} + T_{12}] = \lambda[T_{21}, T_{12}]. \tag{38}$$

By direct calculation we get from Eq. (37):

$$T_{21} = \lambda(E_2 - 2E_1)T_{21}, \quad \text{or,} \quad \lambda = (E_2 - 2E_1)^{-1}.$$

Performing commutations of the Hubbard operators in Eq. (38) according to Eq. (33) we obtain

$$[T_{21}, T_{12}] = \sum_{i\neq j,\sigma} \sum_{l\neq m,\sigma'} 2\sigma t^{21}_{ij} 2\sigma' t^{12}_{lm}[X^{2\bar\sigma}_i X^{0\sigma}_j, X^{\sigma'0}_l X^{\bar\sigma'2}_m]$$

$$= \sum_{i\neq j,\sigma} \sum_{l\neq j} t^{21}_{ij} t^{12}_{jl} \{X^{2\bar\sigma}_i X^{\bar\sigma 2}_l (X^{00}_j + X^{\sigma\sigma}_j) - X^{\sigma 0}_l X^{0\sigma}_i (X^{22}_j + X^{\bar\sigma\bar\sigma}_j)$$

$$- X^{2\bar\sigma}_i X^{\sigma 2}_l X^{\bar\sigma\sigma}_j + X^{2\bar\sigma}_i X^{\bar\sigma 0}_l X^{02}_j + X^{\bar\sigma 0}_l X^{0\sigma}_i X^{\sigma\bar\sigma}_j - X^{\sigma 2}_l X^{0\sigma}_i X^{20}_j\}. \tag{39}$$

In the conventional $t$-$J$ model only the two-site terms are taken into account. They are given by the sums in Eq. (39) with $l = i$ that results in the exchange interaction

$$H_J^{(2)} = \frac{1}{4} \sum_{i\neq j,\sigma} J_{ij} \left(X^{\sigma\bar\sigma}_i X^{\bar\sigma\sigma}_j - X^{\sigma\sigma}_i X^{\bar\sigma\bar\sigma}_j + X^{20}_i X^{02}_j + X^{22}_i X^{00}_j\right), \tag{40}$$

with the exchange energy $J_{ij} = 4(t^{21}_{ij})^2/(E_2 - 2E_1) = 4(t^{21}_{ij})^2/\tilde\Delta$.

The three site term is given by the sums in Eq. (39) with $l \neq i$:

$$H_J^{(3)} = \sum_{i\neq j,\sigma} \sum_{l\neq j, l\neq i} \frac{t^{21}_{ij} t^{12}_{jl}}{(E_2 - 2E_1)} \{X^{2\bar\sigma}_i X^{\bar\sigma 2}_l (X^{00}_j + X^{\sigma\sigma}_j) - X^{\sigma 0}_l X^{0\sigma}_i (X^{22}_j + X^{\bar\sigma\bar\sigma}_j)$$

$$- X^{2\bar\sigma}_i X^{\sigma 2}_l X^{\bar\sigma\sigma}_j + X^{2\bar\sigma}_i X^{\bar\sigma 0}_l X^{02}_j + X^{\bar\sigma 0}_l X^{0\sigma}_i X^{\sigma\bar\sigma}_j - X^{\sigma 2}_l X^{0\sigma}_i X^{20}_j\}. \tag{41}$$

By introducing the number and spin operators

$$N_i = \sum_\sigma X^{\sigma\sigma}_i + 2X^{22}_i, \quad S^\sigma_i = X^{\sigma\bar\sigma}_i, \quad S^z_i = \sum_\sigma \sigma X^{\sigma\sigma}_i, \tag{42}$$

where $\sigma = \pm 1/2$ we can write the exchange interaction in the conventional form

$$H_J^{(2)} = \frac{1}{2} \sum_{i\neq j,\sigma} J_{ij} \left(\mathbf{S}_i \mathbf{S}_j - \frac{1}{4} N_i N_j\right) = \sum_{\langle i,j \rangle} J_{ij} \left(\mathbf{S}_i \mathbf{S}_j - \frac{1}{4} N_i N_j\right), \tag{43}$$

where we omitted two last terms in Eq. (40) with transitions between empty states $|0\rangle$ and doubly occupied states $|2\rangle$ and in the last formula introduced summation over only the nearest neighbor lattice site pairs $\langle i, j \rangle$.

The Hamiltonian (31) after the Schrieffer-Wolf transformation (35) describes one-hole and singlet subbands without hybridization. Neglecting 3-sites term (41) it reads:

$$H_{t-J} = E_1 \sum_{i,\sigma} X_i^{\sigma\sigma} + E_2 \sum_i X_i^{22} + \sum_{i\neq j,\sigma} \{t_{ij}^{11} X_i^{\sigma 0} X_j^{0\sigma} + t_{ij}^{22} X_i^{2\sigma} X_j^{\sigma 2}\}$$

$$+\frac{1}{4} \sum_{i\neq j,\sigma} J_{ij} (X_i^{\sigma\bar\sigma} X_j^{\bar\sigma\sigma} - X_i^{\sigma\sigma} X_j^{\bar\sigma\bar\sigma} + X_i^{20} X_j^{02} + X_i^{22} X_j^{00}). \tag{44}$$

In Sec. III B we study superconducting pairing in the one-subband model (44).

## III  SUPERCONDUCTING PAIRING IN THE t-J MODEL

### A   Low-Doping: Spin Polaron Model

*Spin Polaron Model*

For a small concentration of holes when the long range AFM order is preserved or at least strong AFM correlations for nearest-neighbors still govern the hole motion, the $t - J$ model (5) can be reduced to a more simple spin polaron model as it has been proposed in [80] and [81]. As discussed in the Introduction, a number of studies of this model [82]- [86] predicts that a doped hole dressed by strong AFM spin fluctuations can propagate coherently as a quasiparticle – spin polaron, with weight $Z_k \simeq J/t$. Besides a narrow quasiparticle band of order $J$ there is a broad incoherent band at higher energies. It is important to investigate a possibility of superconducting pairing of the spin polarons by the same spin fluctuations. Below we present the results of a full self-consistent solution of the strong coupling equations both for the normal and superconducting states [66].

To consider superconducting pairing of spin polarons we have to take into account explicitly a two-sublattice structure for the Heisenberg AFM. By introducing two sublattices with spin up ($i \in\uparrow$) and spin down ($i \in\downarrow$) we define the hole spinless fermion operators for two sublattices by the equation:

$$\tilde c_{i\uparrow} = h_i^+, \ \tilde c_{i\downarrow} = h_i^+ S_i^+ \ (i \in\uparrow); \quad \tilde c_{i\downarrow} = f_i^+, \ \tilde c_{i\uparrow} = f_i^+ S_i^- \ (i \in\downarrow) \tag{45}$$

where $S_i^+$, $S_i^-$ are spin operators on the corresponding sublattices. In the linear spin-wave approximation (LSWA) the exchange part of the Hamiltonian (5) can be written as (see, e.g., [83]:

$$H_J = \sum_q \omega_q(\alpha_q^+ \alpha_q + \beta_q^+ \beta_q) + E_0^J \tag{46}$$

where $\alpha_q^+(\alpha_q)$ and $\beta_q^+(\beta_q)$ are the magnon creation (annihilation) operators coupled with the spin lowering operators on two sublattices in LSWA: $S_i^+ \simeq a_i, (i \in \uparrow)$, $S_i^+ \simeq b_i^+, (i \in \downarrow)$ by the Bogoliubov canonical transformation:

$$a_{\mathbf{k}} = v_k \alpha_{\mathbf{k}} + u_k \beta_{-\mathbf{k}}^+, \quad b_{\mathbf{k}} = v_k \beta_{\mathbf{k}} + u_k \alpha_{-\mathbf{k}}^+, \tag{47}$$

$$u_k = \left(\frac{1+\nu_k}{2\nu_k}\right)^{1/2}, \quad v_k = -\mathrm{sign}(\gamma_k)\left(\frac{1-\nu_k}{2\nu_k}\right)^{1/2} \tag{48}$$

with $\nu_k = \sqrt{1-\gamma_k^2}$, $\gamma_k = \frac{1}{2}(\cos ak_x + \cos ak_y)$. The spin-wave energy is given by $\omega_k = SzJ(1-\delta)^2 \nu_k$ with $\delta$ being a hole concentration and $z = 4$ is the number of the nearest neighbors. In derivation of the exchange part of the Hamiltonian (46) the contact interaction between holes was taken into account only in the mean-field approximation that results in the renormalization of the magnon energy proportionally to the factor $(1-\delta)^2$.

By employing the two sublattice representation (45) for holes and the LSWA we get the following expression for the hopping part of the Hamiltonian (5):

$$H_t = \sum_k (\epsilon_k - \mu)(h_k^+ h_k + f_k^+ f_k) + H_{int},$$
$$H_{int} \simeq \sum_{kq}(h_k^+ f_{k-q}[g(k,q)\alpha_q + g(q-k,q)\beta_{-q}^+] + \mathrm{H.c.}), \tag{49}$$

where the next nearest neighbor hopping energy $\epsilon_k = 4t' \cos ak_x \cos ak_y$ and the vertex of magnon-hole interaction is given by

$$g(k,q) = \frac{zt}{\sqrt{N/2}}(u_q \gamma_{k-q} + v_q \gamma_k). \tag{50}$$

The summation over wave-vectors in (46), (49) and below is restricted to $N/2$ points in the AFM Brillouin zone. The chemical potential $\mu$ should be calculated self-consistently as a function of hole concentration $\delta$ and temperature $T$ from the equation:

$$\delta = \langle h_i^+ h_i \rangle + \langle f_i^+ f_i \rangle. \tag{51}$$

## Green Functions

To discuss a singlet superconducting pairing within the spin polaron model (46), (49) we consider the equation of motion method for the matrix GF

$$\hat{G}(k, t-t') = \langle\langle \Psi_k(t) | \Psi_k^+(t') \rangle\rangle, \tag{52}$$

where Zubarev notation [61] is used for the Nambu operators:

$$\Psi_k = \begin{pmatrix} \tilde{c}_{k\uparrow} \\ \tilde{c}^+_{-k\downarrow} \end{pmatrix} = \begin{pmatrix} h^+_k \\ f_{-k} \end{pmatrix}, \quad \Psi^+_k = \begin{pmatrix} \tilde{c}^+_{k\uparrow} & \tilde{c}_{-k\downarrow} \end{pmatrix} = \begin{pmatrix} h_k & f^+_{-k} \end{pmatrix}. \tag{53}$$

By differentiating the GF (52) in respect to two times $t$ and $t'$ we obtain the following Dyson equation as described in [85]:

$$\hat{G}(k,\omega)^{-1} = \omega\hat{\tau}_0 + (\epsilon_k - \mu)\hat{\tau}_3 - \hat{\Sigma}(k,\omega), \tag{54}$$

where $\hat{\tau}_0$ and $\hat{\tau}_3$ are the Pauli matrices. The self–energy operator is given by the many–particle GF:

$$\hat{\Sigma}(k,\omega) = \langle\langle [\Psi_k, H_{int}] \,|\, [H_{int}, \Psi^+_k] \rangle\rangle. \tag{55}$$

The solution of the Dyson equation (54) can be written in the Eliashberg notation as

$$\hat{G}(k,\omega) = \frac{\omega Z_k(\omega)\hat{\tau}_0 + (\chi_k(\omega) - \epsilon_k)\hat{\tau}_3 + \phi_k(\omega)\hat{\tau}_1}{(\omega Z_k(\omega))^2 - (\chi_k(\omega) - \epsilon_k)^2 - \phi_k(\omega)^2}, \tag{56}$$

where

$$\omega(1 - Z_k(\omega)) = \frac{1}{2}[\Sigma_{hh}(k,\omega) + \Sigma_{ff}(k,\omega)],$$

$$\chi_k(\omega) = \frac{1}{2}[\Sigma_{hh}(k,\omega) - \Sigma_{ff}(k,\omega)], \tag{57}$$

$$\phi_k(\omega) = \Sigma_{hf}(k,\omega) = (\Sigma_{fh}(k,\omega))^*,$$

and $\Sigma_{ff}(k,\omega) = -\Sigma_{hh}(k,-\omega)$.

To obtain self–consistent equations for the GF (56) we employ the self–consistent Born approximation (SCBA) (or the noncrossing diagram approximation) which has been proved to be quite reasonable in calculation of the one–hole spectrum in the normal state. In SCBA we get the following equations for the self-energies of the GF (56),

$$\Sigma_{hh}(k,i\omega_n) = -T\sum_q\sum_m G_{hh}(q,i\omega_m)\lambda_{11}(k, k-q \,|\, i\omega_n - i\omega_m), \tag{58}$$

$$\Sigma_{hf}(k,i\omega_n) = -T\sum_q\sum_m G_{hf}(q,i\omega_m)\lambda_{12}(k, k-q \,|\, i\omega_n - i\omega_m), \tag{59}$$

where the Matsubara frequencies $\omega_n = \pi T(2n+1)$. The interaction functions are

$$\lambda_{11}(k, q \,|\, i\omega_\nu) = g^2(k,q)D(q,-i\omega_\nu) + g^2(q-k,q)D(-q,i\omega_\nu), \tag{60}$$

$$\lambda_{12}(k,q \mid i\omega_\nu) = g(k,q)g(q-k,q)\{D(q,-i\omega_\nu) + D(-q,i\omega_\nu)\}. \tag{61}$$

Here the diagonal magnon GF $D(q,\omega) = \langle\langle \alpha_q \mid \alpha_q^+ \rangle\rangle_\omega$ can be written as

$$D(q,\omega) = \frac{\omega + \omega_q + \Pi_{22}(q,\omega)}{[\omega - \omega_q - \Pi_{11}(q,\omega)][\omega + \omega_q + \Pi_{22}(q,\omega)] + \mid \Pi_{12}(q,\omega) \mid^2}. \tag{62}$$

Within the SCBA the polariazation operators are as follows

$$\Pi_{11}(q,i\omega_\nu) = T \sum_k \sum_m \{g^2(k,q) G_{hh}(k,i\omega_m) G_{hh}(k-q, i\omega_\nu + i\omega_m) -$$

$$- g(k,q)g(q-k,q) G_{hf}(k,i\omega_m) G_{hf}(k-q, i\omega_\nu + i\omega_m)\}, \tag{63}$$

$$\Pi_{12}(q,i\omega_\nu) = T \sum_k \sum_m \{g(k,q)g(q-k,q) G_{hh}(k,i\omega_m) G_{hh}(k-q, i\omega_\nu + i\omega_m) -$$

$$- g^2(k,q) G_{hf}(k,i\omega_m) G_{hf}(k-q, i\omega_\nu + i\omega_m)\}, \tag{64}$$

where $\Pi_{22}(q,i\omega_\nu) = \Pi_{11}(-q,-i\omega_\nu)$.

## Numerical Results and Discussion

To calculate superconducting temperature $T_c$ we can study only the linearized system of the equations for the normal GF in (56)

$$G_{hh}(k,i\omega_n) = \frac{1}{i\omega_n + \epsilon_k - \mu - \Sigma_{hh}(k,i\omega_n)}, \tag{65}$$

and for the superconducting gap function (59)

$$\phi(k,i\omega_n) = \sum_{q,m} \lambda_{12}(k, k-q \mid i\omega_n - i\omega_m) \phi(q,i\omega_m) G_{hh}(q,i\omega_m) G_{hh}(q,-i\omega_m). \tag{66}$$

At first a self-consistent calculation of the normal GF (65) with the self-energy operator (58) has been done for a given concentration of holes

$$\delta = \frac{1}{2} + \frac{2T}{N} \sum_k \sum_{n=0}^{\infty} G(k,i\omega_n).$$

Then the gap equation (66) has been solved and the leading eigenvalues for the pairing eigenfuctions $\phi(q,i\omega_n)$ have been obtained. The calculations were performed for the hole concentrations in the range $0.02 \leq \delta \leq 0.35$ and for the parameters of the spin polaron model: $J = 0.4$ and $t' = \pm 0.1$ (all energies here and below are measured in units of $t$).

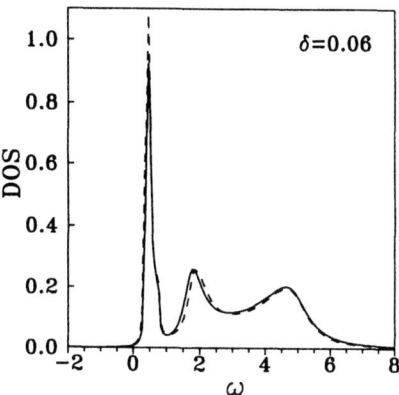

**FIGURE 1.** The DOS for the hole concentration $\delta = 0.06$ with the renormalized (solid line) and unrenirmalized (dotted line) magnon spectra [66].

In the numerical calculations the fast Fourier transformation have been used [107] for a finite mesh of 64×64 **k**-points in the full Brillouin zone ($0 \leq k_x, k_y \leq 1$), in units of $2\pi/a$, and 200-700 points for Matsubara frequencies with a constant cut $\omega_{max} = 10t$ in the summation over it. Usually 10 – 30 iterations were needed to obtain a solution for the self energy with an accuracy of order 0.001. To calculate the hole spectral function

$$A(k,\omega) = -\frac{1}{\pi}\text{Im}\,\langle\langle h_k \mid h_k^+ \rangle\rangle_{\omega+i\epsilon} \qquad (67)$$

and the density of states (DOS)

$$A(\omega) = \frac{1}{N}\sum_k A(k,\omega) \qquad (68)$$

a Padé approximation was used for analytical continuation from Matsubara points on the imaginary axis.

Calculations of the spin polaron quasiparticle spectrum have been done at finite temperature $T = 0.012$ that is slightly higher then the maximal superconducting temperature discussed below. Computations of the hole spectral functions $A(k,\omega)$ at different **k**-points show that for small hole concentrations $\delta \leq 0.10$ there are no much differences for spectral functions calculated with renormalized and unrenormalized magnon energy in the interaction function, Eq. (60). In Fig. 1 [66] we compare the results of calculations for hole density of states $A(\omega)$ with renormalized (solid line) and unrenormalized (dashed line) magnon spectra for $\delta = 0.06$ that demonstrates a small effect of magnon renormalization.

However, at higher hole concentrations a negative contribution to the magnon spectral density appears at $\omega < 0$ due to excitation of electron-hole pairs. That results in negative values for hole spectral functions in the incoherent part of the spectrum. This negative contribution develops at first for long wavelength magnons as has been pointed out already by [84]. Since the main quasiparticle peak at $\mathbf{k} = (\pi/2, \pi/2)$ does not change much in shape with doping the picture of spin polarons as stable quasiparticle seems to be relevant even at moderate hole concentrations. This robust behavior of spin polarons with doping can be explained by a small size of the polarons in comparisons with AFM correlation length due to the large exchange energy.

The quasiparticle energy defined as $E(k,0) = \epsilon(k) + \text{Re } \Sigma(k,0)$ has been calculated for $t' = 0, \pm 0.1$ at different hole concentrations. With doping the hole quasiparticle spectrum does not change much in shape but the band width increases substantially. So, the rigid band approximation adopted by [87] and [88] is inadequate. For $t' = 0$ or $t' = -0.1$ the minimum of the dispersion curves is at the points $(\pm \pi/2, \pm \pi/2)$ in the BZ that results in a 4-pocket like form for the Fermi surface (FS) at low hole concentration. With increasing hole concentration a transition from the 4-pocket like FS to a large one occurs quite sharply. However, for $t' = +0.1$ the minimum of the dispersion curves shifts to points of the type $(0, \pm \pi)$ at the BZ boundary. The corresponding FS appears to be large even at small concentration of doped holes.

Temperature dependence of the momentum distribution for holes in the spin polaron model was investigated in some details in [85] where it was shown that the Fermi surface washed out at some temperature of the order $T_d \simeq 1.5 J \delta \simeq 0.6 t \delta$. So at quite low temperatures $T \approx 0.01 t$ considered here the Fermi surface does not change much with temperature. It should be also pointed out that high density of states in the present calculations (see Fig. 1) results from a narrowing of a free electron band width due to strong correlations (spin polaron formation).

To study the symmetry of the superconducting order parameter and to evaluate the superconducting temperature $T_c$ the linearized equation for the pairing energy $\phi(k, i\omega_n)$ (66) has been considered. Looking for even functions of wave-vector $\mathbf{k}$ that are realized in the singlet pairing one obtains only $d$-type symmetry for the gap function. In Fig. 2 [66] $\mathbf{k}$-dependence of the gap function $\Delta(\mathbf{k}, \omega = 0)$, where $\Delta(\mathbf{k}, \omega) = \phi(\mathbf{k}, \omega)/Z(\mathbf{k}, \omega)$, in the quarter of the full BZ for $\delta = 0.25$, $t' = -0.1$, $J = 0.4t$ and $T/T_c \approx 0.8$ is shown.

It has the typical $d$-wave symmetry with two ridges resulted from sharp changes of the interaction function at the FS. The frequency dependence of Re $\Delta(\mathbf{k}, \omega)$ and Im $\Delta(\mathbf{k}, \omega)$ has the characteristic for the pairing theory cut off energy of order $J \simeq 0.4$ away from the FS [66]. Therefore we have really a strong coupling limit for spin polaron pairing where all quasiparticles are paired contrary to the weak coupling in conventional superconductors. Quite large values of Im $\Delta(k, \omega)$ near the FS [66] also differ from the results for conventional superconductors.

By examining the temperature dependence of the highest egenvalue in the Eq. (66) at different hole concentrations we can find the temperature when it passes through unity for decreasing $T$. At this temperature the normal state becomes unstable due to singlet pairing of quasiparticle – spin polarons on different sublattices.

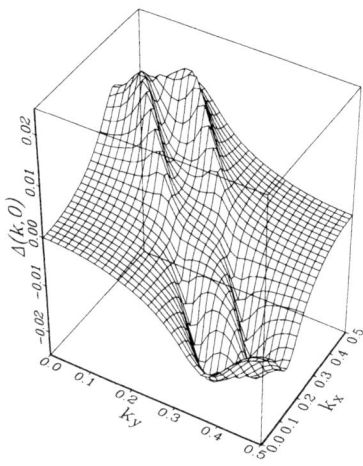

**FIGURE 2.** k-dependence of the gap function $\Delta(\mathbf{k},\omega = 0)$ in the quarter of the full BZ for $\delta = 0.25$ and $T/T_c \approx 0.8$ [66].

In Fig. 3 [66] the dependence of superconducting temperature on hole concentrations is shown for $t' = -0.1$ (solid line), $t' = 0$ (dashed line), and $t' = +0.1$ (solid line with dots). The position of the maximum of $T_c$ at $\delta \simeq 0.25; 0, 20; 0.15$ for $t' = -0.1; 0; +0.1$, respectively, is explained by crossing the maximum of the density of hole states by the Fermi level at given hole concentrations. This results are quite different with the monotone increasing of $T_c$ obtained within the weak coupling limit from the BCS equation in [88].

It was also observed that $T_c$ increases with $J$ but saturates at $T_c \simeq 0.025$ for $J \simeq 3$. However, a large drop of $T_c$ for $J > 3$ observed in small cluster calculations near phase separation [44] was not obtained . But the latter phenomenon is beyond the scope of the discussed theoretical approach. It should be also mentioned calculations done in [89] where superconducting pairing of spin polarons was obtained only by taking into account an additional hole-phonon interaction.

The presented calculations are based on the two sublattice representation, Eq. (45), which can be rigorously proved only for AFM background at low hole concentration  However, it is believed that spin polarons dressed by AFM spin fluctuations continue to be the relevant quasi-particles even at higher hole concentration when the AFM correlation length is much larger then the superconducting correlation length or the polaron size, both of which are of the order of several lattice constants.

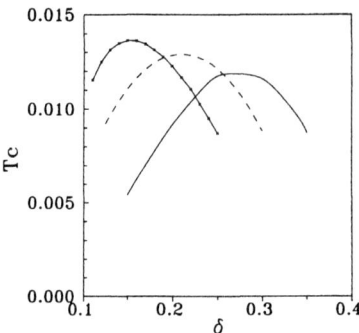

**FIGURE 3.** The superconducting temperature $T_c$ against hole concentration $\delta$ for $J = 0.4$ and $t' = -0.1$ (solid line), $t' = 0$ (dashed line), $t' = +0.1$ (solid line with dots) [66].

## B  Moderate Doping: t-J Model

### Dyson Equation for the t-J Model

In the present Section we consider superconducting pairing in the $t$-$J$ model at moderate doping in paramagnetic phase [65] when the two-sublattice representation considered above becomes less reliable. By using the Hubbard operator representation we write Hamiltonian of the $t - J$ model (44) for electrons in the form

$$H_{t-J} = -\sum_{i \neq j, \sigma} t_{ij} X_i^{\sigma 0} X_j^{0\sigma} - \mu \sum_{i\sigma} X_i^{\sigma\sigma} + \frac{1}{4} \sum_{i \neq j, \sigma} J_{ij} \left( X_i^{\sigma\bar{\sigma}} X_j^{\bar{\sigma}\sigma} - X_i^{\sigma\sigma} X_j^{\bar{\sigma}\bar{\sigma}} \right). \quad (69)$$

Here we introduced the electron hopping energy for the nearest neighbors, $t_{ij} = t$, and the second neighbors, $t_{ij} = t'$, on a 2D square lattice, and the exchange interaction $J_{ij} = J$ for the nearest neighbors. These parameters can be considered as independent ones if starting from a more realistic for copper oxides three-band $p$-$d$ model we reduce it to the $t$-$J$ model as described in Sec. II B. Since we consider here only one singly occupied subband with $X_i^{22} = 0$ the completeness relation Eq. (34) takes the form:

$$X_i^{00} + X_i^{\sigma\sigma} + X_i^{\bar{\sigma}\bar{\sigma}} = 1. \quad (70)$$

The chemical potential $\mu$ is to be calculated from the equation for the average number of electrons

$$n = \langle N_i \rangle = \sum_\sigma \langle X_i^{\sigma\sigma} \rangle. \quad (71)$$

To discuss the superconducting pairing within the model (69) we consider the matrix GF

$$\hat{G}_{ij,\sigma}(t-t') = \langle\langle \Psi_{i\sigma}(t) | \Psi_{j\sigma}^+(t') \rangle\rangle \qquad (72)$$

in terms of the Nambu operators:

$$\Psi_{i\sigma} = \begin{pmatrix} X_i^{0\sigma} \\ X_i^{\bar{\sigma}0} \end{pmatrix}, \qquad \Psi_{i\sigma}^+ = \begin{pmatrix} X_i^{\sigma 0} & X_i^{0\bar{\sigma}} \end{pmatrix},$$

where Zubarev's notation for the anticommutator GF (72) is used [61].

By differentiating the GF (72) over the time $t$ we get the following equation

$$\omega \hat{G}_{ij\sigma}(\omega) = \delta_{ij} \hat{Q}_\sigma + \langle\langle \hat{Z}_{i\sigma} | \Psi_{j\sigma}^+ \rangle\rangle_\omega, \qquad (73)$$

where $\hat{Z}_{i\sigma} = [\Psi_{i\sigma}, H]$. The matrix

$$\hat{Q}_\sigma = \langle\{\Psi_{i\sigma}, \Psi_{j\sigma}^+\}\rangle = \begin{pmatrix} Q_\sigma & 0 \\ 0 & Q_{\bar{\sigma}} \end{pmatrix}$$

in a spin-singlet state does not depend on spin and is given by the correlation function

$$Q_\sigma = \langle X_i^{00} + X_i^{\sigma\sigma} \rangle = Q = 1 - n/2, \qquad (74)$$

which depends only on the average number of electrons (71).

Now, we project the many–particle GF in (73) on the single–electron GF by introducing the *irreducible* part of $\hat{Z}_{i\sigma}$ operator

$$\langle\langle \hat{Z}_{i\sigma} | \Psi_{j\sigma}^+ \rangle\rangle = \sum_l \hat{E}_{il\sigma} \langle\langle \Psi_{l\sigma} | \Psi_{j\sigma}^+ \rangle\rangle + \langle\langle \hat{Z}_{i\sigma}^{(irr)} | \Psi_{j\sigma}^+ \rangle\rangle,$$

$$\langle\{\hat{Z}_{i\sigma}^{(irr)}, \Psi_{j\sigma}^+\}\rangle = \langle \hat{Z}_{i\sigma}^{(irr)} \Psi_{j\sigma}^+ + \Psi_{j\sigma}^+ \hat{Z}_{i\sigma}^{(irr)} \rangle = 0, \qquad (75)$$

that results in the equation for the frequency matrix

$$\hat{E}_{ij\sigma} = \langle\{[\Psi_{i\sigma}, H], \Psi_{j\sigma}^+\}\rangle Q^{-1}. \qquad (76)$$

To calculate the matrix (76) we use the equation of motion for the Hubbard operators:

$$\left(i\frac{d}{dt} + \mu\right) X_i^{0\sigma} = -\sum_l t_{il} B_{i\sigma\sigma'} X_l^{0\sigma'} + \frac{1}{2}\sum_l J_{il}(B_{l\sigma\sigma'} - \delta_{\sigma\sigma'})X_i^{0\sigma'}, \qquad (77)$$

which follows from the commutation relations Eq. (33). Here we introduced the operator

$$B_{i\sigma\sigma'} = (X_i^{00} + X_i^{\sigma\sigma})\delta_{\sigma'\sigma} + X_i^{\bar{\sigma}\sigma}\delta_{\sigma'\bar{\sigma}} = (1 - \frac{1}{2}N_i + \sigma S_i^z)\delta_{\sigma'\sigma} + S_i^{\bar{\sigma}}\delta_{\sigma'\bar{\sigma}}, \qquad (78)$$

where in the second equation we have used the completeness relation, Eq. (70), and definition of number (71) and spin (42) operators. The Bose-like operator (78) describes electron scattering on spin and charge fluctuations caused by the non-fermionic commutation relations for the Hubbard operators (the first term in (77) – the kinematic interaction) and by the exchange spin-spin interaction (the second term in (77)).

The frequency matrix (76) defines the zero–order GF in the generalized MFA

$$\hat{G}^0_{ij\sigma}(\omega) = Q\{\omega\hat{\tau}_0\delta_{ij} - \hat{E}_{ij\sigma}\}^{-1}. \qquad (79)$$

By writing the equation of motion for the irreducible part of the GF in (75) with respect to the second time $t'$ for the right–hand side operator $\Psi^+_{j\sigma}(t')$ and performing the same projection procedure as in (75) we can obtain the Dyson equation for the GF (72) in the form

$$\hat{G}_{ij\sigma}(\omega) = \hat{G}^0_{ij\sigma}(\omega) + \sum_{kl} \hat{G}^0_{ik\sigma}(\omega)\, \hat{\Sigma}_{kl\sigma}(\omega)\, \hat{G}_{lj\sigma}(\omega), \qquad (80)$$

where the self–energy operator $\hat{\Sigma}_{kl\sigma}(\omega)$ is defined by the equation

$$\hat{T}_{ij\sigma}(\omega) = \hat{\Sigma}_{ij\sigma}(\omega) + \sum_{kl} \hat{\Sigma}_{ik\sigma}(\omega)\, \hat{G}^0_{kl\sigma}(\omega)\, \hat{T}_{lj\sigma}(\omega)\,. \qquad (81)$$

Here the scattering matrix is given by

$$\hat{T}_{ij\sigma}(\omega) = Q^{-1} \langle\!\langle \hat{Z}^{(irr)}_{i\sigma} \mid \hat{Z}^{(irr)+}_{j\sigma} \rangle\!\rangle_\omega\, Q^{-1}\,. \qquad (82)$$

From Eq. (81) it follows that the self-energy operator is given by the *proper* part of the scattering matrix (82) that has no parts connected by the single zero-order GF (79):

$$\hat{\Sigma}_{ij\sigma}(\omega) = Q^{-1} \langle\!\langle \hat{Z}^{(irr)}_{i\sigma} \mid \hat{Z}^{(irr)+}_{j\sigma} \rangle\!\rangle^{(prop)}_\omega\, Q^{-1}\,. \qquad (83)$$

Eqs. (79), (80) (83) give an exact representation for the single–electron GF (72). To calculate it, however, one has to introduce an approximation for the many–particle GF in the self-energy matrix (83) which describes inelastic scattering of electrons on spin and charge fluctuations.

## *Self-Consistent Equations*

In the **k**-representation for the GF

$$G^{\alpha\beta}_\sigma(k,\omega) = \sum_j G^{\alpha\beta}_{oj\sigma}(\omega)\, e^{-i\mathbf{kj}},$$

we get for the zero-order GF (79):

$$\hat{G}_\sigma^{(0)}(k,\omega)^{-1} = \{\omega\hat{\tau}_0 - (E_k^\sigma - \tilde{\mu})\hat{\tau}_3 - \Delta_k^\sigma\hat{\tau}_1\}Q^{-1}, \tag{84}$$

where $\hat{\tau}_0$, $\hat{\tau}_1$, $\hat{\tau}_3$ are the Pauli matrices. The energy of QP $E_k^\sigma$, the renormalized chemical potential $\tilde{\mu} = \mu - \delta\mu$ and the gap function $\Delta_k^\sigma$ in the MFA (76) are given by

$$E_k^\sigma = -\epsilon(k)Q - \epsilon_s(k)/Q - \frac{2J}{N}\sum_q \gamma(k-q)N_{q\sigma}, \tag{85}$$

where we have introduced $J(q) = 4J\gamma(q)$ and $\epsilon(k) = t(k) = 4t\gamma(k) + 4t'\gamma'(k)$, $\epsilon_s(k) = 4t\gamma(k)\chi_{1s} + 4t'\gamma'(k)\chi_{2s}$, with $\gamma(k) = (1/2)(\cos a_x q_x + \cos a_y q_y)$ and $\gamma'(k) = \cos a_x q_x \cos a_y q_y$,

$$\delta\mu = \frac{1}{N}\sum_q \epsilon(q)N_{q\sigma} - 2J(\frac{n}{2} - \frac{\chi_{1s}}{Q}), \tag{86}$$

$$\Delta_k^\sigma = -\frac{2}{NQ}\sum_q g(q, k-q)\langle X_{-q}^{0\bar{\sigma}} X_q^{0\sigma}\rangle, \tag{87}$$

where the interaction is given by the function:

$$g(q, k-q) = t(q) - \frac{1}{2}J(k-q). \tag{88}$$

There are two contributions in the gap equation (87): the **k**-independent kinematic interaction $t(q)$ and the exchange interaction $J(k-q)$. The kinematic interaction gives no contribution to the $d$-wave pairing in MFA, Eq. (87)(see [59]), and we disregard it in the following equations. The average number of electrons in Eqs. (85), (86) in the **k**-representation is written in the form:

$$n_{k,\sigma} = \langle X_k^{\sigma 0} X_k^{0\sigma}\rangle = QN_{k\sigma}. \tag{89}$$

In calculation of the normal part of the frequency matrix (85) we have neglected the charge fluctuations and introduced the spin correlation functions:

$$\chi_{1s} = \langle \mathbf{S}_i \mathbf{S}_{i+a_1}\rangle, \quad \chi_{2s} = \langle \mathbf{S}_i \mathbf{S}_{i+a_2}\rangle, \tag{90}$$

for the nearest, $\chi_{1s}$, $(a_1 = (\pm a_x, \pm a_y))$ and the next-nearest, $\chi_{2s}$, $(a_2 = \pm(a_x \pm a_y))$ neighbor lattice sites.

To calculate the self–energy operator we employ the noncrossing approximation (or the self-consistent Born approximation) for the *proper* part of the many–particle GF in (83). It is given by the two-time decoupling for the corresponding correlation functions in (83):

$$\langle X_{j'}^{\sigma'0} B_{j\sigma\sigma'}^+ X_{i'}^{0\sigma'}(t) B_{i\sigma\sigma'}(t)\rangle \simeq \langle X_{j'}^{\sigma'0} X_{i'}^{0\sigma'}(t)\rangle \langle B_{j\sigma\sigma'}^+ B_{i\sigma\sigma'}(t)\rangle. \tag{91}$$

The proposed decoupling does not violate equal time correlations since in Eq. (91) $j \neq j'$ and $i \neq i'$. In the adopted approximation vertex corrections are neglected while the Fermi-like and the Bose-like correlation functions are supposed to be calculated self-consistently from the full GF. As was argued in [67] where the analogous approximation was used, at moderate doping we can consider the spin-charge fluctuations and single-particle excitations as independent modes. Then we can perform the decoupling (91) in the framework of the mode-coupling approximation which has been proved to be quite a reliable one even for systems with strong interactions. As was shown for the spin polaron $t$-$J$ model the vertex corrections to the noncrossing approximation are small and give only numerical renormalization of the model parameters (see, e.g. [83]).

Using the spectral representation for the GF, we obtain in the noncrossing approximation the following expression for the normal and anomalous components of the self-energy:

$$\tilde{\Sigma}^\sigma_{11(12)}(k,\omega) = Q\hat{\Sigma}^\sigma_{11(12)}(k,\omega) =$$

$$= \frac{1}{N}\sum_q \int\!\!\!\int_{-\infty}^{+\infty} dz d\Omega N(\omega,z,\Omega)\lambda_{11(12)}(q,k-q\mid\Omega)A^\sigma_{11(12)}(q,z), \qquad (92)$$

where

$$N(\omega,z,\Omega) = \frac{1}{2}\frac{\tanh(z/2T)+\coth(\Omega/2T)}{\omega-z-\Omega}. \qquad (93)$$

Here we introduce the normalized spectral density:

$$A^\sigma_{11}(q,z) = -\frac{1}{Q\pi}\operatorname{Im}\langle\langle X^{0\sigma}_q \mid X^{\sigma 0}_q\rangle\rangle_{z+i\delta}, \qquad (94)$$

$$A^\sigma_{12}(q,z) = -\frac{1}{Q\pi}\operatorname{Im}\langle\langle X^{0\sigma}_q \mid X^{0\bar{\sigma}}_{-q}\rangle\rangle_{z+i\delta},$$

and the electron - electron interaction functions caused by spin-charge fluctuations

$$\lambda_{11(12)}(q,k-q\mid\Omega) = g^2(q,k-q)[-\frac{1}{\pi}\operatorname{Im}D^\pm(k-q,\Omega+i\delta)], \qquad (95)$$

where the spectral density for the spin-charge fluctuations is defined by the boson-like commutator GF

$$D^\pm(q,\Omega) = \langle\langle \mathbf{S}_q \mid \mathbf{S}_{-q}\rangle\rangle_\Omega \pm \frac{1}{4}\langle\langle n_q \mid n^+_q\rangle\rangle_\Omega. \qquad (96)$$

The solution of the Dyson equation (80) can be written in the Eliashberg notation as

$$\hat{G}^\sigma(k,\omega) = Q\tilde{G}^\sigma(k,\omega) = Q\frac{\omega Z^\sigma_k(\omega)\hat{\tau}_0 + (E^\sigma_k+\xi^\sigma_k(\omega)-\tilde{\mu})\hat{\tau}_3 + \Phi^\sigma_k(\omega)\hat{\tau}_1}{(\omega Z^\sigma_k(\omega))^2 - (E^\sigma_k+\xi^\sigma_k(\omega)-\tilde{\mu})^2 - \mid\Phi^\sigma_k(\omega)\mid^2}, \qquad (97)$$

where

$$\omega(1 - Z_k^\sigma(\omega)) = \frac{1}{2}[\tilde{\Sigma}_{11}^\sigma(k,\omega) + \tilde{\Sigma}_{22}^\sigma(k,\omega)],$$

$$\xi_k^\sigma(\omega) = \frac{1}{2}[\tilde{\Sigma}_{11}^\sigma(k,\omega) - \tilde{\Sigma}_{22}^\sigma(k,\omega)], \qquad (98)$$

$$\Phi_k^\sigma(\omega) = \Delta_k^\sigma + \tilde{\Sigma}_{12}^\sigma(k,\omega),$$

and $\Sigma_{11}^\sigma(k,\omega) = -\Sigma_{22}^{\bar{\sigma}}(-k,-\omega)$. Here we fixed the phase of the gap function by taking it to be real.

## Numerical Results and Discussion

For numerical solution of the system of equations (92)-(98) we have used the imaginary frequency representation for the GF (97) with $\omega = i\omega_n = i\pi T(2n+1)$ and the spin-charge GF (96) with $\Omega = i\omega_n = i\pi T 2n$ where $n = 0, \pm 1, \pm 2, \ldots$. By using the representation for the function (93)

$$N(i\omega_n, z, \Omega) = -T \sum_m \frac{1}{i\omega_m - z} \frac{1}{i(\omega_n - \omega_m) - \Omega}$$

after integration in Eq. (92) we get

$$\tilde{\Sigma}_{11(12)}^\sigma(k, i\omega_n) = -\frac{T}{N} \sum_q \sum_m \tilde{G}_{11(12)}^\sigma(q, i\omega_m)\lambda_{11(12)}(q, k-q \mid i\omega_n - i\omega_m), \qquad (99)$$

The interaction functions are given by

$$\lambda_{11(12)}(q, k-q \mid i\omega_\nu) = g^2(q, k-q)D^\pm(k-q, i\omega_\nu), \qquad (100)$$

To calculate superconducting $T_c$ it is sufficient to study a linearized system of the equations (97) which has the following form

$$\tilde{G}_{11}^\sigma(k, i\Omega_n) = \frac{1}{i\omega_n - E_k + \tilde{\mu} - \tilde{\Sigma}_{11}^\sigma(k, i\omega_n)}, \qquad (101)$$

$$\Phi^\sigma(k, i\omega_n) = \Delta_k^\sigma + \phi^\sigma(k, i\omega_n) = \frac{T}{N} \sum_q \sum_m \{J(k-q)$$

$$+\lambda_{12}(q, k-q \mid i\omega_n - i\omega_m)\}\tilde{G}_{11}^\sigma(q, i\omega_m)\tilde{G}_{11}^{\bar{\sigma}}(q, -i\omega_m)\Phi^\sigma(q, i\omega_m). \qquad (102)$$

At first the system of equations for the normal GF (101) was solved numerically for a given concentration of electrons

$$\frac{n}{1 - n/2} = \frac{1}{N} \sum_{k,\sigma} N_{k\sigma} = 1 + \frac{2T}{N} \sum_k \sum_{n=-\infty}^{\infty} \tilde{G}_{11}(k, i\omega_n). \qquad (103)$$

Then the eigenvalues and eigenfunctions of the gap function (102) were calculated to obtain the superconducting transition temperature $T_c$ and the $(\mathbf{k}, \omega)$-dependence of the gap function.

For numerical calculations we take into account only the spin-fluctuation contribution and write the function $D_s^\pm(q, i\omega_\nu)$ (96) in the form

$$D_s^\pm(q, i\omega_\nu) \simeq \langle\langle \mathbf{S}_q | \mathbf{S}_{-q} \rangle\rangle_{i\omega_\nu} = -\int_0^{+\infty} \frac{2z\,dz}{z^2 + \omega_\nu^2} \chi_s''(q, z). \tag{104}$$

To perform self-consistent calculations one should write an equation for the spin-fluctuation susceptibility (104) in terms of the one-electron Green function (101) as it has been done, e.g. in [108]. However, to make the numerical work tractable we use a model representation for the spin-fluctuation susceptibility suggested in numerical studies [109]

$$\chi_s''(q, \omega) = -\frac{1}{\pi} \mathrm{Im}\, \langle\langle \mathbf{S}_q | \mathbf{S}_{-q} \rangle\rangle_{\omega+i\delta} = \chi_s(q)\, \chi_s''(\omega)$$

$$= \frac{\chi_0}{1 + \xi^2(1 + \gamma(q))} \tanh\frac{\omega}{2T} \frac{1}{1 + (\omega/\omega_s)^2}. \tag{105}$$

The $q$-dependent part has a peak at the AFM wave vector $(\pi, \pi)$ with its intensity defined by the short-range AFM correlation length $\xi$ (measured in lattice constant units). We take quite a small $\xi = 1-3$ to imitate incommensurate character of spin fluctuations with large correlation length $\xi \simeq 5$ observed at finite doping both in La$_{2-x}$Sr$_x$CuO$_4$ and YBa$_2$Cu$_3$O$_{7-x}$ in neutron scattering experiments [110]. For the frequency dependent part we choose the scaling function with an enhanced intensity for frequencies $\omega < T$ and a large cutoff energy $\omega_s \simeq J$ according to neutron scattering experiments in single-layered cuprate La$_{2-x}$Sr$_x$CuO$_4$. It is important that the constant $\chi_0$ in (105) which defines intensity of spin fluctuations at the AFM wave vector is normalized according to the following condition

$$\frac{1}{N} \sum_i \langle \mathbf{S}_i \mathbf{S}_i \rangle = \frac{1}{N} \sum_q \chi_s(q) \int_{-\infty}^{+\infty} \frac{dz}{\exp(z/T) - 1} \chi_s''(z) = \frac{\pi \omega_s}{2N} \sum_q \chi_s(q) = \frac{3}{4} n, \tag{106}$$

which gives

$$\chi_0 = \frac{3n}{2\pi \omega_s C_1}, \quad C_1 = \frac{1}{N} \sum_q \frac{1}{1 + \xi^2(1 + \gamma(q))}.$$

In the approximation (104) we get for the interaction function (100)

$$\lambda_{11}(q, k-q \mid i\omega_\nu) = \lambda_{12}(q, k-q \mid i\omega_\nu) = -g^2(q, k-q) \chi_s(k-q) F_s(i\omega_\nu), \tag{107}$$

where

$$F_s(\omega_\nu) = \int_0^\infty \frac{2x dx}{x^2 + (\omega_\nu/\omega_s)^2} \frac{1}{1+x^2} \tanh\frac{x}{2\tau} \quad (108)$$

is the spectral function, $\tau = T/\omega_s$.
Within the model (105) the static spin correlation functions (90) are calculated by using $\langle \mathbf{S}_q \mathbf{S}_{-q}\rangle = (\pi\omega_s/2)\chi_s(q)$. In the Eq. 109 we give characteristic values for the static spin correlation functions and the constant $\chi_0$ for different values of $\xi$.

| $\xi$ | $\delta$ | $\chi_0$ | $\chi_{1s}$ | $\chi_{2s}$ | $\chi_s(Q)/\chi_s(0)$ |
|---|---|---|---|---|---|
| 1 | 0.30 | 1.56 | $-0.072$ | 0.019 | 3 |
| 3 | 0.10 | 7.40 | $-0.230$ | 0.130 | 19 |
| 5 | 0.05 | 17.08 | $-0.311$ | 0.213 | 51, |

(109)

To analyze a role of different interactions in the electron - spin-fluctuation scattering in (107) we consider the weak coupling approximation for the Eliashberg equation (97). It is given by the following approximation for the interaction (107) [111]:

$$\lambda_{12}(q, k-q \mid i\omega_n - i\omega_m) \simeq -\lambda(q, k-q)\theta(\omega_s - |\omega_n|)\theta(\omega_s - |\omega_m|), \quad (110)$$

where we take $F_s(\omega_\nu) \simeq F_s(\omega_s) \simeq 1$ and introduce

$$\lambda(q, k-q) = g^2(q, k-q)\chi_s(k-q). \quad (111)$$

In the weak coupling limit we have for the anomalous GF:

$$\tilde{G}_{12}^\sigma(k, i\omega_m) \simeq -\frac{\Phi_k^\sigma}{(\omega_m)^2 + (\Omega_k^\sigma)^2}, \quad (112)$$

where $\Omega_k^2 = (E_k^\sigma + \xi_k^\sigma(0) - \tilde{\mu})^2 + |\Phi_k^\sigma|^2$ is the QP energy in the superconducting state with the frequency independent gap function $\Phi_k^\sigma$. By performing summation over $m$ for $\tilde{\Sigma}_{12}^\sigma(k, i\omega_n)$ in Eq. (99), we get the weak coupling BCS equation

$$\Phi^\sigma(k) = \frac{1}{N}\sum_q \{J(k-q) - \lambda(q, k-q)\}\frac{\Phi_q^\sigma}{2\Omega_q} \tanh\frac{\Omega_q}{2T}. \quad (113)$$

In comparison with the results of the diagram technique [57], in Eq. (113) the kinematic interaction is also included in the effective coupling constant of the second order (111). Below we compare results for the superconducting $T_c$ calculated in the weak coupling limit, Eq. (113), and obtained from the equation (102).

The numerical calculations were performed using the fast Fourier transformation [107] for $32\times32$ cluster. In the summation over the Matsubara frequencies we used up to 700 points with the constant cut-off $\omega_{max} = 20\ t$. Usually 10 – 30 iterations were needed to obtain a solution for the self-energy with an accuracy of order 0.1 %. The Padé approximation was used to calculate the one-electron spectral function $A_{11}(k,\omega)$ (94) and the density of states (DOS)

$$A(\omega) = \frac{1}{N}\sum_k A_{11}^\sigma(k,\omega) \qquad (114)$$

on the real frequency axis.

The calculations were performed for several values of the $t-J$ model parameters ($J/t$, $t'/t$), the AFM correlation length $\xi$ in the model function (105) with $\omega_s = J$, and the hole concentration $\delta = 1-n$. The detailed results of the calculations can be found in the original publication [65]. Below we present only several representative results for $\delta = 0.1$ and $\delta = 0.4$ and $\xi = 1, 3$ for the parameters $J = 0.4$, $t' = 0$ if other values are not indicated. All the energies and temperature are measured in units of $t$. To mimic suppression of AFM correlations with doping we usually take $\xi = 3$ for $\delta = 0.1$ and keep $\xi = 1$ for $\delta = 0.2 - 0.4$. Temperature effects are rather small for $T \leq 0.1$ and therefore we present only results for $T = 0.0125$.

*Normal State.* Results for the electron spectral density in the normal state, $A(k,\omega) = A_{11}(k,\omega)$ (94), are shown along the three symmetry directions in the BZ: $\Gamma(0,0) \to X(\pi,0) \to M(\pi,\pi) \to \Gamma$ in Fig. 4 [65] for $\delta = 0.1$, $\xi = 3$ and in Fig. 5 [65] for $\delta = 0.4$, $\xi = 1$.

For small concentration of holes, $\delta = 0.1$, we observe quite narrow QP peaks at the wave vectors crossing the Fermi surface (FS) along $M \to X$ and $M \to \Gamma$ directions. Along $X \to \Gamma$ direction wave vectors are below the FS (see Fig. 6 ) and there are no QP peaks. In addition to the QP dispersion we see also a band of incoherent excitations with large dispersion below the Fermi energy, $\omega < 0$. The incoherent band is caused by the self-energy contribution peaked at the AFM wave vector ("shadow bands"). For $\xi = 3$ in Fig. 4 the incoherent band has a higher intensity due to stronger spin-fluctuations weight at the AFM wave vector .

With increasing hole concentration the dispersion of the QP band also increases and the intensity of QP peaks are enhanced as shown in Fig. 5 for $\delta = 0.4$, $\xi = 1$. At the same time the intensity of the incoherent excitations are suppressed: the "high-energy feature" below the Fermi energy at the $X$ point for $\delta = 0.1$, $\xi = 3$ in Fig. 4 practically disappears for $\delta = 0.4$, $\xi = 1$ in Fig. 5. As was discussed by Shen and Schrieffer [112] (see also [113]), the doping dependence of the spectral line shape near $(\pi,0)$ point can be explained by strong coupling of the QP hole excitations with collective excitations. In our model the latter are spin fluctuations which intensity at $(\pi,\pi)$ point is proportional to $\xi^2$ (see Eq. (105)) resulting in strong suppression of the incoherent excitations with decreasing $\xi$ and increasing $\delta$.

These conclusions has been supported by the doping dependence of the imaginary part of the self-energy $-\text{Im}\Sigma(k,\omega) = -\text{Im}\tilde{\Sigma}_{11}^\sigma(k,\omega + i\epsilon)$. With increasing hole concentration and decreasing AFM correlation length $\xi$ the self-energy decreases due to suppression of electron scattering on spin-fluctuations. It is interesting to

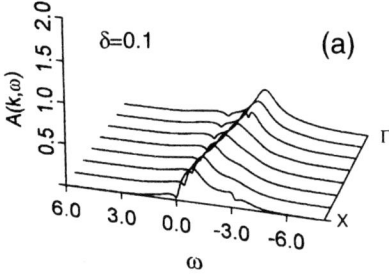
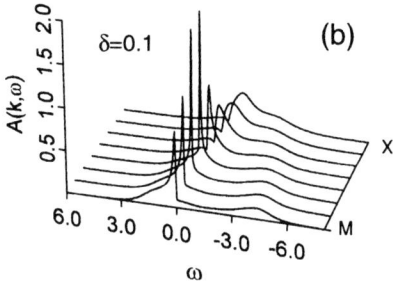
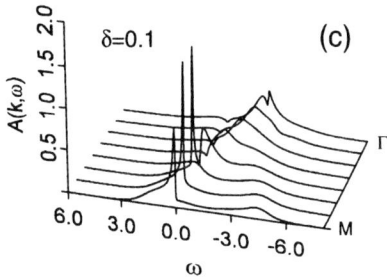

**FIGURE 4.** Electron spectral density $A(k,\omega)$ for $\delta = 0.1$ and $\xi = 3$ along three symmetry directions $\Gamma(0,0) \to X(\pi,0) \to M(\pi,\pi) \to \Gamma$ over the first quadrant of the BZ ($0 \leq k_x, k_y \leq \pi$). Energy is measured in units of $t$ [65].

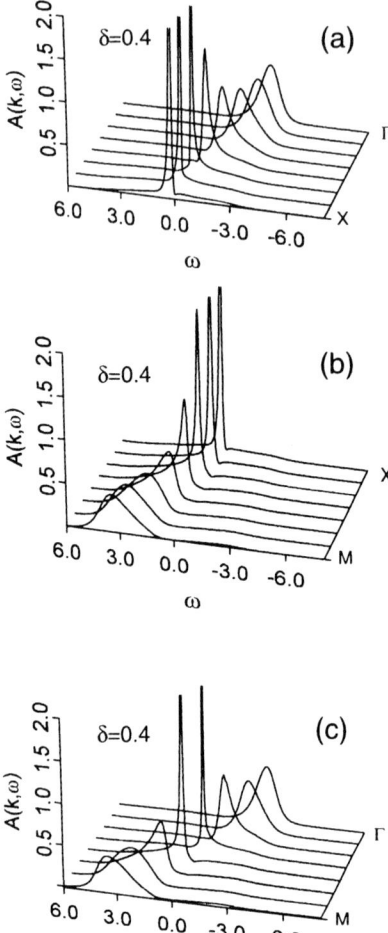

**FIGURE 5.** The same as in Fig. 4 but for $\delta = 0.4$ and $\xi = 1$ [65].

note that for the underdoped region, $\delta \leq 0.1$, $\mathrm{Im}\Sigma(k,\omega)$ for $T \leq \omega \leq J$ is approximately proportional to $\omega$ while for the overdoped region, $\delta \geq 0.3$, for small $\omega$ we have $\mathrm{Im}\Sigma(k,\omega) \propto \omega^2$ (see [65]). However, our $(\mathbf{k},\omega)$ resolution is not high enough to prove a transition from the non-Fermi liquid to the Fermi-liquid behavior with doping.

Our results for electron spectral functions are in a semi-quantitative agreement with the numerical studies of the $t$-$J$ model within the finite-temperature Lanczos method [114]. We observe also a large asymmetry between the photoemission ($\omega < 0$) and the inverse photoemission ($\omega > 0$) spectra. While the spectra below the FS in Figs. 4, 5 are strongly overdamped and show no QP peaks, the spectra for electrons outside the FS, e.g. at $M(\pi,\pi)$-point show QP behaviour. The main disagreement with the results of [114] is a smaller imaginary part of the self-energy which results also in a less pronounced incoherent part of the spectra below the FS. Partly this can be explained by underestimation of the electron scattering on spin fluctuations. Namely, if we make an approximation $(1 + \gamma(q)) \simeq |\mathbf{q} - \mathbf{Q}|^2$ in the spin susceptibility (105) which enhances the peak intensity at $Q = (\pi,\pi)$ we get much stronger scattering and a more intensive incoherent band. To enhance the incoherent contribution and to fulfill the Luttinger theorem it was proposed in [67] to add to the self-energy (92) a momentum-independent part, $\mathrm{Im}\Sigma(\omega) = -c\omega^2$, with quite a large value at maximum of the order of $\simeq 3.5t$. However, this fitting is difficult to justify. We find also a reasonable agreement of our results for the spectral function for $\delta = 0.1$, $\xi = 3$, including both the coherent QP dispersion and incoherent band, with the calculations in [115] done by the exact-diagonalization technique for a finite cluster of 20 lattice sites with 2 holes ($\delta = 0.1$).

The QP dispersion $E(\mathbf{k})$ was calculated from the maxima of spectral density. It appeared that the QP band width strongly increases with doping while the next neighbor hopping $t'$ change the dispersion mostly at $\Gamma(0,0)$ and $X(\pi,0)$ points. These results can be already explained within the spectrum $E_k$ in MFA, Eq. (85). Being written in the form

$$E_k^\sigma = -4t\gamma(k)\,Q[1 + \chi_{1s}/Q^2] - 4t'\gamma'(k)\,Q[1 + \chi_{2s}/Q^2]$$
$$= -4t_{eff}\,\gamma(k) - 4t'_{eff}\,\gamma'(k), \qquad (115)$$

it shows a strong dependence of the effective hopping parameters on the static AFM correlation functions (90): $\chi_{1s} = \langle \mathbf{S}_i \mathbf{S}_{i+a_1} \rangle$, $\chi_{2s} = \langle \mathbf{S}_i \mathbf{S}_{i+a_2} \rangle$. For small hole concentration and large AFM correlation length, e.g., $\delta = 0.1$, $\xi = 3$, we have $\chi_{1s} = -0.23$, $\chi_{2s} = 0.13$ (see Eq. (109)) which strongly reduces the nearest neighbor hopping: $t_{eff} \simeq 0.13t$, while enhances the next nearest neighbor hopping: $t'_{eff} \simeq 0.8t'$. At large hole concentration, e.g., $\delta = 0.4$, $\xi = 1$, we have $\chi_{1s} = -0.06$, $\chi_{2s} = 0.016$ the renormalization is quite small: $t_{eff} \simeq 0.6t$, $t'_{eff} \simeq 0.72t'$.

Here we would like to point out that in the large-$N$ expansion technique, both for the slave boson [50,51] and the Baym-Kadanoff variation GF [78,79], the narrowing of the band due to the discussed above AFM correlations is ignored. In the $1/N$ expansion the static spin correlation functions $\chi_{1s}$, $\chi_{1s}$ appear to be of the higher order in $1/N$ and therefore are omitted. Moreover, the "Hubbard I" factor $Q$ in the spectrum in the MFA in Eq. (85): $t_{eff} = Qt$, is also underestimated. We have in Eq. (85) $Q_\sigma = \langle X_i^{00} + X_i^{\sigma\sigma} \rangle = (1+\delta)/2$, while in the $1/N$ expansion $Q = \langle X_i^{00} \rangle = \delta$ since the correlation function $\langle X_i^{\sigma\sigma} \rangle$ is of the order $1/N$ and is disregarded. These underestimation of the strong kinematic interaction in the large-$N$ expansion changes the doping dependence of the QP spectrum in MFA in comparison with the real situation $N = 2$. In particular, in the limit $\delta \to 0$ in our approximation we have a finite band width, $W = 8Qt = 4t$, while in $1/N$ expansion it collapses to zero if one neglects hopping mediated by the exchange interaction $J$.

The calculated density of states (DOS) $A(\omega)$ (114) has a strongly suppressed incoherent band at large hole concentration ($\delta = 0.4$) and small AFM correlation length ($\xi = 1$). In the latter case DOS has a nearly symmetric form with a broad band width (of the order of $7t$) in comparison with highly asymmetric one for low doping ($\delta = 0.1$) where high density of states below the Fermi level is due to the incoherent band. In that respect DOS at $\delta = 0.1$ agrees quite well with the results of numerical studies by the finite-temperature Lanczos method [114] where the single-particle DOS is also quite asymmetric with large incoherent part below the FS. Moreover, the absolute values of DOS at maximum are close to each other (in [114] DOS $N(\omega) = (1+\delta)A(\omega)$ where $A(\omega)$ is given by Eq. (114)).

Now we consider the results for the electron occupation numbers (89)

$$n_{\mathbf{k},\sigma} = \langle X_k^{\sigma 0} X_k^{0\sigma} \rangle = [(1+\delta)/2] \, N_{\mathbf{k}\sigma}.$$

In Fig. 6 [65] the function $N(\mathbf{k}) = N_{\mathbf{k}\sigma}$ is shown for different hole concentrations: (a) $\delta = 0.1$, $\xi = 3$, and (b)-(d) $\delta = 0.2 - 0.4$, $\xi = 1$. The shape of the FS changes from the hole-like around $M(\pi, \pi)$ point of BZ at small doping to the electron-like one around $\Gamma(0,0)$ point of BZ for large doping. We obtain quite a small drop of $N(\mathbf{k})$ at the FS especially at small doping, which is a specific feature of strongly correlated electronic systems. Large occupation numbers throughout the BZ are due to the incoherent contribution in the spectral density $A(k,\omega)$ under the Fermi level (see Figs. 4, 5). The evolution of the FS with hole concentration is also shown in Fig. 7 [65] by bold solid lines. The maximal occupation numbers for electrons, $n_{\mathbf{k},\sigma} = (1+\delta)N(\mathbf{k})/2 \leq 0.55$ for $\delta = 0.1$, agrees with the results of the exact-diagonalization technique for finite clusters [115]. According to Eq. (103), we have also inequality $n \leq 1$ that results from the restriction of no double occupancy of lattice sites in the $t$-$J$ model.

Concerning the volume of the electron FS, it appears to be proportional at small doping to $(1-\delta)$, e.g., for $\delta = 0.1, 0.2$ the ratio of the BZ part for $k < k_F$ to the whole BZ are close to 90 % and 80 %, respectively, while according to the Luttinger theorem the ratio should be equal to $(1-\delta)/2$. It is possible to obtain a large FS at small doping if one introduce a strong incoherent part in the DOS below the FS as it has been proposed in [67]. However, the one (sub)band $t$-$J$ model could violate the Luttinger theorem since it is derived for the Mott-Hubbard insulating state which may not have an adiabatic connection to a non-interacting

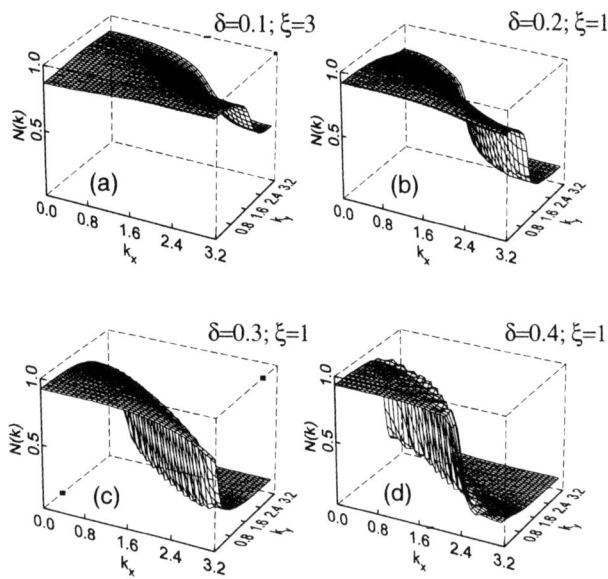

**FIGURE 6.** Electron occupation numbers $N_{\mathbf{k},\sigma}$ (see Eq. (103)) over the first quadrant of the BZ ($0 \leq k_x, k_y \leq \pi$) for different hole concentration $\delta$ and AFM correlation length $\xi$: $\delta = 0.1$, $\xi = 3$ (a), $\delta = 0.2$, $\xi = 1$ (b), $\delta = 0.3$, $\xi = 1$ (c), $\delta = 0.4$, $\xi = 1$ (d) [65].

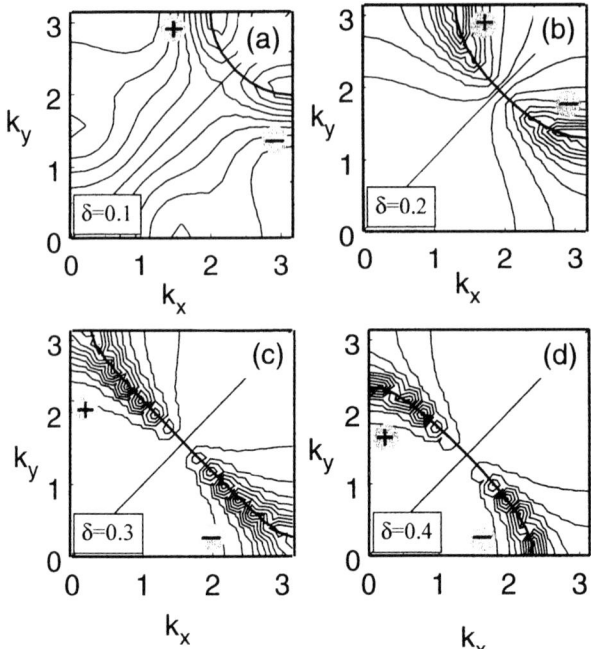

**FIGURE 7.** The Fermi surface (bold solid line) and the gap function $\Phi(\mathbf{k}, \omega = 0)$, over the first quadrant of the BZ ($0 \leq k_x, k_y \leq \pi$) for different hole concentration $\delta$ and AFM correlation length $\xi$: $\delta = 0.1$, $\xi = 3$ (a), $\delta = 0.2$, $\xi = 1$ (b), $\delta = 0.3$, $\xi = 1$ (c), $\delta = 0.4$, $\xi = 1$ (d) [65].

electron gas. It has been recently reported in [116] a violation of the Luttinger theorem in the high-temperature series for the momentum distribution function of the two-dimensional $t - J$.

*Superconducting state.* The results of numerical solution of the linearized equation (102) are presented in Figs. 7, 8 [65]. Figure 7 shows the contour plots in a quarter of BZ, ($0 \leq k_x, k_y \leq \pi$), for the static gap function $\Phi(k) = \Phi(k, \omega = 0)$. At a small doping, $\delta = 0.1$, (Fig. 7 (a)), it has a more complicated $k$-dependence, with two positive and two negative maxima (shown by (+) and (-)) while at $\delta \geq 0.2$ only one positive and one negative maximum survive (Figs. 7 (b)-(d)). The $\Phi(\mathbf{k})$-dependence has a complicated form that cannot be described by simple ($\cos k_x - \cos k_y$) function usually used for the $d$-wave symmetry. However, in all cases the gap function obeys the $B_{1g}$ symmetry: $\Phi(k_x, k_y) = -\Phi(k_y, k_x)$ which breaks the 4-fold symmetry of the FS in $\mathbf{k}$-space. The frequency dependence of the gap function $\Phi(k, \omega)$ is also anomalous. [65]

The superconducting $T_c$ was calculated at different hole concentration $\delta$ for AFM correlation length $\xi = 1, 3$ by direct numerical solution of Eq. (102) [65]. With increasing AFM correlation length $\xi$ effective electron-electron coupling $\lambda_{12}(q, k-q \mid i\omega_\nu)$ mediated by spin fluctuations $\chi_s(k-q)$ also increases which greatly enhances $T_c$. It was observed also that in the weak coupling approximation $T_c$ is much higher in comparison with that one obtained from the frequency-dependent equation (102). The results in the weak coupling approximation, Eq. (113), for $T_c$ at AFM correlation length $\xi = 1$ is presented in Fig. 8 for the full vertex (solid line), the vertex with $t(q) = 0$ (dashed line), and in the MFA with $\lambda(q, k - q) = 0$ (dotted line). The most important contribution in the weak coupling approximation comes from MFA, i.e. $J(k - q)$ in Eq. (113). The second order contribution, $\lambda(q, k - q) = g^2(q, k - q)\chi_s(k - q)$, enhances $T_c$ both due to kinematic, $t(q)$, and exchange, $J(k - q)$, interactions. For larger AFM correlation length superconducting $T_c$ is greatly enhanced in the weak coupling approximation, e.g. $T_c \simeq 0.1$ for $\xi = 3$.

To elucidate the role of AFM short-range fluctuations in the model and particular, strong dependence of $T_c$ on the AFM correlation length $\xi$ we present in Eq. (109) $\xi$-dependence of the static correlation functions, $\chi_{1s}$, $\chi_{2s}$, and the constant $\chi_0$ in Eq. (105). The latter, as well the ratio $\chi_s(Q)/\chi_s(q = 0)$, estimates the electron-spin fluctuation coupling while the static correlation functions $\chi_{1s}$, $\chi_{2s}$, Eq. (90), define the band width in the MFA, Eq. (85) as discussed above. Large increase of these parameters seen in Eq. (109), with increasing $\xi$ from their values at $\xi = 1$, explains changes in the spectral functions $A(k, \omega)$ and strong $T_c$ enhancement.

At the same time the peak position of $T_c(\delta)$ around the hole concentration $\delta \simeq 0.33$ does not change much with $\xi$. As Fig. 7 shows, at this concentration the FS crosses the $(\pm\pi, 0)$, $(0, \pm\pi)$ points of BZ. Since the pairing interaction, Eq. (107), is proportional to the spin susceptibility $\chi_s(k - q)$ with the maximal contribution at $\mathbf{k} - \mathbf{q} = \mathbf{Q}$ the most strong pairing occurs for electrons at the FS with $k = (\pm\pi, 0), (0, \pm\pi)$ coupled by the AFM wave vector $Q = (\pi, \pi)$. This scenario is characteristic for the spin-fluctuation pairing [6] and has been discussed

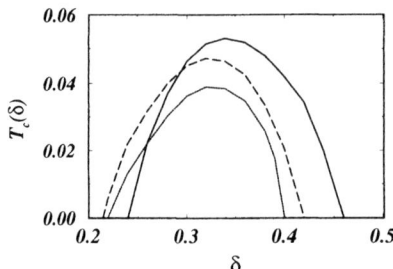

**FIGURE 8.** The superconducting temperature $T_c$ versus hole concentration $\delta$ for AFM correlation length $\xi = 1$ in the weak coupling approximation, Eq. (113), for the full vertex (solid line), the vertex with $t(q) = 0$ (dashed line), and in the MFA with $\lambda(q, k - q) = 0$ (dotted line) [65].

recently by Shen and Schrieffer [112] in connection with the anomalous momentum and temperature dependence of the spectral line-shape in ARPES experiments. Experimentally the highest $T_c$ is observed at the optimal doping of $\delta \simeq 0.16$. To obtain maximal $T_c$ in our calculations at a lower hole concentration the FS should be large and run along the diagonals $(\pm\pi, 0) - (0, \pm\pi)$ at this hole concentration. In the spin polaron $t$-$J$ model [66] $T_c(\delta)$ curve has a narrow peak at the optimal doping $\delta = 0.15 - 0.25$, depending on the next neighbor hopping $t' = \pm 0.1t$, since in that case the FS is close to the AFM BZ along the diagonals $(\pm\pi, 0) - (0, \pm\pi)$ already at a low concentration of doped holes. As discussed by Barabanov et al. [86], a large Fermi surface at low doping in spin-singlet state can be obtained if one properly takes into account a spin polaron formation due to short range AFM order. It appears as a spin-hole bound state given by the products of operators $\sum_{j,\sigma'} X_i^{\sigma\sigma'} X_j^{\sigma'0}$ that should not be decoupled in equation of motion for the GF.

## Conclusions

In the present Section a theory of electron spectrum and superconducting pairing in the $t$-$J$ model (69) in a paramagnetic state is developed by employing the the projection technique for the two-time GF in terms of Hubbard operators. The obtained self-consistent system of equations for the matrix GF (97) and the self-energy (92) in comparison with the diagram technique [57] has an additional contribution proportional to the second order of the kinematic interaction, $t(q)$, which gives important contribution in comparison to the exchange interaction, $J(k-q)/2$, in the vertex (88).

The one-electron spectral functions and the superconducting $T_c$ and the gap function were calculated by numerical solution of the linearized system of Dyson equations (101), (102). To describe kinematic and exchange interactions of electrons with spin fluctuations a model dynamic spin susceptibility (105) with short-range AFM correlation length $\xi$ was used.

The results for the electron spectral density (see Figs. 4, 5) show QP excitations at the FS crossing and a dispersive incoherent band. For small hole concentration and large AFM correlation length $\xi$ the QP dispersion is small while the intensity of the incoherent band is quite large. With doping the QP band width strongly increases and the incoherent band is suppressed. The results for electron spectral functions are in a semi-quantitative agreement with the numerical studies of the $t$-$J$ model within the finite-temperature Lanczos method [114] and also agree quite well with the studies by the exact-diagonalization technique [115]. To perform a quantitative comparison of the numerical results with ARPES investigations [117,118] one has to consider a more general $t$-$t'$-$t''$-$J$ model with the three-site terms as has been discussed recently (see, e.g., [102]).

The occupation numbers $N(\mathbf{k})$ have the characteristic behavior for strongly correlated systems, Fig. 6. Being large throughout the BZ they show only a small drop at the FS. The volume of the FS at small doping is proportional to the hole concentration $\delta$, as shown in Fig. 7, that violates the Luttinger theorem. However, the result for the volume of the FS for $n = 0.8$ is in quantitative agreement with recent numerical results of Putikka et al. [116].

The superconducting pairing due to the exchange and the kinematic interactions (in the second order) has the $d$-wave symmetry, Fig. 7, and high $T_c$, Fig. 8. In the weak coupling approximation, Eq. (113), much larger $T_c$ is observed. The calculations confirm the results of the $d$-wave superconducting pairing obtained within the spin polaron $t$-$J$ model [66] presented in Sec. III A.

Some uncertainty in the interaction is due to the noncrossing approximation (91) where vertex corrections are neglected. They were considered by applying the diagram technique in [58] but no numerical estimations were given. We think that vertex corrections should not change the main conclusions of our calculations. At least, we can argue that in our approach, where the model spin susceptibility (105) with small AFM correlation length, $\xi = 1 - 3$, is used, the vertex renormalization, estimated as $\chi_s(Q)/\chi_s(0)$ (see [43]), should not be large (see Eq. (109)).

# IV  SUPERCONDUCTING PAIRING IN THE HUBBARD MODEL

It is important to compare the results obtained in the previous Section for the $t$-$J$ model with calculations for the more general Hubbard two-band singlet-hole model developed in Sec. II A. It is generally accepted that the superconducting pairing is efficient only close to the Fermi surface and therefore it should be sufficient to study the $t$-$J$ model which is essentially a low energy limit of the Hubbard model. However, the exchange interaction in the $t$-$J$ model is due to strong hybridization with the upper Hubbard subband and therefore it can be studied in-depth only within the two-band Hubbard model. The studies of the Hubbard model [71] show that the exchange interaction couples electrons (holes) in a broad energy shell, of the order of the band width, since for the interband superexchange hopping with a large energy the retardation effects are unimportant as compared with the spin-fluctuation pairing in one subband. That results in high values of superconducting $T_c$ which permits to consider the exchange interaction in copper oxide as the major driving force for superconducting pairing.

Below we present numerical and analytical solutions [71] of a coupled system of equations for superconducting gap functions which show that the spin-fluctuation electron scattering induced by the kinematic interaction in the second order and exchange interaction result in the singlet $d_{x^2-y^2}$-wave pairing of the conventional electron pairs at different lattice sites, both in the lower and the upper Hubbard subbands. $T_c$ is calculated for several hole concentrations within the weak coupling approximation. To treat rigorously the strong correlations, the Hubbard operator technique within the projection method for the GF is used as in the previous Sections. Results of preceding investigations of superconductivity in the Hubbard model are discussed in Sec. I B.

## A  General Formalism

Below we consider a more realistic than the original Hubbard model (4) the asymmetric Hubbard model developed in Sec. II A. We write down the model as follows:

$$H = E_1 \sum_{i,\sigma} X_i^{\sigma\sigma} + E_2 \sum_i X_i^{22} + \sum_{i \neq j, \sigma} \{t_{ij}^{11} X_i^{\sigma 0} X_j^{0\sigma} \\ + t_{ij}^{22} X_i^{2\sigma} X_j^{\sigma 2} + 2\sigma t_{ij}^{12} (X_i^{2\bar\sigma} X_j^{0\sigma} + \text{H.c.})\} \qquad (116)$$

where $X_i^{nm} = |in\rangle\langle im|$ are the Hubbard operators for the four states $n, m = |0\rangle, |\sigma\rangle, |2\rangle = |\uparrow\downarrow\rangle$, $\sigma = \pm 1/2$, $\bar\sigma = -\sigma$. The energy parameters are given by $E_1 = E_D - \mu$ and $E_2 = E_\psi - 2\mu$, respectively, where $E_D$ is a reference (renormalized) energy of a d-hole, $\mu$ is the chemical potential, and $E_\psi$ is the energy of a singlet. The renormalized charge transfer energy $\tilde\Delta = E_\psi - 2E_D$ (see Eqs. (18), (19), (30)). The hopping integrals can be written as $t_{ij}^{\alpha\beta} = K_{\alpha\beta} V_{ij}$, $V_{ij} = 2t\nu_{ij}$, where $t$ is the $p$–$d$ hybridization parameter and $\nu_{ij}$ are the overlapping parameters for the Wannier oxygen states (see Eqs. (14), (25)). The coefficients $K_{\alpha\beta}$ for a realistic value $\Delta = 2t$ are of the order of $|K_{\alpha\beta}| \simeq 0.5 - 0.9$ (see Eq. (29)). Therefore, the effective hopping parameters for the nearest neighbors in the model (116) are given by $t_{eff} \simeq K_{22} 2\nu_1 t \simeq 0.14 t$. In (116), the (renormalized) charge-transfer gap $\tilde\Delta$ plays the role of the Coulomb repulsion $U$ in Eq. (4), while the band width $W = 8 t_{eff}$. Hence the ratio $\tilde\Delta/W \simeq 2$ is quite large and the asymmetric Hubbard model (116) corresponds to the strong correlation limit. The Hubbard operators entering (116) obey the completeness relation, Eq. (34).

To discuss the superconducting pairing within the model Hamiltonian (116), we define the two-band four-component Nambu operators $\hat{X}_{i\sigma}^{\dagger}$ and $\hat{X}_{i\sigma}$, where

$$\hat{X}_{i\sigma}^{\dagger} = (X_i^{2\sigma}\ X_i^{\bar{\sigma}0}\ X_i^{\bar{\sigma}2}\ X_i^{0\sigma}) \tag{117}$$

and $\hat{X}_{i\sigma}$ is the vector Hermitian conjugate of (117).

The pairing correlations of these operators are described within the GF approach. The two-time anticommutator retarded GF associated to the site $(i,j)$ is a $4 \times 4$ matrix. In Zubarev notation [61] the GF can be written as

$$\tilde{G}_{ij\sigma}(t-t') = -i\theta(t-t')\langle\{\hat{X}_{i\sigma}(t), \hat{X}_{j\sigma}^{\dagger}(t')\}\rangle = \langle\langle \hat{X}_{i\sigma}(t)|\hat{X}_{j\sigma}^{\dagger}(t')\rangle\rangle. \tag{118}$$

To get the quasi-particle spectrum of the system, we use the equation of motion method as it has been described in Sec. III B. Differentiation with respect to $t$ of the GF (118) results in the following equation

$$\omega \tilde{G}_{ij\sigma}(\omega) = \delta_{ij}\tilde{\chi} + \langle\langle \hat{Z}_{i\sigma}|\hat{X}_{j\sigma}^{\dagger}\rangle\rangle_\omega, \tag{119}$$

where $\hat{Z}_{i\sigma} = [\hat{X}_{i\sigma}, H]$ and $\tilde{\chi} = \langle\{\hat{X}_{i\sigma}, \hat{X}_{i\sigma}^{\dagger}\}\rangle$.

Assuming that the system is in the paramagnetic state, we get

$$\tilde{\chi} = \hat{\tau}_0 \times \begin{pmatrix} \chi_2 & 0 \\ 0 & \chi_1 \end{pmatrix} \tag{120}$$

where $\hat{\tau}_0$ is a $2 \times 2$ unity matrix. In view of the completeness relation, Eq. (34), the diagonal terms $\chi_2 = \langle X_i^{22} + X_i^{\sigma\sigma}\rangle = \langle X_i^{22} + X_i^{\bar{\sigma}\bar{\sigma}}\rangle$ and $\chi_1 = \langle X_i^{00} + X_i^{\bar{\sigma}\bar{\sigma}}\rangle = \langle X_i^{00} + X_i^{\sigma\sigma}\rangle$ fulfill the relationship $\chi_2 = 1 - \chi_1 = n/2$, where $n$ denotes the hole concentration. In terms of the doping parameter, $\delta = n - 1$, these relationships are

$$\chi_2 = (1+\delta)/2, \quad \chi_1 = (1-\delta)/2. \tag{121}$$

The off-diagonal term $\langle\{X_i^{\sigma 2}, X_i^{0\sigma}\}\rangle$ in the matrix $\tilde{\chi} = \langle\{\hat{X}_{i\sigma}, \hat{X}_{i\sigma}^{\dagger}\}\rangle$ is given by the correlation function $\chi_3 = \langle X_i^{02}\rangle$. In terms of the Fermi annihilation operators $c_{i\sigma}$ it reads: $\langle X_i^{02}\rangle = \langle X_i^{0\downarrow}X_i^{\downarrow 2}\rangle = \langle c_{i\downarrow}c_{i\uparrow}\rangle$ which describes the pairing at the same lattice site. It vanishes for the $d_{x^2-y^2}$-wave symmetry and in what follows we take it to be zero:

$$\chi_3 = \langle X_i^{02}\rangle = 0. \tag{122}$$

The chemical potential $\mu$ is calculated from the equation for the average number of holes,

$$n = \langle N_i\rangle = \sum_\sigma \langle X_i^{\sigma\sigma}\rangle + 2\langle X_i^{22}\rangle. \tag{123}$$

Using the projection technique as described in Sec. III B the operator $\hat{Z}_{i\sigma}$ which enters the last term of Eq. (119) we split it into two parts:

$$\hat{Z}_{i\sigma} = [\hat{X}_{i\sigma}, H] = \sum_l \tilde{E}_{il\sigma} \hat{X}_{l\sigma} + \hat{Z}_{i\sigma}^{(ir)}, \quad \langle\{\hat{Z}_{i\sigma}^{(ir)}, \hat{X}_{j\sigma}^\dagger\}\rangle = 0. \tag{124}$$

The frequency matrix reads:

$$\tilde{E}_{ij\sigma} = \tilde{\mathcal{A}}_{ij\sigma} \tilde{\chi}^{-1}, \tag{125}$$

$$\tilde{\mathcal{A}}_{ij\sigma} = \langle\{[\hat{X}_{i\sigma}, H], \hat{X}_{j\sigma}^\dagger\}\rangle. \tag{126}$$

If the finite lifetime effects due to $\hat{Z}_{i\sigma}^{(ir)}$ in (124) are neglected and the expression (125) of the frequency matrix is used, then Eq. (119) provides the zero-order GF within the generalized MFA. In the $(\mathbf{q}, \omega)$-representation, its expression is given by

$$\tilde{G}_\sigma^0(\mathbf{q}, \omega) = \left(\omega \tilde{\tau}_0 - \tilde{\mathcal{A}}_\sigma(\mathbf{q}) \tilde{\chi}^{-1}\right)^{-1} \tilde{\chi}, \tag{127}$$

where $\tilde{\tau}_0$ is the $4 \times 4$ unity matrix.

Differentiation of the many-particle GF (119) with respect to the second time $t'$ and use of the same projection procedure as in (124) result in the Dyson equation for the GF (118). In the $(\mathbf{q}, \omega)$-representation, the Dyson equation is

$$\left(\tilde{G}_\sigma(\mathbf{q}, \omega)\right)^{-1} = \left(\tilde{G}_\sigma^0(\mathbf{q}, \omega)\right)^{-1} - \tilde{\Sigma}_\sigma(\mathbf{q}, \omega). \tag{128}$$

The self-energy operator is given by the *proper* part of the scattering matrix given by the many-particle GF:

$$\tilde{\Sigma}_\sigma(\mathbf{q}, \omega) = \tilde{\chi}^{-1} \langle\langle \hat{Z}_\sigma^{(ir)} | \hat{Z}_\sigma^{(ir)\dagger} \rangle\rangle_{\mathbf{q}, \omega}^{(prop)} \tilde{\chi}^{-1}. \tag{129}$$

The equations (127), (128), and (129) provide an exact representation for the single-particle GF (118). Its calculation, however, requires the use of some approximations for the many-particle GF in the self-energy matrix (129) which describes the finite lifetime effects (inelastic scattering of electrons on spin and charge fluctuations).

## B  Mean-Field Approximation

*The Frequency Matrix*

In the mean-field approximation, the calculation of the frequency matrix (125) and the zero-order GF (127) require the knowledge of the quantity $\tilde{\mathcal{A}}_{ij\sigma}$, Eq. (126). It can be calculated by using the equations of motion for the Hubbard operators:

$$Z_i^{\sigma 2} = [X_i^{\sigma 2}, H] = (E_1 + \Delta)X_i^{\sigma 2} + \sum_{l \neq i, \sigma'} \left( t_{il}^{22} B_{i\sigma\sigma'}^{22} X_l^{\sigma' 2} - 2\sigma t_{il}^{21} B_{i\sigma\sigma'}^{21} X_l^{0\bar{\sigma}'} \right)$$
$$- \sum_{l \neq i} X_i^{02} \left( t_{il}^{11} X_l^{\sigma 0} + 2\sigma t_{il}^{21} X_l^{2\bar{\sigma}} \right), \tag{130}$$

$$Z_i^{0\bar{\sigma}} = [X_i^{0\bar{\sigma}}, H] = E_1 X_i^{0\bar{\sigma}} + \sum_{l \neq i, \sigma'} \left( t_{il}^{11} B_{i\sigma\sigma'}^{11} X_l^{0\bar{\sigma}'} - 2\sigma t_{il}^{12} B_{i\sigma\sigma'}^{12} X_l^{\sigma' 2} \right)$$
$$- \sum_{l \neq i} X_i^{02} \left( t_{il}^{22} X_l^{2\bar{\sigma}} + 2\sigma t_{il}^{12} X_l^{\sigma 0} \right), \tag{131}$$

$$Z_i^{2\bar{\sigma}} = - \left( Z_i^{\bar{\sigma}2} \right)^{\dagger}, \quad Z_i^{\sigma 0} = - \left( Z_i^{0\sigma} \right)^{\dagger}. \tag{132}$$

Here, $B_{i\sigma\sigma'}^{\alpha\beta}$ are Bose-like operators describing number (charge) and spin fluctuations:

$$B_{i\sigma\sigma'}^{22} = (X_i^{22} + X_i^{\sigma\sigma})\delta_{\sigma'\sigma} + X_i^{\sigma\bar{\sigma}}\delta_{\sigma'\bar{\sigma}}$$
$$= (\frac{1}{2}N_i + S_i^z)\delta_{\sigma'\sigma} + S_i^{\sigma}\delta_{\sigma'\bar{\sigma}}, \tag{133}$$

$$B_{i\sigma\sigma'}^{21} = (\frac{1}{2}N_i + S_i^z)\delta_{\sigma'\sigma} - S_i^{\sigma}\delta_{\sigma'\bar{\sigma}}, \tag{134}$$

$$B_{i\sigma\sigma'}^{11} = \delta_{\sigma'\sigma} - B_{i\sigma\sigma'}^{21}, \quad B_{i\sigma\sigma'}^{12} = \delta_{\sigma'\sigma} - B_{i\sigma\sigma'}^{22}, \tag{135}$$

where we used the completeness relation (34) and introduced the number operator $N_i$, Eq. (123), and the spin operators, Eq. (42). Then by using the commutation relations for the Hubbard operators, Eq. (33), we get for the matrix, Eq. (126),

$$\tilde{A}_{ij\sigma} = \begin{pmatrix} \hat{\omega}_{ij\sigma} & \hat{\Delta}_{ij\sigma} \\ (\hat{\Delta}_{ij\sigma}^*)^{(T)} & -\hat{\omega}_{ij\bar{\sigma}}^{(T)} \end{pmatrix}, \tag{136}$$

where the superscript (T) denotes the $2 \times 2$ transposed matrices which are given by

$$\hat{\omega}_{ij\sigma} = \delta_{ij} \begin{pmatrix} E_1 + \Delta + a_\sigma^{22} & a_\sigma^{21} \\ (a_\sigma^{21})^* & E_1 + a_\sigma^{22} \end{pmatrix} + (1 - \delta_{ij})V_{ij} \begin{pmatrix} K_{ij\sigma}^{22} & K_{ij\sigma}^{21} \\ (K_{ij\sigma}^{21})^* & K_{ij\sigma}^{11} \end{pmatrix}, \tag{137}$$

$$\hat{\Delta}_{ij\sigma} = \delta_{ij} \begin{pmatrix} b_\sigma^{22} & b_\sigma^{21} \\ -b_{\bar{\sigma}}^{21} & b_\sigma^{11} \end{pmatrix} + (1 - \delta_{ij})V_{ij} \begin{pmatrix} L_{ij\sigma}^{22} & L_{ij\sigma}^{21} \\ -L_{ij\bar{\sigma}}^{21} & L_{ij\sigma}^{11} \end{pmatrix}. \tag{138}$$

The components of the matrices determine the energy shift (renormalization of the chemical potential):

$$a_\sigma^{22} = \sum_{m \neq i} V_{im} \left( K_{22} \langle X_i^{2\bar{\sigma}} X_m^{\bar{\sigma}2} \rangle - K_{11} \langle X_m^{\sigma 0} X_i^{0\sigma} \rangle \right), \tag{139}$$

$$a_\sigma^{21} = - \sum_{m \neq i} V_{im} (K_{22} \langle X_i^{\sigma 0} X_m^{\bar{\sigma}2} \rangle + K_{11} \langle X_i^{\bar{\sigma}2} X_m^{\sigma 0} \rangle)$$
$$- 2\sigma \sum_{m \neq i} V_{im} K_{12} (\langle X_i^{\sigma 0} X_m^{0\sigma} \rangle - \langle X_m^{2\bar{\sigma}} X_i^{\bar{\sigma}2} \rangle), \tag{140}$$

and renormalized hopping parameters:

$$K_{ij\sigma}^{22} = K_{22}\chi_{ij}^{cs} - K_{11}\langle X_i^{02} X_j^{20}\rangle, \qquad (141)$$

$$K_{ij\sigma}^{11} = K_{11}(\chi_{ij}^{cs} + 1 - n) - K_{22}\langle X_i^{02} X_j^{20}\rangle, \qquad (142)$$

$$K_{ij\sigma}^{12} = 2\sigma K_{12}(\chi_{ij}^{cs} - \frac{1}{2}n - \langle X_i^{02} X_j^{20}\rangle), \qquad (143)$$

where we introduced static charge and spin correlation functions

$$\chi_{ij}^{cs} = \frac{1}{4}\langle N_i N_j\rangle + \langle \mathbf{S_i S_j}\rangle. \qquad (144)$$

The anomalous correlation functions in Eq. (138) are given by

$$b_\sigma^{22} = \sum_{m\neq i} V_{im}\{K_{22}(\langle X_i^{\bar{\sigma}2} X_m^{\sigma 2}\rangle - \langle X_i^{\sigma 2} X_m^{\bar{\sigma}2}\rangle)$$
$$- 2\sigma K_{12}(\langle X_i^{\sigma 2} X_m^{0\sigma}\rangle + \langle X_i^{\bar{\sigma}2} X_m^{0\bar{\sigma}}\rangle)\}, \qquad (145)$$

$$b_\sigma^{11} = -\sum_{m\neq i} V_{im}\{K_{11}(\langle X_i^{0\sigma} X_m^{0\bar{\sigma}}\rangle - \langle X_i^{0\bar{\sigma}} X_m^{0\sigma}\rangle)$$
$$- 2\sigma K_{12}(\langle X_i^{0\sigma} X_m^{\sigma 2}\rangle + \langle X_i^{0\bar{\sigma}} X_m^{\bar{\sigma}2}\rangle)\}, \qquad (146)$$

$$b_\sigma^{21} = \sum_{m\neq i} V_{im}\{K_{22}(\langle X_i^{0\sigma} X_m^{\sigma 2}\rangle + \langle X_i^{0\bar{\sigma}} X_m^{\bar{\sigma}2}\rangle)$$
$$- 2\sigma K_{12}(\langle X_i^{0\sigma} X_m^{0\bar{\sigma}}\rangle - \langle X_i^{0\bar{\sigma}} X_m^{0\sigma}\rangle)\}, \qquad (147)$$

$$L_{ij\sigma}^{22} = -2\sigma K_{21}\langle X_i^{02} N_j\rangle, \qquad (148)$$

$$L_{ij\sigma}^{11} = 2\sigma K_{21}\langle X_i^{02} N_j\rangle, \qquad (149)$$

$$L_{ij\sigma}^{21} = \frac{1}{2}(K_{22} + K_{11})\langle X_i^{02} N_j\rangle. \qquad (150)$$

In calculation of the anomalous correlation functions in Eqs. (148)-(150) we take into account Eq. (122) and symmetry conditions for the singlet pairing.

## *Anomalous Correlation Functions*

Let us consider anomalous part $\hat{\Delta}_{ij\sigma}$, Eq. (138), of the frequency matrix. Concerning the site independent anomalous correlation functions in Eqs. (145)-(147), we can show that under the assumption of $d_{x^2-y^2}$ pairing they vanish identically. Consider for instance one of the correlation function written in the **q**-representation as follows:

$$\sum_{m\neq i} V_{im}\langle X_i^{\bar{\sigma}2} X_m^{\sigma 2}\rangle = \sum_{\mathbf{q}} V(\mathbf{q})\langle X_\mathbf{q}^{\bar{\sigma}2} X_{-\mathbf{q}}^{\sigma 2}\rangle. \qquad (151)$$

In the tetragonal lattice the permutation of the components $q_x$ and $q_y$ in the sum over $\mathbf{q}$ leaves the interaction $V(\mathbf{q})$ invariant while resulting in the change of sign of the anomalous correlation function $\langle X_{\mathbf{q}}^{\bar{\sigma}2} X_{-\mathbf{q}}^{\sigma 2} \rangle$ having the $d_{x^2-y^2}$-wave symmetry. Thus, the sum over $\mathbf{q}$ equals zero, hence the correlation function at the left-hand side vanishes. Therefore all contributions in Eqs. (145)–(147) equal zero.

To calculate the site dependent contributions to the matrix, Eqs. (148)–(150), we should derive equation for the anomalous correlation function $\langle X_i^{02} N_j \rangle$. For this we apply the equation of motion for the commutator GF

$$L_{ij}(t-t') = \langle\langle X_i^{02}(t) \mid N_j(t') \rangle\rangle.$$

By differentiating it over time $t$ we get the equation:

$$(\omega - E_2) L_{ij}(\omega) \simeq 2\delta_{i,j} \langle X_i^{02} \rangle$$
$$+ \sum_{m,\sigma} 2\sigma t_{im}^{12} \{ \langle\langle X_i^{0\bar{\sigma}} X_m^{0\sigma} | N_j \rangle\rangle_\omega - \langle\langle X_i^{\sigma 2} X_m^{\bar{\sigma}2} | N_j \rangle\rangle_\omega \}. \tag{152}$$

Here we have neglected contributions from the intraband hopping integrals in the Hubbard Hamiltonian (116), while have retained the interband hopping which mediates the exchange interaction. This approximation can be proved if we take into account that for hole doping we have $E_2 - E_1 = \tilde{\Delta} - \mu \simeq 0$ and therefore $E_2 = \tilde{\Delta} - 2\mu \simeq -\tilde{\Delta}$, while for electron doping $\mu \simeq 0$ and $E_2 \simeq \tilde{\Delta}$. So in the both cases $|E_2| = \tilde{\Delta}$ and in the strong correlation limit, $\tilde{\Delta} \gg |t_{ij}^{\alpha\beta}|$, we have $|E_2| \gg |t_{ij}^{\alpha\beta}|$. Now, by using the spectral representation for the GF we obtain for the correlation function for lattice sites $i \neq j$:

$$\langle X_i^{02} N_j \rangle = \int_{-\infty}^{+\infty} \frac{d\omega}{1 - \exp(-\omega/T)} \sum_{m,\sigma} 2\sigma t_{im}^{12}$$
$$\times [-\frac{1}{\pi} \text{Im} \frac{1}{\omega - E_2 + i\delta} (\langle\langle X_i^{0\bar{\sigma}} X_m^{0\bar{\sigma}} | N_j \rangle\rangle_{\omega+i\delta} - \langle\langle X_i^{\sigma 2} X_m^{\bar{\sigma}2} | N_j \rangle\rangle_{\omega+i\delta})]. \tag{153}$$

Below we study a particular case of hole doping when the atomic energy $E_2 \simeq E_1 \simeq -\Delta$. In that case we can neglect the high-energy contribution given by $\delta(\omega - E_2)$ since it is proportional to $\exp(-\Delta/T) \ll 1$. For the one-hole band contribution we can use an estimation

$$-\frac{1}{\pi} \text{Im} \langle\langle X_i^{0\bar{\sigma}} X_m^{0\sigma} | N_j \rangle\rangle_{\omega+i\delta} \simeq \delta_{m,j} \langle X_i^{0\bar{\sigma}} X_j^{0\sigma} \rangle \delta(\omega - 2E_1),$$

which shows that it is proportional to $\exp(-2\Delta/T) \ll 1$ and also can be neglected. Therefore, after integration over $\omega$ of the last term in Eq. (153) we derive the following result for the anomalous correlation function

$$\langle X_i^{02} N_j \rangle = -\frac{1}{\Delta} \sum_{m,\sigma} 2\sigma t_{im}^{12} \langle X_i^{\sigma 2} X_m^{\bar{\sigma}2} N_j \rangle. \tag{154}$$

In treating the exchange interaction the two-site approximation is usually used (see Sec. II B) which is given by the terms $(m = i)$ in Eq. (154). In this approximation after summation over spin $\sigma$ we get:

$$\langle X_i^{02} N_j \rangle = -\frac{4t_{ij}^{12}}{\Delta} 2\sigma \langle X_i^{\sigma 2} X_j^{\bar\sigma 2} \rangle,$$

where we have used the symmetry property of the anomalous averages: $\langle X_i^{\sigma 2} X_j^{\bar\sigma 2} \rangle = -\langle X_i^{\bar\sigma 2} X_j^{\sigma 2} \rangle$. Therefore, we have proved that in the MFA the intersite anomalous correlation function is of the order of $t_{ij}^{12}/\Delta$ and is just proportional to the conventional electron (hole) pairs on neighbor lattice sites, $\langle X_i^{\sigma 2} X_j^{\bar\sigma 2} \rangle$. In the conventional notation we can write:

$$\langle c_{i\downarrow} c_{i\uparrow} N_j \rangle = \langle X_i^{0\downarrow} X_i^{\downarrow 2} N_j \rangle = \langle X_i^{02} N_j \rangle = -\frac{4t_{ij}^{12}}{\Delta} 2\sigma \langle X_i^{\sigma 2} X_j^{\bar\sigma 2} \rangle. \quad (155)$$

Here we have used the definition: $c_{i\sigma} = X_i^{0\sigma} + 2\sigma X_i^{\bar\sigma 2}$ and identities for Hubbard operators: $X_i^{0\sigma} X_i^{0-\sigma} = 0$, $X_i^{\sigma 2} X_i^{-\sigma 2} = 0$, $X_i^{0\downarrow} X_i^{\downarrow 2} = X_i^{02}$. For the anomalous component of frequency matrix, Eq. (138), for lattice sites $i \neq j$ we finally get

$$\Delta_{ij\sigma}^{22} = -\Delta_{ij\sigma}^{11} = J_{ij} \langle X_i^{\sigma 2} X_j^{\bar\sigma 2} \rangle. \quad (156)$$

This is exactly the exchange interaction contribution to the pairing where the exchange energy $J_{ij} = 4(t_{ij}^{12})^2/\Delta$. The off-diagonal anomalous matrix components in Eq. (138) can be neglected since they are of higher order in the hopping energy.

Our calculation proves that the anomalous contributions to the zero-order GF, Eq. (127), are given by the conventional anomalous pairs of quasiparticles and their pairing in MFA is mediated by the exchange interaction which has been studied in the $t$-$J$ model (see Sec. III B). Therefore we can conclude that a nonzero superconducting pairing observed in MFA in Refs. [72–74] is due to the exchange interaction which in the conventional Hubbard model equals to $J_{ij} = 4(t_{ij}^{12})^2/U$. It vanishes in the limit $U \to \infty$ that explains strong suppression of the superconducting $T_c$ observed in Refs. [72–74] for large $U$.

## Spectrum in the Mean-Field Approximation

Let us study at first the zero-order GF (127) in the normal state in MFA when it can be written in the form

$$\tilde{G}_\sigma^0(\mathbf{q}, \omega) = \begin{pmatrix} \hat{G}_\sigma^0(\mathbf{q}, \omega) & 0 \\ 0 & -\hat{G}_\sigma^0(-\mathbf{q}, -\omega)^{(T)} \end{pmatrix}, \quad (157)$$

where the $2 \times 2$ matrix GF reads [69]

$$\hat{G}_\sigma^0(\mathbf{q}\omega) = \{\omega \hat{\tau}_0 - \hat{E}_\sigma(\mathbf{q})\}^{-1} \begin{pmatrix} \chi_2 & 0 \\ 0 & \chi_1 \end{pmatrix}, \quad (158)$$

The **q**–representation for the energy matrix is given by

$$\hat{E}_\sigma(\mathbf{q}) = \begin{pmatrix} \omega_2(\mathbf{q}) & W_\sigma^{21}(\mathbf{q}) \\ W_\sigma^{12}(\mathbf{q}) & \omega_1(\mathbf{q}) \end{pmatrix}. \tag{159}$$

The energy spectra for unhybridized singlets and holes are defined by the equations

$$\omega_2(\mathbf{q}) = E_1 + \Delta + a_\sigma^{22}/\chi_2 + V_\sigma^{22}(\mathbf{q})/\chi_2 ,$$
$$\omega_1(\mathbf{q}) = E_1 + a_\sigma^{22}/\chi_1 + V_\sigma^{11}(\mathbf{q})/\chi_1 , \tag{160}$$

while the hybridization interaction is given by

$$\chi_2 W_\sigma^{21} = a_\sigma^{21} + V_\sigma^{21}(\mathbf{q}), \quad \chi_2 W_\sigma^{21} = \chi_1 W_\sigma^{12}. \tag{161}$$

The effective interaction in (160), (161) has the form

$$V_\sigma^{\alpha\beta}(\mathbf{q}) = \frac{t}{N} \sum_k \nu(\mathbf{k}) \tilde{K}_\sigma^{\alpha\beta}(\mathbf{k\text{-}q}) , \tag{162}$$

where $\tilde{K}_\sigma^{\alpha\beta}(\mathbf{q})$ is the Fourier transform of $\tilde{K}_{ij\sigma}^{\alpha\beta}$, Eqs. (141)-(143), and $\nu(\mathbf{q})$ denoting the geometrical structure factor

$$\nu(\mathbf{q}) = 2\sum_{j\neq 0} \nu_{0j} e^{-i\mathbf{q}\cdot\mathbf{j}} = 8\nu_1 \gamma(\mathbf{q}) + 8\nu_2 \gamma'(\mathbf{q}), \tag{163}$$

for the nearest neighbors, $\gamma(\mathbf{q}) = (1/2)(\cos q_x + \cos q_y))$, and the second neighbors, $\gamma'(\mathbf{q}) = \cos q_x \cos q_y$.

By using the matrix representation (159) the zero–order GF can be written in the diagonal form

$$\hat{G}_\sigma^0(\mathbf{q},\omega) = \begin{pmatrix} \chi_2/[\omega - \Omega_2(\mathbf{q})] & 0 \\ 0 & \chi_1/[\omega - \Omega_1(\mathbf{q})] \end{pmatrix}, \tag{164}$$

if we neglect the off-diagonal GF of the order $|t^{12}|/\Delta$ which appears to give a small contribution to the density of states [69]. The hybridized spectra $\Omega_\alpha(\mathbf{q})$ for singlets ($\alpha=2$) and, respectively, for holes ($\alpha=1$) are given by

$$\Omega_{2,1}(\mathbf{q}) = \frac{1}{2}[\omega_2(\mathbf{q}) + \omega_1(\mathbf{q})] \pm \frac{1}{2}\{[\omega_2(\mathbf{q}) - \omega_1(\mathbf{q})]^2 + 4W_\sigma^{21} W_\sigma^{12}\}^{1/2}. \tag{165}$$

To obtain a closed system of equations we have to calculate self–consistently the correlation functions in (137). The energy shifts $a_\sigma^{\alpha\beta}$ (139), (140) can be readily calculated by using the spectral representation for the GF (164). But to calculate the two–particle correlation functions $K_{ij\sigma}^{\alpha\beta}$ in (141) – (143) we have to adopt some approximations. To calculate the correlation function $\langle X_i^{02} X_j^{20} \rangle$ we can apply the equation of motion method for GF $\langle\langle X_i^{02} | X_j^{20} \rangle\rangle_\omega$ as has been described in the

169

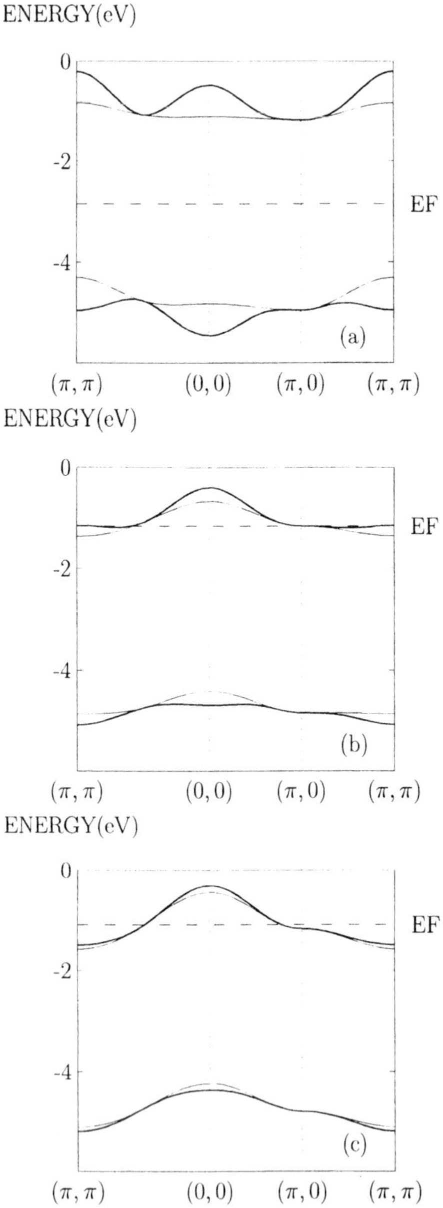

**FIGURE 9.** Energy spectra for one-hole and singlet subbands, Eq. (165), (thick line) in the undoped (a) $n = 1$, optimally doped (b) $n = 1.2$, and overdoped doped (c) $n = 1.4$ cases for parameters: $\Delta = 3eV$ and $t = 1.5eV$. For comparison we show also the result without hybridization (thin line). The zero of energy correspond to the position of the non-bonding oxygen band and the EF denotes the Fermi level [69].

previous section. Due to the large pair excitation energy $|E_2| \simeq \Delta$ the contribution of this correlation function appears to be small, of the order $(t_{ij}^{12}/\Delta)^2$, and can be neglected. Concerning the correlation functions in Eq. (144) we neglect small charge correlations and decouple the product $N_i N_j$ of the numbers operators $N_i$ on different lattice sites $i \neq j$

$$\chi_{ij}^{cs} \simeq (\chi_2)^2 + \langle \mathbf{S_i S_j} \rangle. \tag{166}$$

The results of numerical solution of the self-consistent system of equations for the one-hole and singlet subband specta, Eq. (165), in the adopted approximation is shown in Fig. 9, Ref. [69]. For a spin-singlet state without long-range magnetic order the GF (164) and the one-hole spectrum (165) do not depend on the spin. But short-range AFM spin correlations are very important as can be seen from the position of the excitation energy at $(\pi, \pi)$ point in Fig. 9 (a-c). The spin correlation functions $\langle \mathbf{S_i S_j} \rangle$ in (166) which have been taken as a parameters of the calculation (see [69]) give a considerable contribution to the renormalization of the dispersion relation $\Omega_{2,1}(\mathbf{q})$ (165). For large spin-correlations at undoped case, $(n = 1)$, one finds a next-nearest neighbor dispersion, Fig. 9 (a). With doping, by decreasing the spin correlations, the dispersion changes to an ordinary nearest neighbor one, Fig. 9 (c). It should be also pointed out that for the optimum doping, $n = 1.2$ the dispersion curves are rather flat close to $(\pi, 0)$ points, Fig. 9 (b), which results in a very high density of state at the Fermi level, $N(0) \simeq 10$ eV$^{-1}$ Ref. [69].

By using the results obtained in this section, let us consider the $2 \times 2$ matrix GF for the singlet subband

$$\hat{G}_{ij,\sigma}^{22}(\omega) = \langle\langle \begin{pmatrix} X_i^{\sigma 2} \\ X_i^{2\bar{\sigma}} \end{pmatrix} | (X_j^{2\sigma} \; X_j^{\bar{\sigma}2}) \rangle\rangle_\omega. \tag{167}$$

In the MFA GF (167) for the singlet subband can be written in the same form as for the $t$-$J$ model, Eq. (84):

$$\hat{G}_\sigma^{22}(q,\omega) = \chi_2 \{\omega \hat{\tau}_0 - \Omega_{2,\sigma}(q)\hat{\tau}_3 - \phi_\sigma^{22}(q)\hat{\tau}_1\}^{-1}. \tag{168}$$

The gap function induced by the exchange interaction is defined by the equation:

$$\phi_\sigma^{22}(q) = \frac{1}{N\chi_2} \sum_k J(k-q) \langle X_{-q}^{2\bar{\sigma}} X_q^{2\sigma} \rangle. \tag{169}$$

where we introduced the Fourier component for the nearest neighbor exchange interaction, $J(q) = 4J\gamma(q)$. The anomalous correlation function for the singlet subband is easy to calculate from the anomalous part of the GF (168) that results in the following BCS-type equation for gap function (169) in the MFA:

$$\phi_\sigma^{22}(q) = \frac{1}{N} \sum_k J(k-q) \frac{\phi_\sigma^{22}(k)}{2E_2(k)} \tanh \frac{E_2(k)}{2T}, \tag{170}$$

with the quasiparticle energy $E_2(k) = [\Omega_{2,\sigma}(k)^2 + \phi_\sigma^{22}(k)^2]^{1/2}$. This equation is identical to Eq. (87) derived in MFA for the $t$-$J$ model (see also [59]).

Analogous equations can be obtained for the electron doped case, $n < 1$, with the chemical potential in the one-hole band by considering GF for the Hubbard operators $X_i^{0\sigma}$, $X_i^{\bar\sigma 0}$.

## C  Self-Energy Corrections

In the present section we study the self-energy is given by Eq. (129):

$$\tilde\Sigma_{ij\sigma}(t-t') = \tilde\chi^{-1} \langle\langle \hat Z_{i\sigma}^{(ir)}(t) | \hat Z_{j\sigma}^{(ir)\dagger}(t') \rangle\rangle^{(prop)} \tilde\chi^{-1}. \tag{171}$$

The operators $Z^{(ir)}$ are obtained from Eqs. (130)-(132) by dropping off the average terms which result in reducible diagrams. Below we calculate the self-energy (171) in the self-consistent Born approximation (SCBA) (or the non-crossing approximation) which has already been used in Sec. III B. In SCBA propagation of the Fermi-like and Bose-like excitations in the many-particle GF in (171) are assumed to be independent and therefore the SCBA is given by a decoupling of the corresponding operators in the time-dependent correlation functions. If $X_1(t)$ and $X_2(t')$ are generic notations for Fermi-like operators and $B_{1'}(t)$ and $B_{2'}(t')$ are Bose-like operators, then the decoupling writes for $1' \neq 1$ and $2' \neq 2$:

$$\langle B_{1'}(t) X_1(t) B_{2'}(t') X_2(t') \rangle \simeq \langle X_1(t) X_2(t') \rangle \langle B_{1'}(t) B_{2'}(t') \rangle. \tag{172}$$

Using the spectral theorem, the SCBA results in the following decoupling relation for the emerging GF in the $(\mathbf{r}, \omega)$-representation:

$$\langle\langle B_{1'} X_1 | B_{2'} X_2 \rangle\rangle_\omega \simeq \frac{1}{\pi^2} \int_{-\infty}^{+\infty}\int_{-\infty}^{+\infty} \frac{d\omega_1 d\omega_2 N(\omega,\omega_1,\omega_2)}{\omega - \omega_1 - \omega_2} \operatorname{Im}\langle\langle X_1|X_2\rangle\rangle_{\omega_1} \operatorname{Im}\langle\langle B_{1'}|B_{2'}\rangle\rangle_{\omega_2}. \tag{173}$$

where the function $N(\omega, \omega_1, \omega_2)$ is given by Eq. (93). Within this approximation, the GF associated to the irreducible operators in Eq. (171) can be evaluated and we get the self-energy in the $(\mathbf{r}, \omega)$-representation,

$$\tilde\Sigma_{ij\sigma}(\omega) = \tilde\chi^{-1} \begin{pmatrix} \hat M_{ij\sigma}(\omega) & \hat\Phi_{ij\sigma}(\omega) \\ \hat\Phi_{ij\sigma}^{\dagger(T)}(\omega) & -\hat M_{ij\bar\sigma}^{(T)}(-\omega) \end{pmatrix} \tilde\chi^{-1}, \tag{174}$$

where the $2\times 2$ matrices $\hat M$ and $\hat \Phi$ denote the normal and anomalous contributions to the self-energy respectively:

$$\hat M_{ij\sigma}(\omega) = \begin{pmatrix} M_{ij\sigma}^{22}(\omega) & M_{ij\sigma}^{21}(\omega) \\ M_{ij\sigma}^{12}(\omega) & M_{ij\sigma}^{11}(\omega) \end{pmatrix} = \left\langle\!\left\langle \begin{pmatrix} (Z_i^{\sigma 2})^{(ir)} \\ (Z_i^{0\bar\sigma})^{(ir)} \end{pmatrix} \middle| \left(Z_j^{2\sigma}\right)^{(ir)} \ \left(Z_j^{\bar\sigma 0}\right)^{(ir)} \right\rangle\!\right\rangle_\omega, \tag{175}$$

$$\hat{\Phi}_{ij\sigma}(\omega) = \begin{pmatrix} \Phi^{22}_{ij\sigma}(\omega) & \Phi^{21}_{ij\sigma}(\omega) \\ \Phi^{12}_{ij\sigma}(\omega) & \Phi^{11}_{ij\sigma}(\omega) \end{pmatrix} = \left\langle\!\left\langle\!\left\langle \begin{pmatrix} (Z_i^{\sigma 2})^{(ir)} \\ (Z_i^{0\bar{\sigma}})^{(ir)} \end{pmatrix} \middle| \left(Z_j^{\bar{\sigma}2}\right)^{(ir)} \; \left(Z_j^{0\sigma}\right)^{(ir)} \right\rangle\!\right\rangle\!\right\rangle_\omega. \quad (176)$$

The GF (118) can be also written as a $2 \times 2$ supermatrix of normal, $\hat{G}_{ij\sigma}(\omega)$, and anomalous, $\hat{F}_{ij\sigma}(\omega)$, $2 \times 2$ matrix components:

$$\tilde{G}_{ij\sigma}(\omega) = \begin{pmatrix} \hat{G}_{ij\sigma}(\omega) & \hat{F}_{ij\sigma}(\omega) \\ \hat{F}^\dagger_{ij\sigma}(\omega) & -\hat{G}^{(T)}_{ij\bar{\sigma}}(-\omega) \end{pmatrix}. \quad (177)$$

Eqs. (174) and (177) provide a closed self-consistent system of equations for the self-energy and GF within the SCBA (173).

To get a tractable problem, this system is simplified using diagonal approximations for the GF solution of the normal state, while the matrix elements of $\hat{M}_{ij\sigma}(\omega)$ and $\hat{\Phi}_{ij\sigma}(\omega)$ are calculated keeping the leading terms in $V_{ij}$ (i.e., we assume hybridization ($t^{12}/\Delta$) of the lowest order). Use of Fourier transforms and of the spin reversal symmetries of the GF which allow us to perform the summation over the spin variable, finally results in the $(\mathbf{q}, \omega)$-representation of the contributions to the normal and anomalous parts of the self-energy [70]:

$$\hat{M}_\sigma(\mathbf{q}, \omega) = \frac{1}{N} \sum_\mathbf{k} \int_{-\infty}^{+\infty} d\omega_1 K^{(+)}(\omega, \omega_1 \mid \mathbf{k}, \mathbf{q}-\mathbf{k})$$

$$\times \left\{ -\frac{1}{\pi} \text{Im}[\, \hat{P}^{(+)}_2 G^{22}_\sigma(\mathbf{k}, \omega_1) + \hat{P}^{(+)}_1 G^{11}_\sigma(\mathbf{k}, \omega_1)]\right\}, \quad (178)$$

$$\hat{\Phi}_\sigma(\mathbf{q}, \omega) = \frac{1}{N} \sum_\mathbf{k} \int_{-\infty}^{+\infty} d\omega_1 \, K^{(-)}(\omega, \omega_1 \mid \mathbf{k}, \mathbf{q}-\mathbf{k})$$

$$\times \left\{ -\frac{1}{\pi} \text{Im}[\, \hat{P}^{(-)}_2 F^{22}_\sigma(\mathbf{k}, \omega_1) - \hat{P}^{(-)}_1 F^{11}_\sigma(\mathbf{k}, \omega_1)]\right\}, \quad (179)$$

where

$$\hat{P}^{(\pm)}_2 = \begin{pmatrix} K_{22}^2 & \pm 2\sigma K_{21}K_{22} \\ 2\sigma K_{21}K_{22} & \pm K_{21}^2 \end{pmatrix}, \quad \hat{P}^{(\pm)}_1 = \begin{pmatrix} K_{21}^2 & \pm 2\sigma K_{21}K_{11} \\ 2\sigma K_{21}K_{11} & \pm K_{11}^2 \end{pmatrix} \quad (180)$$

The kernel of the integral equations for the self-energy operators here is defined in the standard way:

$$K^{(\pm)}(\omega, \omega_1 \mid \mathbf{k}, \mathbf{q}-\mathbf{k}) = t^2 \, |\nu(\mathbf{k})|^2 \int_{-\infty}^{+\infty} \frac{d\omega_2}{\omega - \omega_1 - \omega_2}$$

$$\times \quad N(\omega, \omega_1, \omega_2) \, [-\frac{1}{\pi} \text{Im} D^{(\pm)}(\mathbf{q}-\mathbf{k}, \omega_2)], \quad (181)$$

with a spectral density of spin-charge fluctuations given by the commutator GF

$$D^{(\pm)}(\mathbf{q},\omega) = \langle\langle \mathbf{S_q}|\mathbf{S_{-q}}\rangle\rangle_\omega \pm \frac{1}{4}\langle\langle \delta N_\mathbf{q}|\delta N_\mathbf{-q}\rangle\rangle_\omega = \chi^s(\mathbf{q},\omega) \pm \chi^c(\mathbf{q},\omega), \tag{182}$$

and $\nu(\mathbf{q})$ is given by Eq. (163).

The quasi-particle energy $\bar{\Omega}_\alpha(\mathbf{q})$ in the normal state for the diagonal GF in Eqs. (178), (179) can be evaluated as follows

$$\bar{\Omega}_{\alpha,\sigma}(\mathbf{q}) = \Omega_\alpha(\mathbf{q}) - \chi_\alpha^{-1} M_\sigma^{\alpha\alpha}(\mathbf{q},\omega = \Omega_\alpha(\mathbf{q})), \tag{183}$$

where the energy in MFA $\Omega_\alpha(\mathbf{q})$ is defined by Eq. (165).

## D  Weak Coupling Approximation

Once the self-energy corrections (178) and (179) have been calculated in SCBA, we can use them in the Dyson equation (128) to get the one-particle matrix GF. An important simplification of the Dyson equation (128) occurs if we observe that the off-diagonal normal and anomalous matrix elements of the GF, Eqs. (175), (176), are small as compared to the diagonal ones. Neglecting the small matrix elements, the inversion of the GF matrix in (128) results in simple enough analytical formulas. In spite of these simplification, the numerical complexity of the resulting $(q,\omega)$-dependent integral equations which define the critical temperature predicted by the two-band singlet-hole model is enormous.

To get a tractable problem we consider the weak coupling approximation (WCA), which assumes that the behavior of the physical system is dominated by the interactions around the Fermi-level.

In the WCA the interaction kernel (181) at frequencies $(\omega,\omega_1)$ close to the Fermi surface is factorized in the form (compare with Eq. (110))

$$K^{(\pm)}(\omega,\omega_1|\mathbf{k},\mathbf{q}-\mathbf{k}) \simeq -\frac{1}{2}\tanh\left(\frac{\omega_1}{2T}\right)\lambda^{(\pm)}(\mathbf{k},\mathbf{q}-\mathbf{k}), \tag{184}$$

for $|\omega,\omega_1| \leq \omega_s \ll W$ where $\omega_s$ is a characteristic pairing energy and $W$ is the band width. In this approximation the effective interaction is defined by the static susceptibility

$$\lambda^{(\pm)}(\mathbf{k},\mathbf{q}-\mathbf{k}) = t^2 |\nu(\mathbf{k})|^2 \int_{-\infty}^{+\infty} \frac{d\omega_2}{\omega_2}[-\frac{1}{\pi}\mathrm{Im}\, D^{(\pm)}(\mathbf{q}-\mathbf{k},\omega_2)]$$

$$= -t^2 |\nu(\mathbf{k})|^2 \mathrm{Re}\, D^{(\pm)}(\mathbf{q}-\mathbf{k},\omega_2 = 0)]. \tag{185}$$

The WCA can be applied for the band close to the FS while for another band which is far away from the FS, at the energy of the order of the band gap, $\omega_1 \simeq \Delta$, an integration over $\omega_1$ in Eqs. (178), (179) is straightforward.

Let us consider a hole doped system, $n = 1 + \delta \geq 1$. In that case the chemical potential is in the singlet band, $\mu \simeq \Delta$, and we can write the dispersion relations for the two bands in the normal state, Eq. (183), as follow

$$\bar{\Omega}_2(\mathbf{q}) \simeq \Delta - \mu + \epsilon_2(\mathbf{q}) \simeq \epsilon_2(\mathbf{q}),$$

$$\bar{\Omega}_1(\mathbf{q}) \simeq -\mu + \epsilon_1(\mathbf{q}) \simeq -\Delta + \epsilon_1(\mathbf{q}), \qquad (186)$$

where at the Fermi wave-vector $\epsilon_2(\mathbf{q}_F) = 0$. Integration over $\omega_1$ in Eq. (179) gives us the following system of equations for the two diagonal anomalous components of the self-energy, $\Phi^{\alpha\alpha}(\mathbf{q}) = 2\sigma \Phi^{\alpha\alpha}_\sigma(\mathbf{q}, \omega = 0)$,

$$\Phi^{22}(\mathbf{q}) = -K^2_{22}\mathcal{S}_2(\mathbf{q}) + K^2_{12}\mathcal{S}_1(\mathbf{q}),$$
$$\Phi^{11}(\mathbf{q}) = K^2_{12}\mathcal{S}_2(\mathbf{q}) - K^2_{11}\mathcal{S}_1(\mathbf{q}). \qquad (187)$$

The sum $\mathcal{S}_2(\mathbf{q})$ for the singlet band at the FS is given by

$$\mathcal{S}_2(\mathbf{q}) = \frac{1}{N}\sum_{\mathbf{k}} \lambda^{(-)}(\mathbf{k}, \mathbf{q}-\mathbf{k}) \frac{\Phi^{22}(\mathbf{k})}{2E_2(\mathbf{k})} \tanh \frac{E_2(\mathbf{k})}{2T}, \qquad (188)$$

where integration over $\mathbf{k}$ is restricted to an energy shell around the Fermi energy of the order of a characteristic energy $\omega_s$ of spin (charge) fluctuations. The quasi-particle energy here is given by

$$E_2(\mathbf{q}) = [\bar{\Omega}_2^2(\mathbf{q}) + |\chi_2^{-1}\Phi^{22}(\mathbf{q})|^2]^{1/2}, \qquad (189)$$

where $\bar{\Omega}_2(\mathbf{q})$ is the quasi-particle dispersion in the normal state, Eq. (183).

For the one-hole band below the FS at the energy of the order $\Delta \gg W$ we can neglect the dispersion in the one-hole band in Eq. (186), as well as the superconducting gap in the anomalous GF in Eq. (179) in the integration over $\omega_1$ that results in the estimation for the sum

$$\mathcal{S}_1(\mathbf{q}) = \frac{1}{N}\sum_{\mathbf{k}} t^2|\nu(\mathbf{k})|^2 \frac{\Phi^{11}(\mathbf{k})}{\Delta^2} \chi_{sc}^{(-)}(\mathbf{k}-\mathbf{q}). \qquad (190)$$

Here the integration over $\omega_2$ in Eq. (181) results in the spin-charge correlation function

$$\chi_{sc}^{(-)}(\mathbf{q}) = \langle \mathbf{S_q S_{-q}}\rangle - \frac{1}{4}\langle N_{\mathbf{q}} N_{-\mathbf{q}}\rangle. \qquad (191)$$

Simple estimation shows that the sum $\mathcal{S}_1(\mathbf{q})$ gives a small contribution of the order $(t_{eff}/\Delta)^2 \Phi^{11}(\mathbf{q}) \simeq 10^{-2}\Phi^{11}(\mathbf{q})$ and can be neglected in the system of equations (187). By taking into account the contribution due to the exchange interaction, Eq. (170), the equation for the singlet gap in the WCA can written as follows:

$$\Phi^{22}(\mathbf{q}) = \frac{1}{N}\sum_{\mathbf{k}}[J(\mathbf{k}-\mathbf{q}) - K_{22}^2\lambda^{(-)}(\mathbf{k},\mathbf{q}-\mathbf{k})]\frac{\Phi^{22}(\mathbf{k})}{2E_2(\mathbf{k})}\tanh\frac{E_2(\mathbf{k})}{2T}. \quad (192)$$

We note that this equation is identical to Eq. (113), Sec. II B, for the gap function in the WCA within the $t$-$J$ model. However, in the two-band model due to coupling between the subbands the gap $\Phi^{11}(\mathbf{q})$ in the one-hole subbands is also nonzero. From the second equation in (187) we get an estimate for the one-hole band gap $\Phi^{11}(\mathbf{q}) \simeq K_{12}^2 \mathcal{S}_2(\mathbf{q})$.

The same consideration is applicable for an electron doped system, $n = 1+\delta \leq 1$ when $\mu \simeq 0$ in Eq. (186). In that case the gap $\Phi^{11}(\mathbf{q})$ and critical $T_c$ are defined by the second equation in (187) with a proper account of the corresponding exchange interaction as in Eq. (192).

## E  Numerical Results and Discussions

Let us consider at first analytical estimation for superconducting $T_c$ mediated by spin fluctuation coupling $\lambda_s(\mathbf{k},\mathbf{q}-\mathbf{k})$ in Eq. (192). To solve the gap equations, a model for the bosonic GF in the effective interaction (185) is needed. The charge-charge fluctuations are small as compared to the spin-spin fluctuations and they are neglected. To evaluate the spin-spin fluctuations, a model representation of the spin-fluctuation susceptibility is necessary. We use the same model as in Sec. II B suggested in numerical studies [109]. Then the effective static coupling due to spin fluctuations, Eq. (185), reduces to

$$\lambda_s(\mathbf{k},\mathbf{q}-\mathbf{k}) = t^2|\nu(\mathbf{k})|^2\chi_s(\mathbf{q}-\mathbf{k}). \quad (193)$$

For calculation of $T_c$ we can consider a linearized Eq. (192) by neglecting the gap function in (189). For the $d$-wave pairing symmetry we assume the conventional $q$-dependence of the gaps:

$$\Phi^{\alpha\alpha}(\mathbf{q}) = \phi_\alpha(\cos q_x - \cos q_y) = \phi_\alpha \eta(\mathbf{q}). \quad (194)$$

Then integrating over $\mathbf{q}$ both sides of the equation (192) multiplied by $\eta(\mathbf{q}) = (\cos q_x - \cos q_y)$ we get the following equation for $T_c$:

$$\frac{K_{22}^2}{N^2}\sum_{\mathbf{q},\mathbf{k}}\lambda_s(\mathbf{k},\mathbf{q}-\mathbf{k})\frac{\eta(\mathbf{q})\eta(\mathbf{k})}{2\bar{\Omega}_{2,\mathbf{k}}}\tanh\frac{\bar{\Omega}_{2,\mathbf{k}}}{2T} = -1. \quad (195)$$

Let us estimate the sum over $\mathbf{q}$ in this equation. Since the susceptibility term $\chi_s(\mathbf{q}-\mathbf{k})$ in $\lambda_s(\mathbf{k},\mathbf{q}-\mathbf{k})$ is a positive function with a sharp maximum at the AFM wave-vector $\mathbf{q}-\mathbf{k} = \mathbf{Q} = (\pi,\pi)$ the sum over $\mathbf{q}$ in Eq. (195) can be evaluated by the mean value theorem:

$$\frac{1}{N}\sum_{\mathbf{q}}\eta(\mathbf{q})\chi_s(\mathbf{q}-\mathbf{k}) \simeq \eta(\mathbf{k}+\mathbf{Q})\frac{1}{N}\sum_{\mathbf{q}}\chi_s(\mathbf{q}) = -\eta(\mathbf{k})\chi_0 C_1. \quad (196)$$

Using this estimate and averaging over the direction of the wave-vector **k** in Eq. (195) we get the following equation for $T_c$ mediated by spin fluctuation:

$$\frac{\lambda_{s,22}}{N} \sum_{\mathbf{k}} \frac{1}{2\bar{\Omega}_2(\mathbf{k})} \tanh \frac{\bar{\Omega}_2(\mathbf{k})}{2T_c} = 1, \qquad (197)$$

where for the effective spin-fluctuation coupling

$$\lambda_{s,22} \simeq \chi_0 C_1 \frac{K_{22}^2 t^2}{N} \sum_{\mathbf{k}} \eta(\mathbf{k})^2 |\nu(\mathbf{k})|^2 \simeq \frac{t_{eff}^2}{\omega_s}, \qquad (198)$$

where $t_{eff} = K_{22} 2\nu_1 t \simeq 0.14 t$. In the the logarithmic approximation we get from Eq. (197)

$$\frac{\lambda_{s,22}}{N} \sum_{\mathbf{q}} \frac{1}{2\epsilon_2(\mathbf{q})} \tanh \frac{\epsilon_2(\mathbf{q})}{2T_c} \simeq \lambda_{s,22} N(0) \ln(1.13 \frac{\omega_s}{T_c}), \qquad (199)$$

where $N(0)$ is the density of states at the Fermi energy for the singlet band.

To obtain some numerical values we take the parameters of the model given by Eq. (29): $K_{11} \simeq -0.89$, $K_{22} \simeq -0.48$, $K_{12} \simeq 0.83$,. For the density of state we take some average value $N(0) \simeq 1$ (eV · spin)$^{-1}$, while for the spin-fluctuation energy we take $\omega_s = 0.15$ eV - the characteristic exchange energy in cuprates, $\omega_s \ll W \simeq t$. For these parameters we obtain the following estimation $\lambda_{s,22} \simeq 0.27$ eV that results in a weak coupling, $V_{s,22} = \lambda_{s,22} N(0) \simeq 0.27$. However, due to a large spin-fluctuation energy $\omega_s$ we obtain high critical temperature for the singlet band

$$T_c \simeq \omega_s \exp\{-1/V_{s,22}\} \simeq 0.03 \, \omega_s \simeq 50 \text{ K}. \qquad (200)$$

The superconducting gap $\phi_1$ in the one-hole band induced by the gap in the singlet band, according to Eqs. (187), is quite large:

$$\phi_1/\phi_2 \simeq -\frac{K_{12}^2}{K_{22}^2} \simeq -3, \qquad (201)$$

But its contribution to the superconducting density of state is negligible, of the order of $(\phi_1/\Delta)^2$ at the energy $\omega \simeq -\Delta$.

As usually in the BCS-type theory considered here, softening of boson exchange energy leads to the enhancements of $T_c$. Namely, decreasing of the spin-fluctuation energy $\omega_s$ increases the coupling constant $\lambda_{s,22}$, Eq. (198), that results in enhancement of $T_c$. For instance, for $\omega_s = 50$ meV we have $\lambda_{s,22} \simeq 0.81$ eV and $T_c \simeq 200$ K instead of 50 K in Eq. (200). We can speculate here that higher $T_c$ in YBCO compound in comparison with LSCO could be explained by different spin-fluctuation spectra in these materials: a resonance peak around 40 meV in YBCO could give much stronger coupling then a broad featureless spectrum in LSCO [119].

The same estimations can be done for the electronically doped case, $n = 1 + \delta \leq 1$, when $\mu \simeq 0$. In that case the critical $T_c$ is given by the integral $S_1$. For the symmetrical Hubbard model with the equal hopping integrals, $K_{11} = K_{22}$, we will have the electron-hole symmetry and the same $T_c$. In the reduced p-d model considered here the one-hole band has a different coupling constant, $\lambda_{s,11} \simeq 0.95$ eV which is larger then $\lambda_{s,22} \simeq 0.27$ eV for the singlet band. However, due to a lower density of states $N_1(0)$ for a broader one-hole band, the effective coupling constant, $V_{s,11} = \lambda_{s,11} N_1(0)$, can be smaller that results in lower $T_c$, as observed in cuprates. Due to uncertainty in the value of $N_1(0)$, we will not speculate on $T_c$ in this case only pointing out that the gap ratio which does not depend on $N_1(0)$ is close to unity: $\phi_2/\phi_1 \simeq -K_{12}^2/K_{11}^2 \simeq -0.9$.

The analytical estimates given above show a possibility of spin-fluctuation $d_{x^2-y^2}$ symmetry pairing in one subband due to the kinematic interaction in the second order of the intraband hopping matrix elements. This mechanism is absent in the standard spin-fermion models usually considered in phenomenological approach. Superconducting pairing mediated by the exchange interaction which is of the second order in the interband hopping was studied by many authors within the t-J model in MFA (see, e.g., [59,60]).

Below we compare both contributions by presenting the results of numerical solution for the gap equation (192). The critical temperature was calculated from the linearized equation. It corresponds to the temperature at which the maximum eigenvalue of this system equals unity. For a strongly interacting system the symmetry of the order parameter should obey a certain constraint followed from the Hubbard operator algebra as discussed by Plakida et al. [59]. In the present case we have the identities

$$\langle X_i^{\sigma 2} X_i^{\bar{\sigma} 2} \rangle = \langle X_i^{0\bar{\sigma}} X_i^{0\sigma} \rangle = 0,$$

which prohibit double occupancy in the upper and lower Hubbard subbands. The equations are automatically satisfied for the order parameter with $d_{x^2-y^2}$ symmetry for a square lattice. Therefore $T_c$ will be identified with the temperature at which the largest eigenvalue equals one, while the corresponding eigenvector shows $d_{x^2-y^2}$ symmetry.

The dependence of the critical temperature $T_c$ on the doping parameter $\delta$ is shown in Fig. 10 [71], in units of $t_{eff} \simeq K_{22} 2\nu_1 t \simeq 0.14t \simeq 0.2$ eV. The computations were done for the exchange interaction: $J = 0.4 t_{eff}$, usually adopted in the t-J model, and for a short AFM correlation length: $\xi = 3$ which was kept constant for all hole concentrations.

The highest $T_c$ (dotted line) is obtained when both contributions in Eq. (192), the exchange interaction $J$, and the kinematic interaction $\lambda^{(-)}(\mathbf{k}, \mathbf{q}-\mathbf{k})$, are taken into account. The kinematic interaction alone gives the lowest $T_c$ (solid line) while the exchange interaction results in higher $T_c$ (dashed line). The maxima values of $T_c$ are quite high in all cases: from $T_c^{max} \simeq 0.12 t_{eff} \simeq 270$ K at optimum doping $\delta_{opt} \simeq 0.13$ in the highest curve to $T_c^{max} \simeq 0.04 t_{eff} \simeq 90$ K at $\delta_{opt} \simeq 0.07$ in the lowest curve. The latter value is quite close to our crude estimation given above in Eq. (200). The high values of $T_c$ are due to the weak coupling approximation which neglects strong inelastic scattering which results in suppression of $T_c$ (see [65]). We

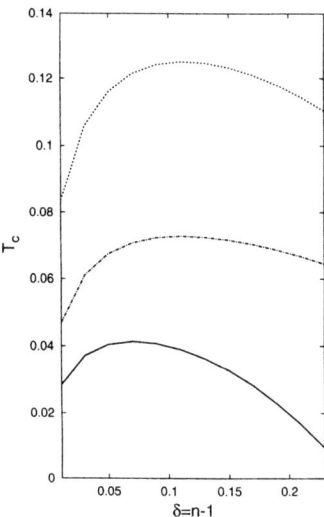

**FIGURE 10.** Variation of $T_c$ (in $t_{eff}$ units) with the doping parameter $\delta$ when in Eq. (192) only the kinematic interaction, $\lambda^{(-)}(\mathbf{k},\mathbf{q}-\mathbf{k})$ with $\xi = 3$, (solid line) or the exchange interaction, $J = 0.4 t_{eff}$, (dashed line), or the sum of them (dotted line) are taken into account [71].

have observed a strong suppression of $T_c$ mediated by kinematic interaction for smaller AFM correlation length which is due to decreasing of $\chi_0$ which defines the electron coupling induced by spin-fluctuation exchange. However, the parabolic behavior of $T_c$ with $\delta$ with a maximum at some optimum doping is similar to the experimental data. This behavior of $T_c(\delta)$ is explained by a strong dependence of the density of states on doping which is high only at optimum value of $\delta_{opt}$ as can be seen from Figs. 9.

We can have more insight in the parabolic dependence observed in Fig. 10 by analyzing the relationship between the gap function and the Fermi surface (FS) in the Brillouin zone (BZ). We have computed the order parameter $\Phi^{22}(\mathbf{k})$ at several temperatures for two cases: in Fig. 11 [71], when in Eq. (192) only the kinematic interaction is taken into account: at the optimum doping $\delta = 0.07$ for $T = 0$ (a); $0.5T_c$ (c); $0.9T_c$ (e) and for the "overdoped" case $\delta = 0.2$ for $T = 0$ (b); $0.5T_c$ (d); $0.9T_c$ (f), and in Fig. 12 [71], when in Eq. (192) both the kinematic and exchange interactions are taken into account: at the optimum doping $\delta = 0.13$ for $T = 0$ (a); $0.5T_c$ (c); $0.9T_c$ (e) and for the "overdoped" case $\delta = 0.2$ for $T = 0$ (b); $0.5T_c$ (d); $0.9T_c$ (f). The occurrence of a superconducting state with $d_{x^2-y^2}$ symmetry is observed. However, the $\mathbf{q}$-dependence of the gap shown by izo-lines is much more complicated then a usually considered simple model in Eq. (194). In the case of only kinematic interaction, Fig. 11, maxima of the gaps are inside the BZ while adding the exchange interaction in Fig. 12 shifts them to the FS close to

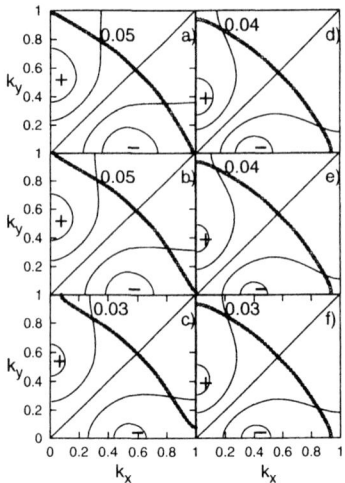

**FIGURE 11.** The temperature dependence of the order parameter $\Phi^{22}(\mathbf{k})$ over the first quadrant of the BZ when in Eq. (192) only the kinematic interaction is taken into account ($J = 0$): at an optimum doping $\delta = 0.07$ for $T = 0$ (a); $0.5T_c$ (c); $0.9T_c$ (e) and at doping $\delta = 0.2$ for $T = 0$ (b); $0.5T_c$ (d); $0.9T_c$ (f). The vector $\mathbf{k}$ is measured in $\pi/a$ units. The circles plot the Fermi surface. The numbers in graphs show the maximum value of the izoline [71].

($\pi, 0$)-type points ("hot spots"). That results also in different values of the optimum doping for the two cases: a smaller value for the kinematic interaction, $\delta = 0.07$, and a larger value for the exchange interaction, $\delta = 0.13$, the latter being close to the experimentally observed in cuprates. That behavior can be explained by a specific wave-vector dependence of the kinematic interaction, $|\nu(\mathbf{k})|^2$ in Eq. (193), where the main contribution due to the nearest neighbor hopping vanishes along the lines $|k_x| + |k_y| = \pi$. Therefore it gives no contribution to pairing for larger doping with FS close to that line. The exchange interaction in Eq. (192), on the contrary, gives the largest contribution close to the "hot spots" and the larger FS at larger hole concentrations results in higher $T_c$. The same nontrivial $\mathbf{q}$-dependence of the gap was obtained for the $t$-$J$ model in Sec. III B, Fig. 7.

Another interesting point in comparison of the exchange and the spin-fluctuation interactions in Eq. (192) is that the former gives twice as large $T_c$ then the latter. The interactions are of the same second order in the hopping matrix elements and in our calculations the exchange interaction $J = 0.4t_{eff} \simeq 0.08$ eV is even smaller then the spin-fluctuation coupling: in Eq. (198) $\lambda_{s,22} \simeq 0.27$ eV. Besides the particular wave-vector dependence of the interactions mentioned above, we should also take into account that the exchange interaction is mediated by the interband hopping, $\propto (t_{ij}^{12})^2$, with large energy transfer. Therefore, the retardation effects become unimportant and the interaction couples the charge carriers in a broad energy

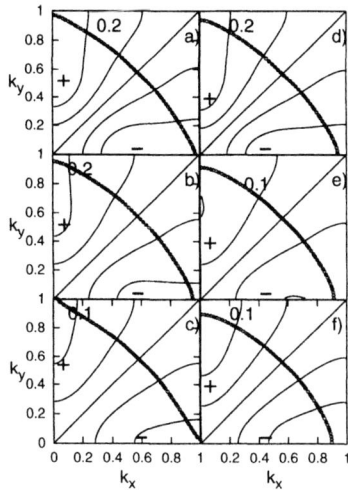

**FIGURE 12.** The temperature dependence of the order parameter $\Phi^{22}(\mathbf{k})$ over the first quadrant of the BZ when in Eq. (192) both the kinematic and exchange interactions are taken into account: at an optimum doping $\delta = 0.13$ for $T = 0\,(a); 0.5T_c\,(c); 0.9T_c\,(e)$ and at doping $\delta = 0.2$ for $T = 0\,(b); 0.5T_c\,(d); 0.9T_c\,(f)$. Other notation are the same as in Fig. 11 [71].

shell, of the order of the band width $W = 8t_{eff}$. The spin-fluctuation interaction, $\propto (t_{ij}^{22})^2$, acting in one subband couples holes (electrons) in a much narrow energy shell, $\omega_s \ll W$ that results in lower $T_c$.

So we can conclude that the most important pairing interaction in the Hubbard model in the strong correlation limit is just the exchange interaction while the spin-fluctuation coupling mediated by the kinematic interaction results only in a moderate enhancement of $T_c$. The same result has been obtained in studies of the $t-J$ model beyond the WCA [65] discussed in Sec. III B. This observation could explain why in numerical simulations the $t-J$ model usually shows much stronger pairing tendency then the original Hubbard model [6,44].

## F  Conclusions

In the present section the microscopic theory of the superconducting pairing within the reduced two-band $p-d$ Hubbard model (116) was formulated. Making use of the projection technique [64] for the two-time GF in terms of the Hubbard operators, we obtained the Dyson equation for the $4 \times 4$ matrix GF (128) with the corresponding matrix self-energy (129). The normal state spectra for the one-hole and singlet subbands were calculated in MFA (see Fig. 9) while the anomalous correlation functions calculated in the second order of the interband hopping resulted in the exchange interaction pairing for holes.

The normal and anomalous components of the self-energy matrix were calculated in the self-consistent Born approximation for the electron – spin-fluctuation scattering, Eqs. (178) and (179). It is to be stressed that, within the present model, the electron – electron coupling is induced by the exchange and kinematic interaction resulting from non-fermionic commutation relations without using additional fitting parameters for that interaction.

The self-consistent system of equations of the two diagonal gap parameters for the lower and upper Hubbard subbands was obtained in the weak coupling approximation, Eq. (187). Analytical and numerical estimates of the superconducting $T_c$ have been derived under the use of a model AFM spin-fluctuation susceptibility, Eq. (105). Simple analytical evaluations proved a possibility of the $d_{x^2-y^2}$-wave pairing with moderate $T_c$ even for a low density of states, Eq. (200), outside the van Hove singularity. It is important to note that the superconducting pairing due to kinematic interaction in the second order occurs in one subband even in the limit $\Delta \to \infty$, which corresponds to $U \to \infty$ in the original Hubbard model (4) when the exchange interaction $J \propto (t_{ij}^{12})^2/\Delta$ vanishes. However, in this limit the spin-fluctuations, Eq. (105), may be suppressed that prevents the quasiparticle formation and the occurrence of the corresponding superconducting pairing.

A more accurate numerical solution of the integral equation (192) for the gap in the singlet subband has been also computed. The obtained $T_c$ dependence on the doping in Fig. 10, showed a maximum value at optimum doping from $T_c^{max} \simeq 0.12 t_{eff} \simeq 270$ K to $T_c^{max} \simeq 0.04 t_{eff} \simeq 90$ K depending on the coupling interactions. The temperature dependence of the order parameters $\Phi^{22}(\mathbf{k})$ over the first quadrant of the BZ, Fig. 11 and Fig. 12, shows a peculiar wave-vector behavior of the $d_{x^2-y^2}$-wave gap depending on the type of involved interactions.

## V  CONCLUSIONS

In this lectures we presented the theory of electron spectrum and superconducting pairing for several microscopical models describing systems with strong electron correlations. At first we derived the models, starting from a more general, two band $p$-$d$ model [91,92] (Sec II A), and then reduced it to the one-band $t$-$J$ model in terms of Hubbard operators (Sec II B). At low doping a further simplification can be done by considering a two-sublattice representation which resulted in the spin polaron model (Sec III A). Since the exchange interaction in the $t$-$J$ model is due to strong hybridization between the Hubbard subband its role can be studied in-depth only within the two-band Hubbard model (Sec IV).

The theory of electron spectrum and superconducting pairing were considered by employing the projection technique for the two-time Green functions [61]. This technique permitted to obtain formally exact Dyson equation with the self-energy in terms of many-particle Green functions. By applying the self-consistent Born approximation (which neglects vertex corrections) a closed system of equations were obtained. They were solved numerically and both the normal state electron spectrum and the superconducting characteristics were studied.

Comparing the calculations presented in the lecture for three different models we can point out the following most important results.

The electron spectral density for the spin polaron, Fig. 1, and $t$-$J$ model, Figs. 4, 5, show QP excitations at the FS crossing and a dispersive incoherent band. For the Hubbard model electron spectra were calculated in the MFA which had no incoherent part but clearly demonstrated a strong dependence on the short-range AFM correlation length and the weight transfer with hole doping from the singly occupied $d$-hole band to the doubly occupied singlet band , Figs. 9(a-c).

Performing a direct numerical solution of the linearized gap equations we obtained for all considered models the $d$-wave superconducting pairing with high superconducting temperature $T_c$ (see Figs. 3, 8, 10). Concerning the superconducting pairing in the Hubbard model, Sec. IV E, in MFA we obtained $d$-wave superconducting pairing mediated by the exchange interaction while the first order of the kinematic interaction failed to produce $d$-wave pairing. So the pairing observed in MFA in Refs. [72–74] is due to the conventional exchange interaction well known in studies within the $t$-$J$ model. As computations showed, just the exchange interaction is the most important pairing interaction in the Hubbard model in the strong correlation limit, while the spin-fluctuation coupling results only in a moderate enhancement of $T_c$. This result is explained by a specific wave-vector dependence of the latter (which vanishes along the lines $|k_x| + |k_y| = \pi$) and a small energy shell, of the order of spin-fluctuation energy $\omega_s \simeq J$ where the pairing occurs. The exchange interaction couples electrons (holes) in a much broader energy shell, of the order of the band width, $W = 8t_{eff} \gg \omega_s$, due to interband hopping where retardation effects are unimportant.

It should be emphasized that our approach is a mean-field type theory in which the phase of the order parameter is fixed - the gaps are taken to be real and the problem of the phase coherence of the superconducting order parameter [120] is beyond the scope of our theory. This remark also explains why we have obtained the pairing long-range order (LRO) for the two-dimensional $t$-$J$ and Hubbard models while it can be rigorously proved, within the Bogoliubov-Mermin-Wagner theory, a nonexistence of the LRO in two dimensions [121]. Numerical results for finite clusters also failed to prove superconducting LRO [48,49] . Small 3D coupling will stabilize the phase fluctuations and implement the LRO. As observed in cuprates, $T_c$ dependence on the inter-layer coupling is rather weak which can occur only for strong intra-layer pairing.

The important advantage of the proposed microscopic theory for the $d$-wave spin-fluctuation superconducting pairing, in comparison with phenomenological approaches based on the Fermi liquid models, is that we have used only few basic parameters of the models. In the $t$-$J$ model they are the hopping energy, $t$, and the (super)exchange energy, $J$, which are characteristic to strongly correlated systems [28,97]. For the Hubbard model we have used the $p$-$d$ hybridization parameter $t$ and the charge-transfer energy $\Delta$. So no additional fitting parameters for the electron – spin-fluctuation interaction were used as in the Pines theory (see Eq. (1) in Sec. I A) since the electron scattering on spin–charge fluctuations is resulted from the kinematic and exchange interactions.

In conclusion, the present investigation points to the existence of singlet $d_{x^2-y^2}$-wave superconducting pairing for holes or electrons in the two-band Hubbard model in the limit of high correlations. It mediated by the exchange interaction and AFM spin-fluctuation scattering induced by the kinematic interaction, characteristic to the Hubbard model. These mechanisms of superconducting pairing are absent in the fermionic models (for a discussion, see Anderson [28]). Therefore we can argue that these specific interactions are the origin of high-temperature superconducting pairing in cuprates, having the largest antiferromagnetic superexchange in transition metal compounds.

# REFERENCES

1. Plakida, N.M., *High Temperature Superconductivity*, Berlin-Heidelberg: Springer-Verlag, 1995.
2. Bourges, Ph., *arXiv:cond-mat*/0009373.
3. Fong, H.F., Bourges, Ph., Sidis, Y., Regnault, L.P., et al., *Phys. Rev. B* **61**, 14773 (2000).
4. Wakimoto, S., Birgeneau, R.J., Lee, Y.S., and Shirane, G., *arXiv:cond-mat*/0011392.
5. Tsuei, C.C. , Kirtley, J.R., Rupp, M. , et al., *Science* **271**, 329 (1996).
6. Scalapino, D.J. , *Phys. Reports* **250**, 329 (1995); *arXiv:cond-mat*/9908287.
7. Van Harlingen, D., *Rev. Mod. Phys.* **67**, 515 (1995).
8. Cyrot, M., *Solid State Comm.* **60**, 253 (1986).
9. Miyake, K., Schmitt-Rink, S., and Varma, C.P., *Phys. Rev. B*, **34**, 6554 (1986).
10. Scalapino, D.J., Loh, E., Jr., and Hirsch, J.E., *Phys. Rev. B*, **34**, 8190 (1986).
11. Bickers, N.E., Scalapino, D.J., and Scalettar, R.T., *Int. J. Mod. Phys. b*, **1**, 687 (1987).
12. Dzyaloshinskii, I.E., *Sov. Phys. JETP*, **66**, 848 (1987).
13. Schrieffer, J.R., Wen, X.-G. , and Zhang, S.-C., *Phys. Rev. Lett.* **60**, 944 (1989).
14. Kampf, A.P., and Scrieffer, J.R., 1990, *Phys. Rev. B*, **41**, 6399 (1990); *Ibid.* **42**, 7967 1990.
15. Schrieffer, J.R. , *Physica C* **185-189**, 17 (1991).
16. Pines, D. , *Physica B* **163**, 78 (1990).
17. Pines, D., *Physica C*, **235-240**, 113 (1994).
18. Monthoux, P., Balatsky, A.V., and Pines, D., *Phys. Rev. Lett.* **67**, 3448 (1991); *Phys. Rev. B* **46**, 14803 (1992).
19. Monthoux, P., and Pines, D., *Phys. Rev. B* **47**, 6069 (1993); *Ibid.* **49**, 4261 (1994).
20. Monthoux, P., *Phys. Rev. B* **55**, 15261 (1997).
21. Moria, T., Takahashi, Y., and Ueda, K., *J. Phys. Soc. Jpn.* **59**, 2905 (1990).
22. Nakamura, S., Moria, T., and Ueda, K., *J. Phys. Soc. Jpn.* **65**, 4026 (1996).

23. Dahm, T., Erdmenger, J., Scharnberg, K., and Rieck, C.T., *Phys. Rev. B* **48**, 3896 (1993).
24. Béal-Monod, M.T., and Maki, K., *Phys. Rev. B* **53**, 5775 (1996).
25. Plakida, N.M., *Phylosophical Magazine B* **76**, 771 (1997).
26. Millis, A., Monien, H., and Pines, D., *Phys. Rev. B*, **42**, 167 (1990).
27. Barzykin, V., and Pines, D., *Phys. Rev. B*, **52**, 13585 (1995).
28. Anderson, P.W., *Advance in Phys.* **46**, 3 (1997).
29. Anderson, P.W., *Science* **235**, 1196 (1987).
30. Anderson, P.W., *The Theory of Superconductivity in the High-$T_c$ Cuprates*, Princeton: Princeton University Press, 1997.
31. Hubbard J., *Proc. Roy. Soc. A (London)* **276** 238 (1963); *Ibid.* **277**, 237 (1963).
32. Pruschke, Th., Jarrell, M., and Freericks, J.K., *Advance in Phys.* **44**, 187 (1995).
33. Georges, A., Kotliar, G., Krauth, W., and Rozenberg, M., *Rev. Mod. Phys.* **68**, 13 (1996).
34. Maier, Th., Jarrel, M., Pruschke, Th., and Keller, J., *Phys. Rev. Lett.* **85**, 1524 (2000).
35. Lichtenstein, A.I., and Katsnelson, M.I., *Phys. Rev. B* **62**, R9283 (2000).
36. Pao Chien-Hua, and Bickers, N.E., *Phys. Rev. Lett.* **72**, 1870 (1994); *Phys. Rev. B*, **51**, 16310 (1995).
37. Monthoux, P., and Scalapino, D.J., *Phys. Rev. Lett.*, **72**, 1874 (1994).
38. Lenck, S.T., Carbotte, J.P., and Dynes, R.C., *Phys. Rev. B*, **50**, 10149 (1994).
39. Dahm, T., and Tewordt, L., *Phys. Rev. Lett.* **74**, 793 (1995); *Phys. Rev. B*, **52**, 1297 (1995).
40. Bickers, N.E., Scalapino, D.J., and White, S.R., *Phys. Rev. Lett.* **62**, 961 (1989).
41. Takimoto, T., and Moria, Y., *J. Phys. Soc. Jpn.* **66**, 2459 (1997).
42. Manske, D., Eremin, I., and Bennemann, K.H., *Phys. Rev. B* **62**, 13922 (2000).
43. Schrieffer, J.R., *J. Low Temp. Phys.* **99**, 397 (1995).
44. Dagotto, E., *Rev. Mod. Phys.* **66**, 763 (1994).
45. Jaklič, J., and Prelovsék, P., *Adv. in Physics* **49**, 1-92 (2000).
46. Scalapino, D.J., and White, S.R., *arXiv:cond-mat*/0007515.
47. Shiwei Zhang, Carlson, J., and Gubernatis, J.E., *Phys. Rev. Lett.* **78**, 4486 (1997).
48. Shih, C.T., Chen, Y.C., Lin, H.Q., and Lee, T.K., *Phys. Rev. Lett.* **81**, 1294 (1998).
49. Suzumura, Y., Hasegawa, Y., and Fukuyama, H., *J. Phys. Soc. Jpn.* **57**, 401 (1988).
50. Kotliar, G., and Liu, J., *Phys. Rev. Lett.* **61**, 1784 (1988).
51. Grilli, M., and Kotliar, G., *Phys. Rev. Lett.* **64**, 1170 (1990).
52. Affleck, I., and Martson, B., *Phys. Rev. B*, **37**, 3774 (1988).
53. Hubbard, J., *Proc. Roy. Soc. A (London)* **285**, 542 (1965).
54. Zaitsev, R.O., and Ivanov, V.A., Soviet Phys. Solid State, **29**, 2554 (1987); *Ibid.* 3111 (1987); *Int. J. Mod. Phys. B* **5**, 153 (1988); *Physica C*, **153-155**, 1295 (1988).
55. Zaitsev, R.O., *Sov. Phys. JETP* **43**, 574 (1976).
56. Izyumov, Yu.A., and Scryabin, Yu. N., *Statistical Mechanics of Magnetically Ordered Systems*, New York: Consultant Bureau, 1989.
57. Izyumov, Yu.A., and Letfulov, B.M., *J. Phys.: Condens. Matter* **3** 5373 (1991).
58. Izyumov, Yu.A., and Letfulov, B.M., *Intern. J. Modern Phys. B* **6**, 3771 (1992).
59. Plakida, N.M., Yushankhai, V.Yu., and Stasyuk, I.V., *Physica C*, **160**, 80 (1989).

60. Yushankhai, V.Yu., Plakida, N.M., and Kalinay, P., *Physcica C*, **174**, 401 (1991).
61. Zubarev, D.N., *Sov. Phys. Usp.* **3**, 320 (1960).
62. Plakida, N.M., and Stasyuk, I.V., *Modern Phys. Lett.* **2** 969, (1988).
63. Bogoliubov, N.N., Aksenov, V.L., and Plakida N.M., *Physica C* **153-155**, 99 (1988).
64. Plakida, N.M., Phys. Lett. A, **43**, 481 (1973).
65. Plakida, N.M., and Oudovenko, V.S., *Phys. Rev. B* **59**, 11949 (1999).
66. Plakida, N.M. , Oudovenko, V.S., Horsch, P., and Liechtenstein, A.I., *Phys. Rev. B* **55**, R11997 (1997).
67. Prelovśek P.M., *Z. Physik B* **103**, 363 (1997).
68. Plakida, N.M., and Hayn, R., *Z. Physik B* **93**, 313 (1994).
69. Plakida, N.M., Hayn, R., and Richard, J.-L., *Phys. Rev. B* **51**, 16599 (1995).
70. Plakida, N.M., *Physcica C* **282-287**, 1737 (1997).
71. Plakida, N.M., Anton, L., Adam, S., and Adam, Gh., *Phys. Rev. B* (submitted).
72. Beenen, J., and Edwards, D.M., *Phys. Rev. B*, **52**, 13636 (1995).
73. Avella, A., Mancini, F., Villani, D., and Matsumoto, H., *Physcica C*, **282-287**, 1757 (1997); Di Matteo, T., Mancini, F., Matsumoto, H., and Oudovenko, V.S., *Physica B* **230 - 232**, 915 (1997).
74. Stanescu, T.D., Martin, I., and Phillips, Ph., *Phys. Rev. B* **62**, 4300 (2000).
75. Kagan, M.Yu., and Rice, T.M., *J. Phys.: Condens. Matter* **3**, 5373 (1994).
76. Heeb, E.S., and Rice, T.M., *Europhys. Lett.*, **27**, 673 (1994).
77. Ruckenstein, A.E., and Schmitt-Rink, S., *Phys. Rev. B* **38**, 7138 (1988).
78. Greco, A., and Zeyher, R., *Europhys. Lett.* **35**, 115 (1996).
79. Greco, A., and Zeyher, R., *Z. Physik B* **104**, 737 (1997).
80. Schmitt-Rink, S., Varma, C.M., and Ruckenstein, A.E., *Phys. Rev. Lett.* **60**, 2793 (1988).
81. Kane, C.L., Lee, P.A., and Read, N., *Phys. Rev. B* **39**, 6880 (1989).
82. Martínez, G., and Horsch, P., *Phys. Rev. B*, **44**, 317 (1991).
83. Liu, Z., and Manousakis, E., *Phys. Rev. B* **45**, 2425 (1992).
84. Scherman, A. and Schrieber, M., *Phys. Rev. B* **48**, 7492 (1993); *Ibid.* **50**, 12887 (1994).
85. Plakida, N.M., Oudovenko, V.S., and Yushankhai, V.Yu., *Phys. Rev. B* **50**, 6431 (1994).
86. Barabanov, A.F., Maksimov, L.A., and Mikheenkov, A.V., "Theory of Spin Polaron for 2D Antiferromagnet", in *Lectures on the Physics oh Highly Correlated Electron Systems, IY*, edited by F. Mancini, AIP Conference Proceedings 527, New York, 2000, pp. 1-117.
87. Dagotto, E., Nazarenko, A., and Moreo, A., *Phys. Rev. Lett.* **74**, 310 (1995).
88. Belinicher, V.I., Chernyshov, A.L., Dotsenko, A.V., and Sushkov, O.P., *Phys. Rev. B* **51**, 6076 (1995).
89. Scherman, A. and Schrieber, M., *Phys. Rev. B*, **52**, 10621 (1995).
90. Zhang , F.C., and Rice, T.M., *Phys. Rev.B* **37**, 3759 (1988).
91. Emery, V.J., Phys. Rev. Lett. **58**, 2794 (1987);
92. Varma, C.M., Schmitt-Rink, S., and Abrahams, E., *Solid State Commun.* **62**, 681 (1987).
93. Horsch, P., Stephan, W., in: *"Dynamics of Magnetic Fluctuations in High Temper-*

*ature Superconductors"*, Proceedings of the NATO Advanced Research Workshop, edited by G. Reiter, P. Horsch, and G. Psaltakis, New York: Plenum, 1991.

94. Eskes, H., Tjeng, L.H., and Sawatzky, G.A., *Phys. Rev.B* **41**, 288 (1990); Eskes, H., Sawatzky, G.A., and Feiner, L.F., *Physica C* **160**, 424 (1989); Eskes, H., and Sawatzky, G.A., *Phys. Rev. B* **43**, 119 (1991).
95. Tohyama, T., and Maekawa, S., *Physica C* **191**, 193 (1992).
96. Unger, P., and Fulde, P., *Phys. Rev. B* **47**, 8947 (1993);*Ibid.* **48**, 16607 (1993).
97. Fulde, P., *Electron Correlations in Molecules and Solids*, Berlin-Heidelberg: Springer-Verlag, 1991, ch. 14, pp. 351-375.
98. Pickett, W.E., *Rev. Mod. Phys.* **61**, 433 (1989).
99. Lovtsov, S.V., and Yushankhai, V.Yu., *Physica C* **179**, 159 (1991).
100. Hayn R., Yushankhai, V.Yu., and Lovtsov, S.V., *Phys. Rev.B* **47**, 5253 (1993).
101. Feiner, L.F., Jefferson, J.H., and Raimondi, R., *Phys. Rev. B* **53**, 8751 (1996); Raimondi, R., Jefferson, J.H., and Feiner, L.F., *Phys. Rev. B* **53**, 8774 (1996).
102. Yushankhai, V. Yu., Oudovenko, V.S., and Hayn, R., *Phys. Rev. B* **55**, 15 562 (1997).
103. Eskes, H., Meinders, M.B.J., and Sawatzky, G.A., *Phys. Rev. Lett.* **67**, 1035 (1991).
104. Eskes, H., and Oleś, A.M., *Phys. Rev. Lett.* **73**, 1279 (1994).
105. Krüger, E., phys. stat. sol. (b) **52**, 215, 512 (1972); Kohn, W. *Phys. Rev. B***7**, 4388 (1997).
106. Schrieffer, J.R., and Wolf, P.A., *Phys. Rev.* **149**, 491 (1966).
107. Serene, J.W., and Hess, D.W., *Phys. Rev.* **B 44**, 3391 (1991).
108. Jackeli, G., and Plakida, N.M., *Theor. and Math. Phys.* **114**, 426 (1998).
109. Jaklič, J., and Prelovsék, P., *Phys. Rev. Lett* **74**, 3411 (1995); *Ibid.* **75**, 1340 (1995).
110. Yamada, K., Lee, C.H., Kurahashi, K., Wada, J., Wakimoto, S., Ueki, S., Kamura, H., and Endoh, Y., *Phys.Rev. B*, **57**, 6165 (1998).
111. Allen, Ph., and Mitrović, B., *Solid State Physics*, vol. **37**, 1-92 (1982).
112. Shen, Z.-X., and Schrieffer, J.R., *Phys. Rev. Lett.* **78**, 1771 (1997).
113. Schmalian, J., Pines, D., and Stojković, B., *Phys. Rev. Lett.* **80**, 3839 (1998).
114. Jaklič J., and Prelovsek P., *Phys.Rev. B*, **55**, R7307 (1997).
115. Stephan, W., and Horsch, P., Phys. Rev. Lett. **66**, 2258 (1991).
116. Putikka, W.O., Luchini, M.U., and Singh, R.R.P., *Phys. Rev. Lett.* **81**, 2966 (1998).
117. Shen, Z.-X., and Dessau, D.S., *Phys. Rep.* **253**, 1-161 (1995).
118. Wells, B.O., Shen, Z.-X., Matsuura, A., King, D.M., Kastner, M.A., Greven, M., and Birgeneau, R.J., *Phys. Rev. Lett.* **74**, 964 (1995).
119. Yamada, K., Endoch, Y., Chul-Ho Lee, *et al.*, *J. Phys. Soc. Jpn.* **64**, 1626 (1995).
120. Deutscher, G., *Nature* **397**, 410 (1999).
121. Gang Su, and Suzuki, M., *Phys. Rev. B* **58**, 117 (1998).

# Non Abelian bosonization and WZNW models

## A. M. Tsvelik*

*Department of Physics, University of Oxford
Brasenose College, 1 Keble Road, Oxford OX1 3NP, UK*

## R.Citro[†]

†*Dipartimento di Fisica "E.R. Caianiello", University of Salerno
Via S. Allende, I-84081 Baronissi, Italy*

**Abstract.** In these Lectures we discuss some aspects of theory of strongly correlated low-dimensional systems. In (1+1) and two dimensions availability of non-perturbative techniques makes it possible to obtain solutions of non-trivial models. One of such techniques is called bosonization. In this course we discuss a generalization of this technique for systems with isotopic symmetry.

## I INTRODUCTION

In these lectures we discuss the general properties of Wess-Zumino-Novikov-Witten (WZWN) models. These models have critical points where they are described by conformal field theory. To start with, let us discuss the related physical background. There are at least two areas of condensed matter physics where WZWN model emerge:

- (1+1)-D strongly correlated systems including quantum impurities (Kondo effect);

- 2D-disordered critical models.

As known, in one-dimensional interacting electrons systems single-particle excitations are not *coherent*; coherent modes are collective excitations, such as spin or charge density waves.

Collective modes are conveniently described by the so-called *current* operators. Let us first introduce such operators for 1D spinless fermions. The Fermi surface consists of two points. For energies much less than the Fermi energy we can linearize the spectrum around the Fermi points, i.e. $\epsilon(k \pm k_F) = \pm v_F k$, where $v_F$ is the Fermi velocity. The energy of a particle-hole excitation at the right Fermi point is

$$E = V_F(k+q) - V_F k = V_F q. \tag{1}$$

Since it does not depend on the individual wave-vectors, but only on the transferred momentum, all particle-hole pairs with the same $q$ will propagate coherently provided their energy is much less than $E_F$ (otherwise approximation (1) is violated). Their propagation is described by the following operator:

$$J_q = \sum_{|k|<\Lambda} a^\dagger_{k+q} a_k, \tag{2}$$

where the cutoff $\Lambda \simeq E_F$. It creates particle-hole excitations with energy $E = V_F q$ when acting on the occupied Fermi sea:

$$J_q|FS> = \sum_{|k|<\Lambda} a^\dagger_{k+q} a_k |FS> = \sum_{k=-q}^{0} a^\dagger_{k+q} a_k |FS>. \tag{3}$$

Now, let us consider a 1D gas of spinless fermions interacting via the Coulomb-like potential:

$$V(q) = \int_0^\infty e^{iqx} \frac{e^2}{|x|} dx \simeq 2e^2 \ln \frac{1}{|q|d}, \tag{4}$$

where $d$ is a short-distance cutoff.

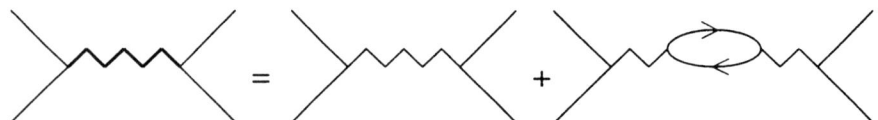

**FIGURE 1.** Renormalized interaction.

The screening of such interaction is represented by the diagrams shown in figure 1 which can be summed up in a Random Phase Approximation (RPA) leading to the following expression of the renormalized interaction:

$$U(q,\omega) = \frac{V(q)}{1 - V(q)\Pi(q,\omega)}, \qquad (5)$$

where $\Pi(q,\omega)$ is the polarization. The one-loop corrections to the polarization, represented by diagrams shown in Fig.2, vanish and this result is exact. It can be rederived using *bosonization*.

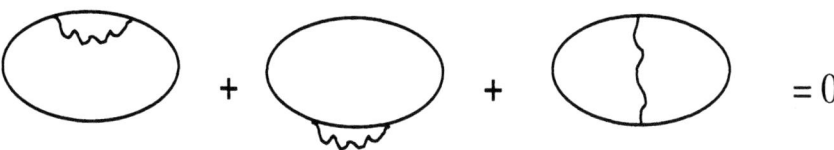

**FIGURE 2.** One-loop corrections to polarization.

The spectrum of the collective excitation (plasmons) is given by the poles of the above correlation function. To determine the spectrum, let us first calculate the polarization loop:

$$\Pi(q,\omega) - \frac{1}{2\pi}\left(\frac{v_F q^2}{\omega^2 - v_F^2 q^2}\right). \qquad (6)$$

Inserting it in (5) we derive the spectrum of the collective excitations:

$$\omega^2 = v_F^2 q^2 \left(1 + \frac{1}{2\pi v_F} v(q)\right). \qquad (7)$$

Let us note that one-dimensional plasmons have no spectral gap, contrary to what happens in 2- or 3D.

# A  Bosonization

The approach known as *bosonization* is a way to reformulate complicated interacting models in such a way that they become weakly interacting. The idea of bosonization was pioneered by Jordan and Wigner in 1928 [1] when they established the equivalence between the spin $S = 1/2$ anisotropic Heisenberg chain and the model of interacting fermions. The transformation from fermions to bosons was conceived in 1975 indipendently by two particle and two condensed matter physicists-Coleman [3] and Mandelstam [5], and Mattis and Luther [4], respectively. The approach was based on the properties of the Dirac fermions in (1+1)-dimensions. They established that correlation functions of such fermions can be expressed in terms of correlation functions of free bosonic field. In the bosonic representation the fermion forward scattering became trivial, thus leading to a simple solution of the Tomonaga-Luttinger model.

As we have previously discussed, when we consider one dimensional spinless fermions, we can reduce the degrees of freedom to those living near Fermi points, introducing left and right fermion operators. The Hamiltonian $\hat{H} = \sum_k \epsilon(k) a_k^\dagger a_k$ is written as

$$\hat{H} = \sum_{|k| \ll k_F} (k R_k^\dagger R_k - k L_k^\dagger L_k), \tag{8}$$

where $R_k = a_{k+k_F}$ and $L_k = a_{k-k_F}$ with $|k| \ll k_F$.

The Hamiltonian (8) is equivalent to that of relativistic particles which is invariant under 1+1 Lorents transformation. This is a first surprising feature in the physics of one dimension. It turns out that the important symmetry connected to Lorents invariance exists not only in high-enrgy physics but may appear for low energy states in a one dimensional metal. Of course the light speed in this case is replaced by the Fermi velocity. The spectrum of (8) is similar to that of sound waves in 1D, but the analogy goes much further. Infact, given the Hamiltonian of a bosonic system:

$$H_B = \sum_{|q| \ll k_F} |q| a_q^\dagger a_q, \tag{9}$$

the calculation of the free energy for the bosons:

$$\frac{F}{L} = \frac{T}{2\pi} \int dk \ln(1 - e^{-|k|/T}), \tag{10}$$

where $L$ is the size of the system and $T$ is the temperature, yields the same result for that of left and right fermions. This simple calculation shows that left and right fermions are equivalent to a *single boson*. This equivalence is known as *bosonization*. To render much explicit this equivalence, it is convenient to deal with the following correlation functions in real space:

$$\ll R(x,\tau)R^\dagger(0,0) \gg = \frac{1}{2\pi(\tau - ix)} = \frac{1}{2\pi z},$$

$$\ll L(x,\tau)L^\dagger(0,0) \gg = \frac{1}{2\pi(\tau + ix)} = \frac{1}{2\pi \bar{z}}, \tag{11}$$

where we have used complex coordinates $z = (\tau + ix)$ and $\bar{z} = (\tau - ix)$. The polarization loop for left fermions will simply yield $-\frac{1}{4\pi^2 z^2}$.

Let us consider a theory of free bosonic massless scalar fields in a two dimensional Euclidean space whose action is

$$S = \frac{1}{L}\int_A (\nabla\phi)^2 d\tau dx. \tag{12}$$

where $A$ is some area of the infinite plane, $A = (0 < \tau < \beta; 0 < x < L)$. In momentum space this action becomes

$$S = \frac{1}{2}\int \frac{d\omega}{dq}\phi(-\omega,q)(\omega^2 + q^2)\phi(\omega,q). \tag{13}$$

The correlation function for the bosonic field $\phi$ gives:

$$<\phi(-q,\omega)\phi(q,\omega)> = \frac{1}{\omega^2 + q^2}. \tag{14}$$

When we perform the same calculation in real space and polar cooordinates, the correlation function will be calculated by the integral:

$$<\phi(x,\tau)\phi(0,0)> = \int_{|r|>1/L}\frac{drd\alpha}{r^2}e^{-(i\tau\cos\alpha + ix\sin\alpha)r} = \frac{1}{2\pi}\int_{1/L}^\infty \frac{dr}{r}e^{-\sqrt{\tau^2 + x^2}}$$

$$\simeq \frac{1}{2\pi}\ln\frac{L}{\sqrt{\tau^2 + x^2}} = \frac{1}{4\pi}\ln\frac{L^2}{z\bar{z}}, \tag{15}$$

where $r = \sqrt{\tau^2 + x^2}\cos(\alpha - \phi)$ and $\tan\phi = x/\tau$.

This result shows that the left fermion polarization loop can be understood in terms of the bosonic correlation functions $1/4\pi^2 z^2 \iff -\frac{1}{\pi}<\partial_z\phi(x,\tau)\bar{\partial}_z\phi(0,0)>$. The connection with bosonic fields is completed by introducing the *current* operators

$$J(x,\tau) =: L^\dagger(x,\tau)L^\dagger(x,\tau) := \frac{i}{\sqrt{\pi}}\partial_z\phi,$$

$$\bar{J}(x,\tau) =: R^\dagger(x,\tau)R^\dagger(x,\tau) := -\frac{i}{\sqrt{\pi}}\bar{\partial}_z\phi, \tag{16}$$

where $:\ldots:$ is the normal ordering and $J$ and $\bar{J}$ are bosonic operators. Within the previous procedure, known as *bosonization*, the two-point correlation functions for fermions coincide with the correlation function for the bosonic field $\phi$. Since this fact is crucial for the transmutation of statistics, let us say that the apparent simplicity that brings us to identify the fermionic currents with the derivatives of the Bose field hides highly non-trivial features. One of this is the *chiral anomaly* that we are going to discuss in the next section. The second one is that the model (12) has quite remarkable correlation functions.

# II THE CHIRAL ANOMALY

The physics of one dimensional systems is not a trivial subject and even non-interacting fermions may reveal some surprising features. One of these is the so called *chiral anomaly*, a subject of great fascination for many years. It is related to the screening of the electromagnetic field in presence of massless electrons. This field excites electron-hole pairs through the Fermi points and becomes screened. Mathematically the anomaly appears as a paradox. Let us consider massless fermions in Minkovsky space-time in an external potential which can be parametrised as follows:

$$A_0 + A_1 = \partial_x \phi(x), A_0 - A_1 = \partial_x \theta(x), \tag{17}$$

where $\phi$ and $\theta$ are some functions of $x$. Classically, the above potential appears to have no effect on fermions at all becouse it can be removed from the action by the canonical transformation

$$R(x) \to e^{-i\phi(x)} R(x), \ L(x) \to e^{-i\theta(x)} L(x). \tag{18}$$

Anyway this result is a nonsense. Infact, since the potential is not a pure gauge, the corresponding electric field does not vanish:

$$E_{01} = \partial_t A_0 + \partial_x A_1 = \frac{1}{2} \partial_x^2 (\phi + \theta) \neq 0. \tag{19}$$

Thus it is unthinkable that charged particles do not respond to an electric field. The answer to this paradox is that, from the quantum point of view, the chiral symmetry with respect to transformation (18) is broken. The transformation (18) does not leave the partition function invariant and therefore modifies the measure in the path integral. The latter fact means that single-fermion wave functions:

$$\begin{pmatrix} f_{R,E}(x) \\ f_{F,E}(x) \end{pmatrix}, \begin{pmatrix} f_{R,E'}(x) \\ f_{F,E'}(x) \end{pmatrix} \tag{20}$$

with energies $E, E'$, being orthogonal before the transformation are not orthogonal after it. In general, *anomalies* appear in theories which allow transformations affecting only the measure of integration and not the action itself. In the particular case of $(1 + 1)$-dimensional massless Quantum Electrodynamics, the anomaly leads to the illusion that the fermions decouple from the vector potential. They do not, as we have just found out. As we shall demonstrate later, the anomaly allows us to integrate over the fermions and obtain an explicit expression for the effective action of the gauge field.

# III  ANOMALOUS COMMUTATORS

The chiral anomaly can be rephrased in terms of commutators. Let us introduce the density operator for left-moving particles:

$$J(q) = \sum_p L^+(p+q)L(p)$$

Adopting the terminology of field theory we shall call this operator the left *current*. Let us consider a commutator of two currents. The straightforward calculation yields

$$[J(q), J(p)] = 0 \tag{21}$$

with the same result for the right-moving currents $\bar{J}(p)$.

The above result is also absurd because it suggests that the pair correlation function of currents vanishes. On the other hand we can calculate this correlation function using Feynman diagrams. It is more convenient to do this in real space-time. The current–current correlation function is given by the diagram in Fig. 3.

**FIGURE 3.** Current–current correlation function.

The single electron Green's functions are given by

$$\langle\langle R(1)R^+(2)\rangle\rangle = G_R(\bar{z}_{12}) = \frac{1}{2\pi \bar{z}_{12}}, \tag{22}$$

$$\langle\langle L(1)L^+(2)\rangle\rangle = G_L(z_{12}) = \frac{1}{2\pi z_{12}} \tag{23}$$

Thus Green's function of the left-moving particles depends only on $z = \tau + ix$ and the one for right-moving particles contains only $\bar{z} = \tau - ix$. For this reason the corresponding components of currents can also be treated as analytical or antianalytical functions. Substituting these expressions into the diagram in Fig. 2.2, we get

$$\langle\langle J(z_1)J(z_2)\rangle\rangle = \frac{1}{4\pi^2 z_{12}^2} \tag{24}$$

We can use this expression to calculate the commutator. In fact, the procedure described below is quite general and will be used throughout the book. Therefore we discuss it in detail. Let $A$ and $B$ be two Bose operators whose correlation function is known. Then one can calculate their commutator at equal times according to the following rule:

$$\langle\langle [A(x), B(y)]\rangle\rangle = \lim_{\tau\to+0}[\langle\langle A(\tau,x)B(0,y)\rangle\rangle - \langle\langle A(-\tau,x)B(0,y)\rangle\rangle] \tag{25}$$

Let us show that it follows from the definition of the time-ordered correlation function. Let us imagine that the operators $\hat{A}(z)$ and $\hat{B}(z_1)$ stand inside of some correlation function, perhaps surrounded by other operators which we denote as $X(\{z_i\})$:

$$G(z, z_1, \{z_i\}) = \langle\langle A(z)B(z_1)X(\{z_i\})\rangle\rangle \tag{26}$$

Recall, that in our notations $z = \tau + ix$. Let us consider the situation when $\Re e z = \Re e z_1 \pm \delta$ where $\delta$ is a positive infinitesimal (see Fig.4).

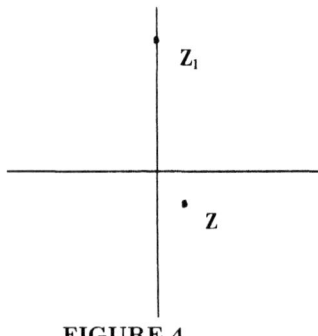

FIGURE 4.

Without loss of generality we can put $z = ix \pm 0$, $z_1 = iy$, where $x, y$ are real. Then due to the time-ordering present in the correlation function the following is valid:

$$G(\tau - \tau' = +0; x, y) - G(\tau - \tau' = -0; x, y) = \langle\langle [A(x), B(y)] X(\{z_i\}) \rangle\rangle \quad (27)$$

which leads to Eq. (25).

Substituting (24) into Eq. (25), we get

$$\langle\langle [J(x), J(y)] \rangle\rangle = \frac{1}{4\pi^2} \left\{ \frac{1}{[0 + i(x - y)]^2} - \frac{1}{[-0 + i(x - y)]^2} \right\}$$

$$= \frac{1}{4\pi^2} \partial_x \left[ \frac{1}{(-i0 + x - y)} - \frac{1}{(i0 + x - y)} \right] = \frac{i}{2\pi} \partial_x \delta(x - y) \quad (28)$$

A similar procedure for the right currents yields

$$\langle\langle [\bar{J}(x), \bar{J}(y)] \rangle\rangle = -\frac{i}{2\pi} \partial_x \delta(x - y) \quad (29)$$

One can say that there is a difference in definitions. The original commutator does not include any averaging and the one which we have just calculated does. Here we have an example of a commutator which is nonzero only for a system of infinite number of particles and vanishes for any finite system. Such commutators are called *anomalous*. There is a nice and clear discussion of the anomalous commutator of currents in Subsection 4.3.1 of the Fradkin's book.

An **important lesson** which follows from the present discussion is the following. In order not to miss anomalous commutators, it is always better not to calculate them straightforwardly, but rather use a diagram expansion and the identity (25). Sometimes it is more convenient to use instead of commutators the *Wilson operator product expansion* (OPE):

$$J(z) J(z') = \frac{1}{4\pi^2 (z - z')^2} + \ldots \quad (30)$$

where dots denote nonsingular terms. This expansion means that one can substitute the product of currents *inside of a correlation function* by the right-hand side of Eq. (30).

Now consider a massless Bose field $\Phi$ governed by the action

$$S_B = \frac{1}{2} \int d^2 x (\nabla \Phi)^2 \quad (31)$$

and a model of free massless Dirac fermions

$$S_F = \int d^2 x \, \bar{\psi} \gamma_\mu \partial_\mu \psi \quad (32)$$

with a Noether vector current $J_\mu = \bar{\psi} \gamma_\mu \psi$. Conservation of this current $\partial_\mu J_\mu = 0$ will follow automatically, if we express it in terms of derivatives of the field $\Phi$:

$$J_\mu(x) \equiv \frac{1}{\sqrt{\pi}} \epsilon_{\mu\nu} \partial_\nu \Phi \quad (33)$$

**TABLE 1.** Bosonization dictionary. From [13], reprinted with the permission of Cambridge University Press.

| Massless bosons | Massless fermions |
|---|---|
| Action | Action |
| $\frac{1}{2}\int d^2x(\nabla\phi)^2$ | $2\int d^2x(\psi_R^\dagger \partial_{\bar z}\psi_R + \psi_L^\dagger \partial_z \psi_L)$ |
| Operators | Operators |
| $(1/\sqrt{2\pi a})\exp[\pm i\sqrt{4\pi}\phi(z)]$ | $\psi_R, \psi_R^\dagger$ |
| $(1/\sqrt{2\pi a})\exp[\mp i\sqrt{4\pi}\bar\phi(\bar z)]$ | $\psi_L, \psi_L^\dagger$ |
| $\frac{1}{\pi a}\cos[\sqrt{4\pi}\phi(z,\bar z)]$ | $\bar\psi\psi$ |
| $\frac{i}{\sqrt{\pi}}\partial\phi$ | $:\psi_R^\dagger\psi_R:$ |
| $-\frac{i}{\sqrt{\pi}}\bar\partial\phi$ | $:\psi_L^\dagger\psi_L:$ |

According to (33), the chiral components of the fermionic current, $J$ and $\bar J$, are then represented as

$$J = \frac{1}{\sqrt{4\pi}}(\partial_x\Phi + \Pi), \quad \bar J = \frac{1}{\sqrt{4\pi}}(\partial_x\Phi - \Pi) \tag{34}$$

where $\Pi$ is the momentum conjugate to the field $\Phi$. It is readily seen that, at quantum level, the anomalous current algebra (28) and (29) is correctly reproduced, if $\Phi$ and $\Pi$ satisfy the canonical commutation relation

$$[\Phi(x), \Pi(x')] = i\delta(x - x') \tag{35}$$

Thus, the identification (33) gives us an example of *bosonization*. Therefore for a fermionic theory with a linear spectrum, the fermionic operators can be represented in terms of a massless bosonic field and its dual field. This procedure called *bosonization* was first suggested by Coleman (1975) [3] and Mandelstam (1975) [5]. We can establish a correspondence between operators of the theory of massless bosons and massless fermions with the same correlation functions as reported in Table1.

Given the bosonization dictionary, one can solve many highly non-trivial models. The simplest example is given by the following action:

$$S = \int d^2x[\frac{1}{2}(\nabla\phi)^2 + \frac{M}{\pi a}\cos(\sqrt{4\pi}\phi)] \equiv \int d^2x(\bar\psi\gamma^\mu\partial_\mu\psi + M\bar\psi\psi). \tag{36}$$

The equivalence follows from the fact that in the two theories the perturbation expansions in M coincide.

For a review on bosonization approach to strongly correlated systems see [6].

# IV KAC-MOODY ALGEBRAS

As we have demonstrated in the first lecture, the Hamiltonian of spinless fermions can be written in terms of currents which leads to the Gaussian model and bosonization. A similar procedure exists for fermions with spin and is called *non-Abelian bosonization*. To be adequately equipped for it, we have to discuss algebraic properties of the corresponding currents.

Let $R^+_{\alpha,n}, R_{\alpha,n}$ ($L^+_{\alpha,n}, L_{\alpha,n}$) be the right- (left-) moving components of *free* massless fermions, with $\alpha$ and $n$ ($n = 1, 2, ...k$) being the spin and 'flavour' indices, respectively. It is assumed that the fermionic fields transform according to some Lie groups G and F, operating in the spin and flavour spaces. The reason for introducing the second index $n$ will become clear later. The current operators on group G are defined as follows:

$$J^a(z) = \sum_{n=1}^{k} L^+_{\alpha,n}(z)\tau^a_{\alpha\beta}L_{\beta,n}(z)$$

$$\bar{J}^a(\bar{z}) = \sum_{n=1}^{k} R^+_{\alpha,n}(\bar{z})\tau^a_{\alpha\beta}R_{\beta,n}(\bar{z}) \tag{37}$$

where $\tau^a$ are matrices – generators of the Lie algebra $\mathcal{G}$. They satisfy the following relations:

$$[\tau^a, \tau^b] = if^{abc}\tau^c$$

$$\mathrm{Tr}\tau^a\tau^b = \frac{1}{2}\delta^{ab} \tag{38}$$

For the case of the SU(2) group $\tau^a = S^a$ are the matrices of spin $S = 1/2$ and $f^{abc} = \epsilon^{abc}$. In a similar way one can define the current operators on the group F, using for this purpose its generators $T^j$.

Since the fermionic operators with different chiralities commute, so do the currents with different chiralities; currents with the same chirality satisfy the following Wilson operator expansion:

$$J^a(z)J^b(z') = \frac{k}{8\pi^2(z-z')^2}\delta^{ab} + if^{abc}\frac{J^c(z')}{2\pi(z-z')} + ... \tag{39}$$

To check the validity of this very important identity, one should consider two diagrams: one for the current-current correlation function depicted on Fig. 2.2 and another for the correlation function of three currents.

It is also very convenient to rewrite the operator expansion (39) in terms of commutators:

$$[J^a(x), J^b(y)] = \frac{ik}{4\pi}\delta^{ab}\delta'(x-y) + if^{abc}J^c(y)\delta(x-y) \tag{40}$$

In this form it is called a *Kac-Moody* algebra.

As for the spinless currents, the term with a derivative of the delta function is anomalous; it appears only in infinite systems. The second term in the right-hand side of Eq. (40) being absent in the spinless case is an ordinary commutator and can be obtained by a straightforward calculation.

Often the Kac–Moody algebra is written for the Fourier components of current operators. In this case we assume that the system is placed on the strip $0 < x < L$ with periodic boundary conditions and expand the current operators into the series

$$J^a(x) = \frac{1}{L}\sum_n e^{-2i\pi nx/L} J_n^a$$

Substituting this expansion into Eq. (40) we get the Kac–Moody algebra for the Fourier components:

$$[J_n^a, J_m^b] = \frac{nk}{2}\delta^{ab}\delta_{n+m,0} + if^{abc} J_{n+m}^c \tag{41}$$

The Kac–Moody algebra includes the original algebra $\mathcal{G}$ as its subalgebra; this is the algebra of the zeroth components of the currents:

$$[J_0^a, J_0^b] = if^{abc} J_0^c \tag{42}$$

Therefore one can think of these zeroth components as about matrices – generators of the group G.

Let us now follow the example of spinless fermions and consider the operators defined as G-invariant quadratic forms of chiral currents.

$$T(z) = \frac{1}{k+c_v} : J^a(z)J^a(z) :$$

$$\bar{T}(\bar{z}) = \frac{1}{k+c_v} : \bar{J}^a(\bar{z})\bar{J}^a(\bar{z}) : \tag{43}$$

The numerical value of the coefficient $1/(k+c_v)$, where $c_v$ is the Casimir operator in the adjoint representation, i.e., is defined by the identity

$$f_{abc}f_{\bar{a}bc} = c_v\delta_{a\bar{a}}$$

We leave it to the reader to show that the operators such defined are indeed chiral components of a stress energy tensor, i.e. they satisfy the Virasoro algebra with the central charge

$$C = \frac{\dim G}{k+c_v} \tag{44}$$

The easiest way to prove it would be to calculate the two- and three-point correlation functions of $T$ making use of the free fermion representation of current operators (37).

An extensive review of affine Kac-Moody algebras, may be found in [7].

# A   Example: some free fields representation

Free systems can also be used to realize particular representations of Kac-Moody algebras. For example we take N free fermions $\psi^i$ with operator product algebra

$$\psi^i(z)\psi^j(z') = -\frac{\delta^{ij}}{(z-z')}. \tag{45}$$

Consider these fermions to transform in the vector representation of the $SO(N)$, with matrices $t^a$. For $N \geq 4$, the currents

$$J^a(z) = \psi(z)t^a\psi(z), \tag{46}$$

are easily verified to satisfy (30) for SO(N) at level $k = 1$. We also verify that the central charge is given by

$$c_{SO(N)_1} = \frac{1 \cdot 1/2N(N-1)}{1+(N-2)} = N/2, \tag{47}$$

consistent with the central charge for N free fermions. We could in the same way use N complex fermions taken to transform in the vector representation of SU(N), and construct currents $J^a(z) = \psi^\star(z)t^a\psi(z)^\star$ analogous to (46). These realize affine $SU(N) \times U(1)$, with the SU(N) at level $k$. The central charge comes out to be

$$c_{U(1)} + c_{SU(N)_k} = 1 + \frac{1(N^2-1)}{(1+N)} = N, \tag{48}$$

consistent with the result for N free complex fermions.

An other example is to take $|G|$ free fermions to transform in the adjoint representation of some group $G$. Then the currents

$$J^a(z) = \frac{i}{2}f^{abc}\psi^b(z)\psi^c(z), \tag{49}$$

give a realization of affine G at level $k = h_G$. The central charge is $c = h_G|G|/(h_G + h_G) = |G|/2$. This case of dimG is also known as super-affine $G$ algebra [21]. In general, a super-affine algebra has, in addition to the structure (30) a spin-3/2 superstress tensor given by

$$T_F = -f^{abc}\psi^a\psi^b\psi^c/12\sqrt{2C}. \tag{50}$$

# V  SUGAWARA HAMILTONIAN FOR WESS–ZUMINO–NOVIKOV–WITTEN MODEL

According to the general property of conformal theories a sum of the zeroth components of traceless stress-energy tensors $T_0$, $\bar{T}_0$ gives a Hamiltonian. From Eq. (43) we extract the corresponding Hamiltonian which traditionally carries the name of Sugawara:

$$\hat{H}(k,G) = \frac{2\pi v}{L}(L_0 + \bar{L}_0)$$
$$= \frac{2\pi v}{(k+c_v)L}\left[J_0^a J_0^a + 2\sum_{n>0} J_{-n}^a J_n^a + (J \to \bar{J})\right], \quad (51)$$

where $J_n^a$ satisfy the Katz-Moody algebra. The Sugawara Hamiltonian defines the Hamiltonian of the Wess–Zumino–Novikov–Witten (WZNW) model.

Since we have introduced the Hamiltonian (51) axiomatically, the reader may remain perplexed about its physical meaning. It is clear that the WZNW model represents some conformal theory, but we have not explained relations between this theory and more conventional models. These relations are established by the following identity which we give only for the case G = SU(N):

$$\int dx[-iR_{\alpha,n}^+ \partial_x R_{\alpha,n} + iL_{\alpha,n}^+ \partial_x L_{\alpha,n}] = H[U(1)] + H[k, \mathrm{SU}(N)] + H[N, \mathrm{SU}(k)] \quad (52)$$

$$H[U(1)] = 2\pi \int dx[: J(x)J(x): + : \bar{J}(x)\bar{J}(x):] \quad (53)$$

$$H[k, \mathrm{SU}(N)] = \frac{2\pi}{(N+k)}\sum_{a=1}^{D_N} \int dx[: J^a(x)J^a(x): + : \bar{J}^a(x)\bar{J}^a(x):] \quad (54)$$

$$H[N, \mathrm{SU}(k)] = \frac{2\pi}{(N+k)}\sum_{\lambda=1}^{D_k} \int dx[: \mathcal{J}^\lambda(x)\mathcal{J}^\lambda(x): + : \bar{\mathcal{J}}^\lambda(x)\bar{\mathcal{J}}^\lambda(x):] \quad (55)$$

where $\mathcal{J}^\lambda$ are currents of the SU(k) group. This representation corresponds to the decomposition of the nonsimple symmetry group of the free fermions into the product of simple groups U(1)× SU(N)× SU(k). It has been used here that for the SU(N) group $c_v = N$ and dim SU(N) = $N^2 - 1$. The decomposition (51) is a generalization of the decomposition for a case of fermions with U(1)×SU(2) symmetry. It would be too complicated to give a complete rigorous proof of the validity of this identity. A simple argument in favor of this identity is equality of the central charges. The central charge of the theory free Dirac fermions is equal to the number of fermion species, so it is $Nk$. According to (44) the central charge of the SU(N)-invariant WZNW model is given by

$$C = \frac{k(N^2-1)}{k+N} \quad (56)$$

where the number of generators is $D_N = N^2 - 1$. Thus we have

$$Nk = 1 + \frac{k(N^2-1)}{k+N} + \frac{N(k^2-1)}{k+N}$$

which is identically satisfied.

From Eq. (51) we see that the WZNW model on group G can be understood as a G-invariant subsector of the model of free fermions. So, take free fermions and project some states out. If you are careful and do this projection intelligently preserving the conformal symmetry, you will obtain a model with interesting properties. Many interactions do this job perfectly thus leaving the WZNW model to be an effective theory for low-lying excitations.

Now we have to discuss the WZNW theory in more detail. It is instructive to construct the full basis of its eigenstates. One should do this using the current operators *only*, because only in this way can one be sure that all eigenstates are G-invariant. To fulfill this task, we need to know how the current operators commute with the Hamiltonian. The corresponding commutation relations follow from the definition of the stress-energy tensor (43):

$$[L_n, J_m^a] = -m J_{n+m}^a \tag{57}$$

Since $H = 2\pi v(L_0 + \bar{L}_0)/L$, we get

$$[H, J_m^a] = -\frac{2\pi v m}{L} J_m^a \tag{58}$$

Now we have everything we need to construct the eigenstates. Let us define the vacuum vectors $|h\rangle$ as states annihilated by positive Fourier components of the currents:

$$J_m^a |h\rangle = 0; \quad \bar{J}_m^a |h\rangle = 0 \tag{59}$$

The Hamiltonian (51) acting on these states is reduced to

$$H_{\text{reduced}} = \frac{2\pi v}{(k+c_v)L}(J_0^a J_0^a + \bar{J}_0^a \bar{J}_0^a) \tag{60}$$

Therefore $|h\rangle$ are eigenstates of the Hamiltonian if they realize irreducible representations of the left and right G groups:

$$J_0^a J_0^a |h\rangle = C_{\text{rep}} |h\rangle; \quad \bar{J}_0^a \bar{J}_0^a |h\rangle = \bar{C}_{\text{rep}} |h\rangle \tag{61}$$

Now we can use the negative components of the currents as creation operators. According to Eq. (58), the following vectors

$$\bar{J}_{-m_1}^{a_1} ... \bar{J}_{-m_p}^{a_p} J_{-n_1}^{a_1} ... J_{-n_q}^{a_q} |h\rangle \tag{62}$$

are eigenvectors of the Hamiltonian (51) with the energies

$$E_{pq}(h) = \frac{2\pi v}{L}\left(\frac{C_{\text{rep}}}{k+c_v} + \sum_{i=1}^{q} n_i + \frac{\bar{C}_{\text{rep}}}{k+c_v} + \sum_{i=1}^{p} m_i\right) \tag{63}$$

For every vacuum state we have towers of states, where $\sum_i n_i$ is the total momentum of the system. As known, each eigenstate in conformal field theories is associated with some conformal field, whose conformal dimensions are related to the eigenvalues of energy and momentum. In the given case the eigenstates of the Hamiltonian are eigenstates of the Cazimir operators of the group G. The conformal dimensions are

$$\Delta = \frac{C_{\text{rep}}}{k+c_v} + \sum_{i=1}^{q} n_i$$
$$\bar{\Delta} = \frac{\bar{C}_{\text{rep}}}{k+c_v} + \sum_{i=1}^{p} m_i \tag{64}$$

It turns out that the basis of states (62) is overcomplete. To make it complete one has to restrict the number of vacuum states $|h\rangle$ choosing a finite number of irreducible representations of G. For example, in the case G = SU(2) where the irreducible representations are representations of spin operators with $C_{\text{rep}} = S(S+1), S = 1/2, 1, ...$, the basis is composed by $S = 1/2, 1, ...k/2$ (Fateev and Zamolodchikov (1986) [8]). Each vacuum vector $|h\rangle$ can be considered as a state created from the lowest vacuum $|0\rangle$ by a corresponding primary field $g_h(\tau, x)$. Indeed, we have

$$|h\rangle = \lim_{\tau \to +\infty} e^{\tau(E_h - E_0)}(L/2\pi)^{(2\Delta_h + 2\bar{\Delta}_h)} g_h(\tau, x=0)|0\rangle \tag{65}$$

Thus the primary fields are matrices $g_{ab}$ – tensors realizing irreducible representations of the group G. There is one more restriction on the vacuum states, namely a requirement that physical fields must have integer or half-integer conformal spins. If these fields are built exclusively from the WZNW fields, this means that

$$\frac{C_{\text{rep}}}{k+c_v} - \frac{\bar{C}_{\text{rep}}}{k+c_v} = n/2 \tag{66}$$

where $n$ is an integer. In general, this requirement is difficult to satisfy except for $n = 0$. In this latter case $C_{\text{rep}} = \bar{C}_{\text{rep}}$ and the primary fields are dim $C_{\text{rep}} \times$ dim $\bar{C}_{\text{rep}}$. This is the case considered in the original paper by Knizhnik and Zamolodchikov (1984) [17]. There are other cases, however, where physical fields are products of fields of several WZNW models (the corresponding example is the spin–charge separation). Some details can be found in the paper of Affleck and Ludwig (1991) [10].Then the restriction on right and left representations is different.

# VI KNIZHNIK–ZAMOLODCHIKOV (KZ) EQUATIONS

Now we shall derive differential equations for multi-point correlation functions of the WZNW primary fields. The derivation is based on the fact that the WZNW stress energy tensor is quadratic in currents (43). Writing this expression in components, we get

$$L_n = \frac{1}{c_v + k} \sum_m : J^a_m J^a_{n-m} : \tag{67}$$

where the normal ordering assumes that the operators $J^a_m$ with a positive subscript (annihilation operators) stay on the right. Among the Virasoro generators there is one whose action on operators is particularly simple – it is $L_{-1} \equiv \partial_z$. For $n = -1$ we have

$$\partial_z \equiv L_{-1} = \frac{2}{c_v + k} [J^a_{-1} J^a_0 + \sum_{m=1}^{\infty} J^a_{-m-1} J^a_m] \tag{68}$$

Since primary fields are vacuum states, they are annihilated by positive components of current operators (59). Therefore from (68) it follows that any primary field satisfies the identity

$$\left(\partial_z - \frac{2}{c_v + k} J^a_{-1} J^a_0\right) \phi_h(z) = 0 \tag{69}$$

We already know that $J^a_0$ acts simply as the generator of the group (see Eq. (61)). To proceed further we need to know the action of $J_{-1}$. This can be extracted from the Ward identity

$$\langle J^a(z)\phi(1)...\phi(N)\rangle = \sum_j \frac{T^a_j}{z - z_j} \langle \phi(1)...\phi(N)\rangle \tag{70}$$

Thus we have

$$\langle J^a_{-1}\phi(1)...\phi(N)\rangle = \frac{1}{2\pi i} \int_C dz(z_1 - z)^{-1} \langle J^a(z)\phi(z_1)...\phi(N)\rangle$$

$$= \sum_{j \neq 1} \frac{T^a_j}{z_1 - z_j} \langle \phi(1)...\phi(N)\rangle \tag{71}$$

where the contour $C$ encircles $z$.

Combining these results we conclude that the $N$-point function of primary fields satisfy the following system of equations:

$$\left[\frac{1}{2}(c_v + k)\partial_{z_i} - \sum_{j \neq i} \frac{T^a_i T^a_j}{z_i - z_j}\right] \langle \phi(1)...\phi(N)\rangle = 0 \tag{72}$$

where a matrix $\tau_i^a$ acts on the indices of the $i$-th operator.

Let us consider this equation for the case $N = 2$. Then we have

$$\left[\frac{1}{2}(c_v + k)\partial_{z_1} - \frac{\tau_1^a \tau_2^a}{z_1 - z_2}\right] \langle \phi(1)\phi(2) \rangle = 0$$
$$\left[\frac{1}{2}(c_v + k)\partial_{z_2} + \frac{\tau_1^a \tau_2^a}{z_1 - z_2}\right] \langle \phi(1)\phi(2) \rangle = 0 \qquad (73)$$

¿From these two equations it follows that $\langle \phi(1)\phi(2) \rangle = G(z_{12})$ and the function $G(z)$ satisfy the following equation:

$$\partial_z G_{\alpha,\alpha'}^{\beta,\beta'}(z) - \frac{2\tau_{\alpha,\gamma}^a \tau_{\beta,\delta}^a}{(c_v + k)z} G_{\gamma,\alpha'}^{\delta,\beta'}(z) = 0 \qquad (74)$$

It is natural to suggest that a nonzero solution exists only if the second field is the Hermitian conjugate of the first one: $\phi(2) = \phi^+(1)$. It is essential that $\tau_2^a$ acting on this operator gives minus sign. Then $\tau_1^a \tau_2^a$ becomes a Kazimir operator:

$$\tau_1^a \tau_2^a = -(\tau^a)^2 = -C_{\text{rep}}$$

and we get

$$\langle \phi(1)\phi^+(2) \rangle = z_{12}^{-2\Delta_h} \qquad (75)$$

with the correct conformal dimension (64). Solutions of KZ equations for four-point correlation functions of primary fields in the fundamental representation of the SU(N) group were found by Knizhnik and Zamolodchikov (1984) [17]. Solutions for four-point functions of all primary fields of the $SU_k(2)$ WZNW model were found by Fateev and Zamolodchikov (1986) [8]. There is a regular procedure for finding multi-point correlation functions, called the Wakimoto construction, which is based on the representation of the WZWN operators in terms of free bosonic fields (Wakimoto (1986) [11]). The details of this procedure can be found in Dotsenko (1990) [2].

# VII WZNW MODEL IN THE LAGRANGIAN FORMULATION AND NON-ABELIAN BOSONIZATION

In many cases it is necessary to have Wess-Zumino-Novikov-Witten model in its Lagrangian form, whose derivation is given in the original paper of Polyakov and Wiegmann (1983) [15]. Let us start from the following **theorem**:
The Euclidean action for the Sugawara Hamiltonian

$$H = \frac{2\pi}{k+c_v} \sum_{a=1}^{D} \int dx [: J^a(x)J^a(x) : + : \bar{J}^a(x)\bar{J}^a(x) :] \tag{76}$$

where the currents $J^a$, $\bar{J}^a$ satisfy the Kac-Moody algebras for the group G with the central extension $k$, is given by $S = kW(U)$, where $U$ is a matrix from the fundamental representation of the group G and $W(U)$ is

$$W(U) = \frac{1}{16\pi} \int d^2x Tr(\partial_\mu U^{-1} \partial_\mu U) + \Gamma[U]$$
$$\Gamma[U] = -\frac{i}{24\pi} \int_0^\infty d\xi \int d^2x \epsilon^{\alpha\beta\gamma} Tr(U^{-1}\partial_\alpha U U^{-1}\partial_\beta U U^{-1}\partial_\gamma U) \tag{77}$$

It is supposed that the field $U(\xi, x)$ in Eq.(77) is defined on a three-dimensional hemisphere whose boundary coincides with the two-dimensional plane where the original theory is defined, so that $U(\xi = 0, x) = U(x)$. The first remarkable fact about the Lagrangian $W(U)$ is that it has local equations of motion:

$$\bar{\partial}\bar{J} = 0, \bar{J} = \frac{k}{2\pi} U \partial U^{-1},$$
$$\partial J = 0, J = -\frac{k}{2\pi} U \bar{\partial} U^{-1}. \tag{78}$$

The currents appearing in (78) have the same correlation functions as the currents obeying the Sugawara Hamiltonian of the Theorem. The easiest way to obtain the equation of motions (78) is to use the following identity:

$$W(gU) = W(U) + W(g) + \frac{1}{2\pi} \int d^2x Tr(g^{-1}\partial g U \bar{\partial} U^{-1}). \tag{79}$$

The meaning of Eq.(79) is easy to grasp when considering its Abelian limit. Let $U = \exp(i\phi)$, $g = \exp(i\eta)$, where $\phi$ and $\eta$ are scalar fields. In this case the $\Gamma$ term in (77) vanishes identically and the WZNW action $W(U)$ becomes:

$$W[\exp(i\phi)] = \frac{1}{16\pi} \int d^2x (\partial_\mu \phi \partial_\mu \phi). \tag{80}$$

Substituting it into the identity (79) we get

$$\frac{1}{16\pi}\int d^2x[\partial_\mu(\phi+\eta)\partial_\mu(\phi+\eta)] =$$
$$\frac{1}{16\pi}\int d^2x(\partial_\mu\phi\partial_\mu\phi) + \frac{1}{16\pi}\int d^2x(\partial_\mu\eta\partial_\mu\eta) + \frac{1}{8\pi}\int d^2x(\partial_\mu\eta\partial_\mu\phi). \tag{81}$$

An important thing is that the WZNW action W(U) is a *multivalued* functional. Indeed, the fact that the action W(U) is defined modulo $2\pi i$ does not prevent the equations of motion (78) from being local. Neither does it prevent the partition function from being well defined, provided $k$ is an integer. This means that the $\Gamma$-functional can become a part of an action of a physical theory only if it has an integer coefficient $k$. In other words, the fact that the action is multivalued yields the quantization condition for $k$.

From the previous sections, we are familiar with primary fields of the WZNW model. In the Lagrangian formulation we identify them with the following matrix fields:

- U-field: a primary field transforming according to the fundamental representation of the group G. If G=SU(N) ($c_v = N$) its scaling dimension is equal to
$$\Delta_U = \bar\Delta_U = \frac{N^2-1}{2N(N+k)}; \tag{82}$$

- $\phi_1^{ab} = Tr(U^{-1}t^a U T^b)$ where $t^a$ are the generators of $G$. This field is also a primary one and belongs to the adjoint representation; its scaling dimensions are equal to
$$\Delta_1 = \bar\Delta_1 = \frac{c_v}{c_v+k} \tag{83}$$

- The 'wrong currents' $K^a \simeq Tr(t^a U^{-1}\partial U)$, $\bar K^a \simeq Tr(t^a \bar\partial U U^{-1})$, whose scaling dimensions are equal to $(\Delta_1+1, \Delta_1)$ and $(\Delta_1, \Delta_1+1)$, respectively. These fields are not primary

- The Lagrangian density $Tr(\partial_\mu U^{-1}\partial_\mu U)$ is also not a primary field and has scaling dimensions equal to $(\Delta_1+1, \Delta_1+1)$. As we see, the Lagrangian density is an irrelevant operator. Therefore the general theory with the action
$$W(\lambda; U) = \frac{1}{2\lambda}\int d^2x Tr(\partial_\mu U^{-1}\partial_\mu U) + k\Gamma[U], \tag{84}$$

scales to the critical point $\lambda = 8\pi/k$. The crossover was described exactly by Polyakov and Wiegmann who solved this model by the Bethe-Ansatz (1984) [16]. In fact, it would be proper to call this model the Wess-Zumino-Novikov-Witten model; then the theory described by the action W(U) appearing in (77) is the WZNW model at the 'critical' point. The details on the derivation of the Lagrangian can be found in the paper of Polyakov and Wiegmann (1983) [15]. Using the above results we can compile the following Table

**TABLE 2.** Non-Abelian bosonization.

| Critical U(N)WZNW with $k=1$ | Massless Dirac fermions |
|---|---|
| Action | Action |
| $1/2 \int d^2x (\partial_\mu \phi)^2 + W[g]$; g$\in SU(N)$ | $2 \int d^2x (R_\alpha^\dagger \partial_{\bar z} R_\alpha + L_\alpha^\dagger \partial_z L_\alpha)$ |
| Operators | Operators |
| $1/(2\pi a)\exp[i(N/4\pi)^{1/2}\phi]g_{\alpha\beta}$ | $R_\alpha^\dagger L_\beta$ |
| $i(N/4\pi)^{1/2}\partial\phi$ | $1/2: R_\alpha^\dagger R_\alpha:$ |
| $-i(N/4\pi)^{1/2}\bar\partial\phi$ | $1/2: L_\alpha^\dagger L_\alpha:$ |
| $1/2\pi Tr(\tau^a g \partial g^{-1})$ | $:R_\alpha^\dagger \tau^a_{\alpha\beta} R_\beta:$ |
| $-1/2\pi Tr(\tau^a g \bar\partial g^{-1})$ | $:L_\alpha^\dagger \tau^a_{\alpha\beta} L_\beta:$ |

More details on other critical WZWN models could be found in the book [13].

As we have seen from the previous discussion certain combinations of fermionic fields governed by the Dirac Hamiltonian can be written in terms of fields of the WZNW model. In particular, there exist an explicit equivalence for currents. In particular, the slow components of the $R^\dagger$ L field in the $U(N) \times SU(k)$-model with the current-current interaction are proportional to the Wess-Zumino matrix $U$. The latter property holds also for noninteracting Dirac fermions with symmetry U(N) (see the paper of Knizhnik and Zamolodchikov (1984) [17]) and for the noninteracting Majorana fermions with the symmetry O(N) (Witten (1984) [14]). Thus we can compose the previous bosonization Table2 for non-Abelian theories. Why non-Abelian bosonization can help? Let us consider for example a dimerized spin chain, whose Hamiltonian is

$$\sum_n \mathbf{S}_n \mathbf{S}_{n+1}[1 + \delta(-1)^n]. \tag{85}$$

In non-Abelian bosonization its action is given by

$$\delta Tr(g + g^\dagger) + W[g], \tag{86}$$

while in the Abelian bosonization it is

$$\frac{1}{2}(\nabla\phi)^2 + \delta \cos\sqrt{2\pi}\phi. \tag{87}$$

It turns immediately out that non-Abelian bosonization is the most adequate approach in this case because it explicitly preserves $SU(2)$ symmetry.

## VIII  A NON-TRIVIAL DETERMINANT

Using the previous Table and the properties of the WZNW-action we can calculate the following fermionic determinant (Tsvelik (1994) [12]):

$$D[U] = Trln[\gamma_\mu \partial_\mu + (1+\gamma_5)mU/2 + (1-\gamma_5)mU^\dagger/2] \tag{88}$$

where $U$ is an external matrix field from the $SU(N)$ group and $m$ is a parameter of the dimension of mass. We assume that the field $U(x,\tau)$ varies slowly on the scale $1/m$ and will be interested in the expansion of $D[U]$ in powers $1/m$. Such problems appear frequently in applications. Using the Table we can rewrite the determinant as path integral over the matrix field $g$ with the WZNW action:

$$e^{D[U]} = \int DgD\phi\, e^{-S}, \tag{89}$$

$$S = \int d^2x [\frac{m}{2\pi a} Tr(Ug^\dagger e^{-i\gamma\phi} + gU^\dagger e^{i\gamma\phi})] + W(g) + \frac{1}{2}\int d^2x (\partial_\mu \phi)^2, \tag{90}$$

where $\gamma = (4\pi/N)^{1/2}$. Now we make a shift of variables in the path integral introducing a new variable $G$, $g = UG$. This shift leaves the measure of integration unchanged. Using the identity (79) we get the following result:

$$e^{D[U]} = \int DGD\phi \exp\{-S[G,\phi] - S_{int}[U,G] - W[U]\}$$

$$S[G,\phi] = \int d^2x [\frac{m}{2\pi a} Tr(G^\dagger e^{-i\gamma\phi} + e^{i\gamma\phi}G)] + W(G) + \frac{1}{2}\int d^2x (\partial_\mu \phi)^2,$$

$$S_{int} = \frac{1}{2\pi}\int d^2x Tr(U^{-1}\partial U G \bar{\partial} G^{-1}). \tag{91}$$

The only term in the action connecting $G$ and $U$ is the interaction term $S_{int}$. At small energies this term is irrelevant and from Eq.(91) we find

$$D[U]/D[I] = -W(U) + O(m^{-2}). \tag{92}$$

The latter result is not-trivial since it includes the topological term which is difficult to get by more conventional methods.

Let us conclude saying that there are many applications of the bosonization technique to physical models in (1+1)-dimensions, ranging from spin-1/2 Tomonaga-Luttinger liquid to spin-1/2 Heisenberg chain with alternating exchange , and single impurity problems. In the following we are goind to discuss some specific example.

# IX $SU(2)_1$ WZNW MODEL: A GAUSSIAN MODEL

As an example where Abelian bosonization applies, let us consider the $SU(2)_1$ WZNW model. The possibility of an Abelian bosonization is due to the fact that its conformal charge is equal to 1, $C_{SU(2)_1} = C_{boson} = 1$. Moreover, since $1/3 \mathbf{J}_{R(L)} \mathbf{J}_{R(L)} = J^z_{R(L)} J^z_{R(L)}$, the noninteracting Hamiltonian $H_{SU(2)}$ can be expressed in terms of $J^z$ currents only. Introducing the canonical variables $\phi_S$ and $\Pi_S$ via

$$J^z + \bar{J}^z = \frac{1}{\sqrt{2\pi}} \partial_x \phi_S,$$

$$J^z - \bar{J}^z = \frac{1}{\sqrt{2\pi}} \Pi_S. \tag{93}$$

One finds

$$H_{SU(2)} \equiv H_B = \frac{v_s}{2} \int dx [\Pi_S^2 + (\partial_x \phi_S(x))^2]. \tag{94}$$

The Hamiltonian thus corresponds to that of a gaussian model. The price for such semplification is the loss of the spin rotational invariance of the currents (i.e. $J^x$, $J^y$ cannot be expressed as $J^z$) and we need to bosonize the left and right fermion fields in the following way:

$$\psi_{R,L,\alpha} \simeq \frac{1}{\sqrt{2\pi a}} e^{\pm i\sqrt{4\pi} \phi_{R,L,\alpha}(x)}, \tag{95}$$

where $\alpha$ is a spin index. We now introduce the linear combinations:

$$\Phi_\alpha = \phi_{R\alpha} + \phi_{L\alpha}$$

$$\Theta_\alpha = \phi_{L\alpha} - \phi_{R\alpha}. \tag{96}$$

The spin and charge degrees of freedom are described by the following fields:

$$\Phi_c = \frac{\Phi_\uparrow + \Phi_\downarrow}{\sqrt{2}}, \quad \Theta_c = \frac{\Theta_\uparrow + \Theta_\downarrow}{\sqrt{2}}$$

$$\Phi_s = \frac{\Phi_\uparrow - \Phi_\downarrow}{\sqrt{2}}, \quad \Theta_s = \frac{\Theta_\uparrow - \Theta_\downarrow}{\sqrt{2}}, \tag{97}$$

where the conjugate fields of $\Phi_{c,s}$ are $\partial_x \Theta_{c,s} = \Pi_{c,s}$.

To bosonize the currents we use Table1 and the result gives:

$$J^+ = \psi^\dagger_{R\uparrow} \psi_{R\downarrow} = \frac{1}{2\pi a} e^{-i\sqrt{2\pi}(\Phi_s - \Theta_s)}$$

$$\bar{J}^+ = \psi^\dagger_{L\uparrow} \psi_{L\downarrow} = \frac{1}{2\pi a} e^{i\sqrt{2\pi}(\Phi_s + \Theta_s)}. \tag{98}$$

As expected the charge part does not contribute to the spin SU(2) currents. The SU(2) currents $\mathbf{J}, \bar{\mathbf{J}}$ permits to determine the smooth part of the spin operators in the continuum limit (i.e. at $a \to 0$). We have

$$\mathbf{S}_i \to a\mathbf{S}(x), \quad \mathbf{S}(x) = \mathbf{J}(x) + \bar{\mathbf{J}}(x) + (-)^i \mathbf{n}(x), \tag{99}$$

where

$$\mathbf{n}(x) = \psi_{R\alpha}^\dagger \frac{\sigma^{\alpha\beta}}{2} \psi_{L\beta}(x) + h.c., \tag{100}$$

is the staggered part of the local spin density. Let us note that since the off-diagonal terms $\psi_{R(L)}^\dagger \psi_{L(R)}$ describe particle-hole excitations with transfer momentum $2k_F$, charge excitations emerge. Infact,

$$\begin{aligned} n^z &= -\frac{1}{2\pi a} \cos(\sqrt{2\pi}\Phi_c) \sin(\sqrt{2\pi}\Phi_s) \\ n^\pm &= \frac{1}{2\pi a} \cos(\sqrt{2\pi}\Phi_c) \exp(\pm\sqrt{2\pi}\theta_s). \end{aligned} \tag{101}$$

Assuming that the charge excitations are massive with a large energy gap $m_c$, since we are interested in the low-energy sector $E \ll m_c$, we can replace $\cos(\sqrt{2\pi}\Phi_c)$ by its nonzero expectation value, denoted by $\lambda$. The final bosonization formulas for $\mathbf{n}$ is:

$$\begin{aligned} n^z &= -\frac{\lambda}{2\pi a} \sin(\sqrt{2\pi}\Phi_s) \\ n^\pm &= \frac{\lambda}{2\pi a} \exp(\pm\sqrt{2\pi}\theta_s). \end{aligned} \tag{102}$$

Since the $SU(2)_1$ WZNW model describes the continuous limit of the isotropic spin 1/2 Heisenberg chain, the previous procedure complets the bosonization of the spin operators for the isotropic spin chain.

# X  NON-ABELIAN BOSONIZATION OF INTERACTING FERMIONS WITH SPIN: THE HUBBARD MODEL

In this section we discuss a model of itinerant electrons interacting with a short range interaction. Apart the intrinsic interest towards this model, we shall employ non-Abelian bosonization which makes the SU(2) symmetry more manifest. The Hamiltonian is given by

$$H = \sum_{<ij>}(t\psi_{i,\alpha}^\dagger \psi_{j,\alpha} + h.c.) + U\sum_i n_{i\uparrow}n_{j\downarrow}, \qquad (103)$$

where $n_{i\alpha}$ is the electron number $\psi_{i\alpha}^\dagger \psi_{i\alpha}$, $\alpha$ is a spin index. The first term describes the hopping between nearest neighbor sites and $U$ is a highly screened Coulomb repulsion between electrons. We take the continuum limit as for the spinless fermion model, defining left and right moving fermions for each spin. The hopping term gives a Lorentz-invariant free Dirac fermion theory with two flavors. The continuum free fermion theory, as we have seen, has $U(1)$ and $SU(2)$ symmetries. Corresponding to these symmetries we have conserved currents, that can be written as

$$J_{R,L} \equiv\, :\psi_{R,L}^{\dagger i\alpha}\psi_{L,R,i\alpha}:,$$
$$\mathbf{J}_{R,L} \equiv \psi_{R,L}^{\dagger i\alpha}\frac{\sigma_\alpha^\beta}{2}\psi_{L,R,i\alpha}. \qquad (104)$$

They satisfy $SU(2)_1$ Kac-Moody algebra. The energy-momentum tensor can be written in terms of the currents as:

$$T_L = \frac{\pi}{2}v J_L J_L + \frac{2\pi v_F}{3}\mathbf{J}_L \mathbf{J}_L, \qquad (105)$$

and similarly for the right part, where $v_F$ is the velocity of light given by

$$v_F = 2a|t|. \qquad (106)$$

We now consider the Hubbard interaction $U$ in the continuum limit. Writing the couplings in terms of the left and right movers, we get terms with power four in $\psi_{R(L)}$ and terms with two $\psi_R$ and two $\psi_L$. The completely left-moving (or completely right-moving) terms can be written quadratically in currents and thus simply renormalize the speed of light (106). This renormalization is different for terms involving charge and spin degrees of freedom. Of course, a theory with two different velocities is not Lorents invariant. However, in the low energy limit only the spin part of the energy tensor can be retained, so the problem does not arise. This is due to the fact that the charge sector has a gap with a large mass $mc$. The mixed left-right terms correspond, instead, to Lorents invariant interactions in the Lagrangian formalism. There are only three different Lorents-invariant interaction permitted by symmetries of the Hubbard model. Using the continuum representation of the Fermi fields operators and omitting highly oscillatory terms, the three permitted interactions can be written as

$$H_{int(LR)} = \int dx[g_4(J_R^2 + J_L^2) + g_C J_R J_L + g_S \mathbf{J}_R \cdot \mathbf{J}_L], \qquad (107)$$

where

$$g_4 = u/4, g_S = -2U, g_C = U. \qquad (108)$$

The resulting Hamiltonian naturally splits into two commuting parts describing charge and spin degrees of freedom, respectively:

$$H = H_c + H_s, \qquad (109)$$

where

$$H_c = \frac{\pi v_c}{2} \int dx(: J_R J_R : + : J_L J_L :) + g_C \int dx J_R J_L,$$
$$H_s = \frac{2\pi v_s}{3} \int dx(: \mathbf{J}_R \mathbf{J}_R : + : \mathbf{J}_L \mathbf{J}_L :) + g_S \int \mathbf{J}_R \mathbf{J}_L. \qquad (110)$$

Using the Abelian bosonization we can write the Hamiltonian for $H_c$ in the gaussian form. Namely,

$$H_c = \frac{v_c}{2} \int dx[K_c(\partial_x \theta_c)^2 + K_c^{-1}(\partial_x \phi_c)^2], \qquad (111)$$

where $v_c$ and $K_c$ depend on the coupling constants.

The most elegant way of doing this is to use non-Abelian bosonization [14]. The version that is most useful in this case is to rewrite the theory in terms of an SU(2) Wess-Zumino-Witten matrix $g(\mathbf{x})$ representing the SU(2) degrees of freedom, and a single scalar field $\eta(\mathbf{x})$ representing the U(1) degrees of freedom. The free fermion theory is equivalent to decoupled theories for $g$ and $\eta$. Namely,

$$L(\eta) = \frac{1}{2}\partial_\mu \eta \partial^\mu \eta,$$
$$S_{WZNW}(g) = \frac{1}{8\pi} \int d^2x tr \partial_m u g^\dagger \partial^\mu g$$
$$+ \frac{1}{12\pi} \int d^3x \epsilon^{\mu\nu\lambda} tr g^\dagger \partial_\mu g g^\dagger \partial_\nu g g^\dagger \partial_\lambda g, \qquad (112)$$

where the second WZ term is defined by extending the two-dimensional space (or space-time) to a three dimensional half-space ($x_3 < 0$). The boundary value determines the WZ term up $2\pi n$ where $n$ is an integer. This implies that the path integral is well defined if the coefficient in front of this term is an integer. The currents are written as

$$J_L = -\frac{1}{\sqrt{4\pi}}\partial_+\eta, \quad J_L = \frac{1}{\sqrt{4\pi}}\partial_-\eta,$$
$$\mathbf{J}_L = -\frac{i}{4\pi}tr\partial_+ g\partial_- g^\dagger \sigma, \quad \mathbf{J}_R = \frac{i}{4\pi}tr\partial_+ g\partial_- g^\dagger \sigma, \tag{113}$$

while the left-right fermion product is

$$\psi_L^{\dagger\alpha}\psi_{R\beta} \propto g_\beta^\alpha \exp i\sqrt{2\pi}\eta. \tag{114}$$

Thus the Lagrangian decouples into two parts involving $\eta$ or $g$ only:

$$L(\eta) = \frac{1}{2}\partial_\mu \eta \partial^\mu \eta (1 - \frac{g_4}{2\pi}) + \text{constant } g_C \cos\sqrt{8\pi}\eta,$$
$$S(g) = S(g)_{WZW} + g_S \int d^2x \mathbf{J}_L \cdot \mathbf{J}_R. \tag{115}$$

The bosonization procedure has allowed a complete spin-charge separation in the continuum limit. The charge sector should be massive for any finite $U > 0$ since we get a relevant sine-Gordon interaction, while, concerning the spin sector we have a marginal interaction $g_S$ that can be marginally relevant or irrelevant depending on the sign of the interaction. When $g_S > 0$ the coupling between currents becomes relevant, and the spin sector acquires a spectral gap. Meanwhile the charge sector remains gapless away from half-filling. This situation at low temperature is characterized by an exponentially small Pauli susceptibility, indicating that the material remains still a good conductor.

The followed derivation serves for relating the parameters of the effective theory to the bare interactions. These relations in general holds only for small values of the bare coupling constants, but in the present case, the general form of the low-energy Hamiltonian being determined by the symmetry requirements, survives even if the interactions are strong.

The parameters $v_c$, $v_s$ and $g_{S(C)}$ are known exactly from Bethe-ansatz solution [22,23]. In particular, for $U/t \ll 1$ we have

$$v_s/v_c \simeq (t/U)\sin\pi\nu. \tag{116}$$

where $\nu$ is the band filling, i.e. the average number of electrons per site divided by 2. The example of the Hubbard model demonstrates that $v_s/v_c$ may change with the interaction, while $K_c$ is always greater than $1/2$ (as shown in Figure5).

The fact that charge and spin sectors have different velocities, $v_c \neq v_s$, is a peculiarity of one-dimensional systems. This phenomenon is known as *spin-charge separation*.

To conclude let us summarize the possible states of a general system in one dimension characterized by a spin-charge separation:
*away from half-filling*

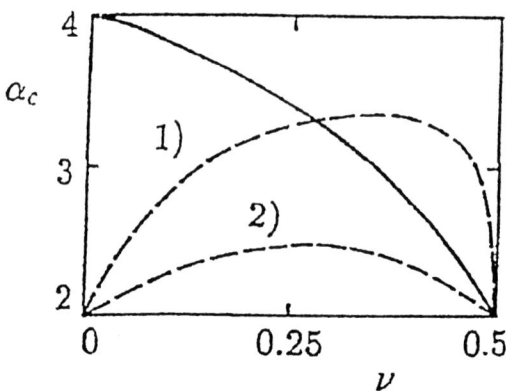

**FIGURE 5.** $K_c$ (in figure labeled by $\alpha_c$) for the Hubbard model (dashed lines) as a function of the filling for 1) $U/t = 2$ and 2) $U/t = 8$ from Ref. [24].

a) if $g_s < 0$ both spin and charge excitations are gapless with different velocities, and the single electron excitations are ill-defined;

b) if $g_s > 0$ charge excitations are gapless whyle the spin excitations acquire a gap.

*at half-filling*

c) if $g_s < 0$ charge excitations have a spectral gap whyle spin excitation are gapless. This case corresponds to Mott-Hubbard insulator or spin density wave state;

d) if $g_s > 0$ both charge and spin excitations are gapfull. Such situation can arise in a system with an 1/2 filled band and non-point like interaction. The ground state is a commensurate charge-density wave with a true long-range order.

# XI THE KONDO CHAIN

To conclude our Lectures let us discuss, as an example of the physics in one dimension, the Kondo chain at half-filling ($\mu = 0$). The following model Hamiltonian describes an M-fold degenerate band of conduction electrons interacting with a periodic arrangement of local spins $S$:

$$H = -\sum_{j=1}^{M}\sum_{n,\alpha} c^\dagger_{n+1,\alpha,j} c_{n,\alpha,j} + J \sum_n c^\dagger_{n,\alpha,j}\sigma_{\alpha\beta} c_{n,\beta,j} \mathbf{S}_n + J' \sum_n \mathbf{S}_n \mathbf{S}_{n+1}, \qquad (117)$$

In the following we shall use the path integral formalism. As known in path integral formulation spins are treated as classical vectors $\mathbf{S} = S\mathbf{M}(\mathbf{M}^2 = 1)$ whose action includes the Berry phase. We have the following partition function:

$$Z = \int \mathcal{D}c^*_n \mathcal{D}c_n \mathcal{D}\mathbf{n}(\tau) e^{A[c^*,c,S\mathbf{M}]}, \qquad (118)$$

where $\mathbf{n}$ is a unit vector on the lattice ($\mathbf{n}^2 = 1$).

The Euclidean action is:

$$A = \int d\tau \sum_n \{c^*_{n,\alpha}\partial_\tau c_{n,\alpha} - H(c^*,c,S\mathbf{M}) + iS\int_0^\infty d\xi (\mathbf{M}_n[\partial_\xi \mathbf{M} \times \partial_\tau \mathbf{M}])\} \qquad (119)$$

where the last term represents the Berry phase responsible for the correct quantization of the local spins. In the following we shall employ a semiclassical approach assuming that all the fields can be separated in the slow and fast components. Integrating out the fast components, we shall obtain an effective action for the slowly varying part. If the gap is much smaller than the bandwidth we go first to the continuum limit, i.e. we write the fermions in terms of the right and left components:

$$c_{n,\alpha} = R_\alpha e^{-i\pi/2n} + L_\alpha e^{i\pi/2n}, \qquad (120)$$

Thus we get

$$c^\dagger_{n,\alpha} c_{n,\beta} = (R^\dagger_\alpha R_\beta + L^\dagger_\alpha L_\beta + (-1)^n (R^\dagger_\alpha L_\beta + L^\dagger_\alpha R_\beta)), \qquad (121)$$

where the second term in (121) represents the staggered (fasting oscillating) part. The suggested decomposition for the field $\mathbf{M}_n$ is instead the following:

$$\mathbf{M}_n = a\mathbf{m}(x) + (-1)^n\sqrt{1-\mathbf{m}^2}\mathbf{N}(\tau,x), \qquad (122)$$

where $a$ is the lattice spacing and $|\mathbf{m}|a \ll 1$ is a rapidly varying ferromagnetic component of the local magnetization. Substituting Eq.(121) and Eq.(122) into (119) and keeping only the slowly varying part we obtain:

$$A = \int d\tau dx L,$$
$$L = iS(\mathbf{m}[\mathbf{N} \times \partial_\tau \mathbf{N}] + (2\pi S \times top.term.) + R_\alpha^\dagger(\partial_\tau - i\partial_x)R_\alpha +$$
$$L_\alpha^\dagger(\partial_\tau + i\partial_x)L_\alpha + J\mathbf{m}(R_\alpha^\dagger \sigma_{\alpha\beta} R_\beta + L_\alpha^\dagger \sigma_{\alpha\beta} L_\beta) + J\sqrt{1-\mathbf{m}^2}\mathbf{N}(R_\alpha^\dagger \sigma_{\alpha\beta} L_\beta + L_\alpha^\dagger \sigma_{\alpha\beta} R_\beta), \quad (123)$$

where the first term comes from the integration $\int d\xi \partial_\xi \mathbf{m}$, and there appears a topological term given by

$$iS \int d\tau dx (M[\partial_\tau \mathbf{N} \times \partial_x \mathbf{N}]). \quad (124)$$

We employ now the non-Abelian bosonization and write the fermionic determinant as a path integral over the matrix field $g$:

$$A = W[g] + \lambda \mathbf{m}(\mathbf{J}_R + \mathbf{J}_L) + \lambda\sqrt{1-\mathbf{m}^2} Tr\{\mathbf{N}\sigma(g^\dagger e^{i\sqrt{2\pi}\phi} + g e^{-i\sqrt{2\pi}\phi})\}$$
$$+ iS\mathbf{m}[\mathbf{N} \times \partial_\tau \mathbf{N}] + iS(M[\partial_x \mathbf{N} \times \partial_\tau \mathbf{N}]), \quad (125)$$

where we have called $\lambda = JS$.

Now we introduce the $SU(2)$ matrix $h = i\sigma\mathbf{N} \in SU(2)$ and define the new variable $G = gh^\dagger$. The metric is invariant under the transformation, thus

$$A = W[Gh] + \lambda(1 - \frac{\mathbf{m}^2}{2}) Tr(G^\dagger e^{i\sqrt{2\pi}\phi} + G e^{-i\sqrt{2\pi}\phi}) \quad (126)$$

where we have expanded in $\mathbf{m}$ and neglected the terms in $\mathbf{N} \cdot \mathbf{m}$. Separating the part depending on $G$ and $h$ Eq.(126) becomes

$$A = W[G] + W[h] + \frac{1}{2\pi} Tr(G^{-1}\partial G h^{-1}\partial h) + \lambda(1 - \frac{\mathbf{m}^2}{2}) Tr(G^\dagger e^{i\sqrt{2\pi}\phi} + G e^{-i\sqrt{2\pi}\phi}). \quad (127)$$

The part in $G$ can be refermionized and the transformed action will be

$$A = R_\alpha^\dagger(\partial_\tau - i\partial_x)R_\alpha + L_\alpha^\dagger(\partial_\tau + i\partial_x)L_\alpha + \lambda(1 - \frac{\mathbf{m}^2}{2})(R_\alpha^\dagger L_\alpha + L_\alpha^\dagger R_\alpha)$$
$$+ W[i\sigma\mathbf{N}] + iS\mathbf{m}[\mathbf{N} \times \partial_\tau \mathbf{N}] + iS(M[\partial_x \mathbf{N} \times \partial_\tau \mathbf{N}]) +$$
$$+ \frac{1}{2\pi}(\mathbf{J}_L \cdot h\bar\partial h^{-1}). \quad (128)$$

The new fermions are rotated ($R_\alpha \to iR_\alpha$ and $L_\alpha \to L_\alpha$) and the term of the interaction with the magnetization generates a mass term. The fermion spectrum will have a gap of the order $\lambda$:

$$\epsilon^2(q) = q^2 + \lambda^2. \quad (129)$$

Now we are in position to integrate over the fast ferromagnetic fluctuations described by $\mathbf{m}$, and rescaling time and space, as a final result, we get:

$$A = \frac{M}{2\pi} \int d\tau dx [v^{-2}(\partial_\tau \mathbf{N})^2 + (\partial_x \mathbf{N})^2] + [\pi(2S - M) \times top.terms.], \tag{130}$$

where

$$v^{-2} = 1 + \frac{2\pi}{\lambda^2 \ln(1/\lambda)}. \tag{131}$$

This is the action of the $O(3)$-nonlinear sigma model with the dimensionless coupling constant

$$g = \frac{\pi v}{M}. \tag{132}$$

Now we have to distinguish two cases:
If $|M - 2S|$ is *odd* the topological term is essential. In this case the action will be of the form

$$A = \int d\tau dx [A(\partial_\tau \mathbf{N})^2 + B(\partial_x \mathbf{N})^2], \tag{133}$$

where $A$ and $B$ are constants, $\mathbf{N}^2 = 1$. This model is a spin liquid with a spectral gap. After the rescaling, the action can be written as

$$A = \frac{1}{2g} \int d\tau dx (\partial_\mu \mathbf{N})^2, \tag{134}$$

where

$$g \simeq \frac{1}{2\pi} \frac{1}{\sqrt{\lambda^2 \ln(t/\lambda)}}. \tag{135}$$

The model becomes critical and its low-energy behavior is the same as for the spin-1/2 antiferromagnetic Heisenberg chain. In this case the spectral gap for the charge excitations is $\Delta_c \simeq \lambda$, while the energy scale for the spin sector is

$$\Delta_s(J) = \pi v/\xi \simeq \epsilon g^{-1} \exp(-2\pi/g), \tag{136}$$

It marks the crossover to the critical regime where $\mathbf{N}$ has the same correlation lenght as the staggered magnetization of the Heisenberg chain. Let us remember that the expression for the Kondo temperature is

$$T_K = \sqrt{J} \exp(-2\pi/J), \tag{137}$$

It follows that $m(J)$ is always larger due to the presence of the large logarithm. Therefore at small $J$ the antiferromagnetic exchange induced by the conduction electrons (the well known RKKY) interaction plays a stronger role than the Kondo screening. This justifies why we neglect the term $\mathbf{m}(R\sigma R + L\sigma L)$ in the evaluation of the determinant. This implies that the leading contribution to the low-energy dynamics come from antiferromagnetic fluctuations in agreement with the results of Tsunetsugu et al. (1992) [18], Yu and White (1993) [19]. Let us note that an important point in the previous derivation consists in the fact that the fermionic determinant (123) contains the topological term. This term cannot be obtained semiclassically. Infact, a semiclassical expansion assumes adiabaticity, i.e. the fasting oscillating terms adjust themselves to a slowly varying external field. This adiabacity is violated when an energy level crossing occurs, giving rise to a topological term. This happens when the field $\mathbf{N}(x,\tau)$ has a non-zero topological charge.

If $|M - 2S|=even$, the topological term does not contribute to the partition function, since it is given by $2\pi i k$, where $k$ is an integer. In this case the action (130) corresponds to that of the O(3)-nonlinear sigma model. This model has a disordered ground state with a correlation lenght:

$$\xi = a^{-1}g\exp(2\pi/g). \qquad (138)$$

It follows from the previous analysis that, at half-filling, the Kondo chain has very different properties in the charge and spin sectors.

# XII CONFORMAL SYMMETRY AND GENERAL PROPERTIES OF CORRELATION FUNCTIONS FOR WZNW MODELS

An important property of a gaussian model is that it possesses a special hidden symmetry, known as the conformal symmetry. The conformal symmetry manifests in the special properties that correlation functions possess under the transformation of the area $A$ of the complex plane on which the field $\phi$ is defined. Given an area $A$ in the complex plane, the multipoint correlation function of the bosonic exponent is defined by:

$$\langle \exp[i\beta_1\phi(\xi_1)]\ldots\exp[i\beta_n\phi(\xi_n)]\rangle = e^{-\sum_{i>j}\beta_i\beta_j G(\xi_i;\xi_j)} e^{-\frac{1}{2}\sum_i \beta_i^2 G(\xi_i;\xi_i)}, \tag{139}$$

where $G$ is the Green's function of the Laplace operator on $A$. The explicit expression for the Green's function is obtained from the transformation $z(\xi)$ which maps $A$ onto the infinite plane. Explicitly, this Green's function is given by

$$G(\xi_1,\xi_2) = -\frac{1}{2\pi}\ln|z(\xi_1) - z(\xi_2)| - \frac{1}{4\pi}\ln|\partial_{\xi_1} z(\xi_1)\partial_{\xi_2} z(\xi_2)|. \tag{140}$$

Let us consider the case of two exponents. Remembering that the correlation functions (139) are products of the analytical and antianalytical part, substituting Eq.(140) into Eq.(??) with N=2 and $\beta_1 = \beta$, $\beta_2 = -\beta$, one obtains the following expression for the analytical part of the pair correlation function:

$$D(\xi_1,\xi_2) = \frac{1}{[z(\xi_1) - z(\xi_2)]^{2\Delta}}[\partial_{\xi_1} z(\xi_1)\partial_{\xi_2} z(\xi_2)], \tag{141}$$

where $\Delta = \beta^2/8\pi$. Eq.(141) shows that correlation functions transform locally under analytical coordinate transformations. Therefore one can assign these transformations to the corresponding operators. i.e. the bosonic exponents $A_\Delta(z) \equiv \exp[i\beta\phi(z)]$. From Eq.(141) it's clear that these operators transform as tensors of rank $(\Delta, 0)$, i.e.:

$$A_\Delta(\xi) = A_\Delta[z(\xi)](dz/d\xi)^\Delta. \tag{142}$$

In the same way the antianalytical exponent $A_{\bar\Delta}(\bar z)$ transforms as a tensor of rank $(0, \bar\Delta)$. From the previous expressions we can immediately write the following correlation function:

$$\langle A_\Delta(\xi_1)A_\Delta(\xi_2)\rangle = \frac{\delta_{\Delta_1\Delta_2}}{[z(\xi_1) - z(\xi_2)]^{2\Delta}}. \tag{143}$$

Let us postulate now, that there are models besides the Gaussian model that possess a linear spectrum, whose correlation functions factorize into products of analytical (antianalytical) parts and which have operators transforming under analytical (or antianalytical) transformations as (142). Such operator will be called *primary* fields. This defines what a conformal field theory is. To make the definition more rigorous we have to add one more property, namely, to postulate that the three-point correlation functions of the primary fields have the following form:

$$\langle A_{\Delta_1}(z_1)A_{\Delta_2}(z_2)A_{\Delta_3}(z_3)A_{\Delta_4}(z_4)\rangle = \frac{C_{123}}{(z_{12})^{\Delta_1+\Delta_2-\Delta_3}(z_{13})^{\Delta_1+\Delta_3-\Delta_1}(z_{23})^{\Delta_2+\Delta_3-\Delta_1}}, \tag{144}$$

where $C_{123}$ are some constants, and we have used subindices for $\xi_i$. Since the eigenvalues of the Hamiltonian and momentum are related to the zeroth components of the stress-energy tensor, this suggests that the generator of the transformation (142) are Fourier components of the stress-energy tensor. From the commutation relations between these components and the primary fields we are able to derive the **Ward identity** for the correlation functions of stress energy-tensor with primary fields:

$$\langle T(z)A_{\Delta_1}(z_1)\ldots A_{\Delta_N}(z_N)\rangle = \sum_{i=1}^{N}[\frac{\Delta_i}{(z-z_i)^2} + \frac{1}{(z-z_i)}\partial_{z_i}]\langle A_{\Delta_1}(z_1)\ldots A_{\Delta_N}(z_N)\rangle. \tag{145}$$

## A  An example, free bosons

We shall use the formalism developed so far to illustrate an example, the case of a single massless free boson, also known as gaussian model. The action is written as

$$S = \frac{1}{2\pi}\int \partial X \bar{\partial} X, \tag{146}$$

where the field $X(z,\bar{z})$ has propagator $<X(z,\bar{z})X(z',\bar{z}')> = -\frac{1}{2}\log|z-z'|$. For solutions of the equations of motion, we find that $X(z,\bar{z}) = \frac{1}{2}(x(z) + \bar{x}(\bar{z}))$, splits into two pieces with holomorphic and anti-holomorphic dependence respectively. These are the left and right-movers. These pieces have propagators

$$<x(z)x(z')> = -\log(z-z'), \quad <\bar{x}(\bar{z})\bar{x}(\bar{z}')> = -\log(\bar{z}-\bar{w}). \tag{147}$$

Note that the field $x(z)$ is not itself a conformal field, but its derivative, $\partial x(z)$, has leading short distance expansion

$$\partial x(z)\partial x(z') = -\frac{1}{(z-z')^2} + \ldots, \tag{148}$$

obtained by taking two derivatives of (147). We see from the scaling properties of the right-hand side of (148) that $\partial x(z)$ behaves as a (1,0) conformal field. We can define the stress-energy tensor $T(w)$ via the normal ordering prescription

$$T(w) = -\frac{1}{2} : \partial x(z)\partial x(w) := -\frac{1}{2} lim_{z \to w}[\partial x(z)\partial x(w) + \frac{1}{(z-w)^2}]. \qquad (149)$$

Using Wick's theorem and performing a Taylor expansion in the limit $z \to w$, we can compute the singular part:

$$\begin{aligned} T(z)\partial x(w) &= -\frac{1}{2} : \partial x(z)\partial x(z) : \partial x(w) \\ &= -\frac{1}{2}\partial x(z) < \partial x(z)\partial x(w) > 2 + \cdots \\ &= \partial x(z)\frac{1}{(z-w)^2} + \cdots \\ &= (\partial x(w) + (z-w)\partial^2 x(w))\frac{1}{(z-w)^2} + \cdots . \end{aligned} \qquad (150)$$

Thus, we find

$$T(z)\partial x(w) \simeq \frac{\partial x(w)}{(z-w)^2} + \frac{1}{(z-w)}\partial^2 x(w) + \cdots , \qquad (151)$$

which is just the definition of a (1,0) primary field. Identical considerations apply to the anti-holomorphic operator $\bar{\partial}\bar{x}\bar{z}$. Its product with $\bar{T}(\bar{z})$ shows it to have conformal dimension (0,1).

## B  Finite temperature formalism

Now let us turn to describe a transformation which is particularly important for applications:

$$z(\xi) = \exp(2\pi \xi / L). \qquad (152)$$

This transforms a strip of width $L$ in the $x$-direction into the infinite complex plane (see figure 6. This transformation relates correlation functions at zero temperature with those at finite $T$ (where we identify $T = i/L$). Substituting $z(\xi)$ into Eq.(142) we get

$$D(\xi_1, \xi_2) = \langle A_\Delta(z_1) A_\Delta(z_2) \rangle = \{\frac{\pi/L}{\sinh[\pi(\xi_1 - \xi_2)/L]}\}^{2\Delta} \qquad (153)$$

$$\bar{D}(\xi_1, \xi_2) = \langle \bar{A}_{\bar{\Delta}}(\bar{z}_1) \bar{A}_{\bar{\Delta}}(\bar{z}_2) \rangle = \{\frac{\pi/L}{\sinh[\pi(\xi_1 - \xi_2)/L]}\}^{2\bar{\Delta}}. \qquad (154)$$

Let us remind that while considering an arbitrary area of the complex plane the correlation functions decay power-law, since there is no energy scale, the transformation (152) generates a scale, i.e. the width of the strip, and the correlation lenght becomes finite. There is no reason for the correlation functions to be power-law.

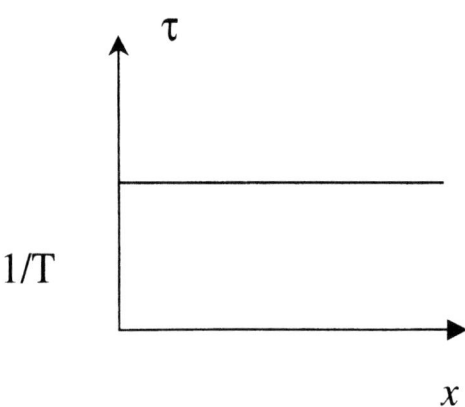

**FIGURE 6.** Tranformation of strip to the infinite plane

## C  Energy-momentum eigenvalues and conformal dimensions

A general property that we will describe here is the one to one correspondence between the scaling dimension of the correlation functions and the energy levels. This correspondence is true for any (1+1)-dimensional theory and is of great help in calculating correlation functions. The intimate relation between CFT and the conventional bosonization has become manifest when Dotsenko and Fateev represented the CFT correlation functions in terms of correlators of bosonic exponents (1984) [20].

Let us consider $T = 0$ and the system be a circle of a finite lenght $L$:

$$\xi = \tau + ix, \quad 0 < x < L \tag{155}$$

Let us expand the expression of $D,\bar{D}$ for $\tau \ll (L/v_F)$. The result is

$$D(\xi_1, \xi_2)\bar{D}(\xi_1, \xi_2) =$$
$$= \sum_{n,m=0} (\pi/L)^d \exp[-\frac{2\pi}{L}(d+n)\tau_{12}] \exp[-ix_{12}\frac{2\pi}{L}(S+m)], \tag{156}$$

where $C_{nm}$ are universal numerical coefficients (with $n$ and $m \geq 0$). On the other hand we can use a Lehman representation:

$$\langle A_\Delta(1) A_\Delta^\dagger(2) \rangle = \sum_m | <0|A\Delta(0,0)|m>|^2 e^{-\tau(E_m - E_0) - iX(P_m - P_0)}, \tag{157}$$

where $|m>$ labels the eigenstates of the Hamiltonian and $E_n, P_n$ are the eigenvalues of the energy and momenta of the state $m$. Comparing the two expansions we find

$$E_n = \frac{2\pi}{L}(d+n)$$
$$P_n = \frac{2\pi}{L}(S+m) \tag{158}$$

These two equations relate conformal dimensions of the correlation functions to the eigenvalues of energy and momentum operators. We can also write:

$$E_n - E_0 = \frac{2\pi}{L}(\Delta_n + \bar{\Delta}_n + integer)$$
$$P_n - P_0 = \frac{2\pi}{L}(\Delta_n - \bar{\Delta}_n + integer) \tag{159}$$

For example, for the $SU_k(2)$ WZW model we already know the energy eigenvalues. From their knowledge we identify

$$\Delta_n = \frac{l(l+1)}{k+2} + integer, \tag{160}$$

where l=1/2,3/2 and the same for $\bar{\Delta}_n$.

In general, one solves first the problem of low-lying energy levels in a finite size system (which can be done numerically or even exactly) and then, using the relationships (158)-(159) calculates the correlation functions. From Eqs.(158)-(159) we see that the problem of conformal dimensions can be formulated as an eigenvalue problem for the following operators:

$$\Delta = \frac{L}{4\pi}(\hat{H} + \hat{P}) \equiv \hat{T}_0$$
$$\bar{\Delta} = \frac{L}{4\pi}(\hat{H} - \hat{P}) \equiv \hat{\bar{T}}_0. \qquad (161)$$

It can be shown that the operators $\hat{T}_0$, $\hat{\bar{T}}_0$ are related to the analytical and anti-analytical components of the stress energy tensor defined as

$$T_{ab}(x) = \frac{\delta S}{\delta g^{ab}(x)}, \qquad (162)$$

where $g^{ab}$ is the inverse metric tensor. Indeed the scaling dimensions are related to coordinate transformations and such transformations change the metric on the surface. The relations between scaling dimensions and the stress-energy tensor have a general character that hold for all conformal theories.

From the previous relations, suppose to consider the $SU_k(2)$ WZNW model, we are able to write down the correlation function for the group tensor $g_{\alpha,\bar{\alpha}}$

$$\langle g_{\alpha,\bar{\alpha}}(z_1,\bar{z}_1) g_{\beta,\bar{\beta}}(z_2,\bar{z}_2) \rangle = \delta_{\alpha\beta} \delta_{\bar{\alpha}\bar{\beta}} \frac{1}{|z_{12}|^{\frac{l(l+1)}{k+2}}}. \qquad (163)$$

## XIII  CONCLUSIONS

In these Lectures we have focused on some aspects of the theory of strongly correlated low-dimensional systems. We have discussed some non-perturbative solutions related to (1+1)-dimensional quantum or two-dimensional classic models based on the bosonization approach, with a generalization to systems with isotopic symmetry. We have shown that non-Abelian bosonization is very convenient when there are spin degrees of freedom in the problem. In particular, we have discussed its application to the Kondo and Hubbard model. Partially, we have devoted to Conformal Field Theory describing why it provides a unified approach to all models with gapless linear spectrum in (1+1)-dimension and discussed its relation to conventional bosonization.

# REFERENCES

1. See I. Affleck, *Fields, String and Critical Phenomena*, Edited by E. Brezin and J. Zinn-Justin (Elsevier, Amsterdam, 1988)
2. Vl. S. Dotsenko, *Nucl. Phys.* **B338**, 747 (1990); **358**, 547 (1990).
3. S. Coleman, *Phys. Rev.* **D11**, 2088 (1975).
4. A. Luther and I. Peschel, *Phys. Rev. B* **12**, 3906 (1975).
5. S. Mandelstam, *Phys. Rev.* **D11**, 3026 (1975).
6. A.O. Gogolin, A.A. Nersesyan and A.M. Tsvelik, *The bosonization approach to strongly correlated systems*, (Cambridge University Press, Cambridge 1998).
7. P. Goddard and D. Olive, *Int. J. Mod. Phys. A* **1**, 303 (1986).
8. V. A. Fateev and A. B. Zamolodchikov, *Yad. Fiz. (Sov. Nucl. Phys.)* **43**, 657 (1986).
9. V. G. Knizhnik and A. B. Zamolodchikov, *Nucl. Phys.* **B247**, 83 (1984).
10. I. Affleck and A. W. W. Ludwig, *Nucl. Phys.* **B352**, 849 (1991); *Phys. Rev. Lett.* **67**, 3160 (1991).
11. M. Wakimoto, *Commun. Math. Phys.* **104**, 605 (1986).
12. A.M. Tsvelik, *Phys. Rev. Lett.* **72**, 1048 (1984).
13. A.M. Tsvelik, *Quantum Field Theory in Condensed Matter Physics*, Cambridge University Press (1995)
14. E. Witten, *Commun. Math. Phys.* **92**, 455 (1984).
15. A.M. Polyakov and P.B. Wiegmann, *Phys. Lett.* **B 131**, 121 (1983).
16. A.M. Polyakov and P.B. Wiegmann, *Phys. Lett.* **B 141**, 223 (1984).
17. V.G. Knizhnik and A.B. Zamolodchikov, *Nucl. Phys. B* **247**, 83 (1984).
18. H. Tsunetsugu, Y. Hatsugai, K. Ueda and M. Sigrist, *Phys. Rev. B* **46**, 3175 (1992).
19. C.C. Yu and S.R. White, *Phys. Rev. Lett.* **71**, 3866 (1993).
20. Vl. S. Dotsenko and V.A. Fateev, *Nucl. Phys. B* **240**, 312 (1984).
21. P. Goddard, W. Nahm, and D. Olive, *Phys. Lett.* **160**, 111 (1985).
22. N. Kawakami and S.K. Yang, *Phys. Lett. A* **148**, 359 (1990).
23. H. Frahm and V. E. Korepin, *Phys. Rev. B* **42**, 10553 (1990).
24. N. Kawakami and S.K. Yang, *Phys. Rev. Lett.* **65**, 2309 (1990).

# AUTHOR INDEX

## A
Alexandrov, A. S., 1

## C
Citro, R., 189

## M
Maritato, L., 97

## P
Plakida, N. M., 121

## T
Tsvelik, A. M., 189